Time Series Analysis with Long Memory in View

WILEY SERIES IN PROBABILITY AND STATISTICS

Time Series Analysis with Long Memory in View

Uwe Hassler
Goethe University Frankfurt

Registered Office(s)

John Wiley & Sons, Inc., 111 River Street, Hoboken, NJ 07030, USA

Editorial Office

111 River Street, Hoboken, NJ 07030, USA

For details of our global editorial offices, customer services, and more information about Wiley products visit us at www.wiley.com.

Wiley also publishes its books in a variety of electronic formats and by print-on-demand. Some content that appears in standard print versions of this book may not be available in other formats.

Library of Congress Cataloging-in-Publication Data

Names: Hassler, Uwe, author.
Title: Time series analysis with long memory in view / Uwe Hassler.
Description: 1st edition. | Hoboken, NJ : John Wiley & Sons, 2019. | Series:
   Wiley series in probability and statistics |
Identifiers: LCCN 2018023760 (print) | LCCN 2018036671 (ebook) | ISBN
   9781119470281 (Adobe PDF) | ISBN 9781119470427 (ePub) | ISBN 9781119470403
   (hardcover)
Subjects: LCSH: Time-series analysis.
Classification: LCC QA280 (ebook) | LCC QA280 .H385 2019 (print) | DDC
   519.5/5—dc23
LC record available at https://lccn.loc.gov/2018023760

Cover Design: Wiley
Cover Image: Courtesy of Uwe Hassler

Set in 10/12pt WarnockPro by SPi Global, Chennai, India

Printed in the United States of America

C10004343_092718

PROMETHEUS

> "Then I invented arithmetic for them,
> the most ingenious acquired skill,
> and joining letters to write down words,
> so they could store all things in Memory,
> the working mother of the Muses' arts."

AESCHYLUS, Prometheus Bound

Quoted from the translation by Ian Johnston, *Richer Resources Publications,* 2012

Contents

List of Figures

Preface

Scope of the Book

Since the book by Box and Jenkins (1970), autoregressive moving average (ARMA) models integrated of order d are a standard tool for time series analysis, where typically $d \in \{0, 1, 2\}$. The integrated ARMA (ARIMA) model of order d means that a time series has to be differenced d times in order to obtain a stationary and invertible ARMA representation. The papers by Granger and Joyeux (1980) and Hosking (1981) extended the ARIMA model with integer d to the so-called fractionally integrated model, where d takes on noninteger values, often restricted to $|d| < 1/2$. In particular, the case of $0 < d < 1/2$ corresponds to a stationary model with long memory, where the latter means that the autocorrelations die out so slowly that they are not absolutely summable. For $1/2 \leq d < 1$, the fractionally integrated model bridges the gap from stationarity to the so-called unit root behavior ($d = 1$), where past shocks have a permanent effect on the present and values of $d > 1$ allow for even more extreme persistence.

This book grew out of lecture notes from which I taught PhD courses on time series analysis and in particular on time series with long memory. Long memory and fractional integration have become key concepts in time series analysis over the last decades. For instance, the updated edition of Box and Jenkins (1970), i.e. Box et al. (2015), contains a section on long memory and fractional integration, and so do Kirchgässner et al. (2013), Pesaran (2015), or Palma (2016). Also, previous textbooks like Brockwell and Davis (1991, Section 13.2) and Fuller (1996, Section 2.11) include short sections on this topic. Contrary to these books on general times series analysis containing only short digressions into the realm of long memory, there are nowadays specialized monographs dedicated to this topic exclusively, most recently by Giraitis et al. (2012) and Beran et al. (2013); see also the earlier books by Beran (1994) and Palma (2007). The approach of the present book differs from both routes, from the general interest track and from the specialized long memory track. I rather attempt to introduce into the theory of univariate time series analysis, and the foundations thereof, in such a way that long memory and fractional integration arise as a special case, naturally embedded into the general theory. This is reflected by the title: Time Series Analysis with Long Memory in View. This view is largely directed by the author's research agenda in this field over the last 25 years.

Acknowledgment

Twenty-five years ago I wrote my doctoral thesis on time series with long memory under the supervision of Professor Wolters at the Freie Universität Berlin. Jürgen Wolters passed away in November 2015. I take this opportunity to commemorate his enthusiasm, generosity, and open-mindedness from which I profited so much not only during my doctoral studies but also later on as his coauthor. Since my thesis, I have written a sequence of papers on long memory. I am indebted to many anonymous referees for writing, in many cases, critical and constructive reports on my papers before publication. Most papers were written with coauthors. I thank them for sharing their knowledge and endurance with me. All of them I owe insights that influenced my research agenda and hence this book. In particular, I wish to mention Matei Demetrescu and Mehdi Hosseinkouchack with whom the collaboration was especially fruitful. The intense discussions we had on a daily basis when they held postdoc positions at Goethe University Frankfurt shaped my view not only on how to address long memory but also on time series analysis in general. Christoph Hanck, Paulo Rodrigues, and Verena Werkmann have proofread an earlier draft of this book, and their many comments and corrections are gratefully acknowledged. Finally, I am grateful to the Volkswagen Stiftung for financing a year of sabbatical leave in 2014/2015 by an *opus magnum* grant; without this support it would not have been possible to write this book.

October 2017   *Uwe Hassler*

List of Notation

$\mathbb{C}$	set of complex numbers
$\mathbb{N}$	set of natural numbers
$\mathbb{N}_0$	set of natural numbers including 0
$\mathbb{R}$	set of real numbers
$\mathbb{Z}$	set of integers
$\lfloor x \rfloor$	largest integer smaller or equal to $x \geq 0$, $x \in \mathbb{R}$
$\ln x$	natural logarithm of $x > 0$, also $\ln(x)$
I_k	identity matrix of dimension k
γ	Euler's constant
$\mathrm{P}(\cdot)$	probability
$\mathrm{E}(\cdot)$	expectation operator
$\mathrm{Var}(\cdot)$	variance operator
$\mathrm{Cov}(\cdot, \cdot)$	covariance operator
$\gamma(h)$	autocovariance at lag h
$\rho(h)$	autocorrelation at lag h
ω^2	long-run variance
L	lag operator
Δ	difference operator
$\mathcal{EXP}(\theta)$	exponential distribution with parameter θ
$\mathcal{N}(\mu, \sigma^2)$	normal distribution with mean μ and variance σ^2
$\chi^2(n)$	chi-square distribution with n degrees of freedom
$\rightarrow$	convergence
$\xrightarrow{a.s.}$	almost sure convergence
$\xrightarrow{2}$	convergence in mean square
$\xrightarrow{p}$	convergence in probability
$\xrightarrow{D}$	convergence in distribution
$\Rightarrow$	weak convergence
$\approx$	approximately equal

$a := b$	a is defined to equal b
$A \Rightarrow B$	A implies B
$A \Leftrightarrow B$	A and B are equivalent
$a_n \sim b_n$	a_n/b_n converges to 1
$y_t \sim I(d)$	y_t is integrated of order d
$\sim \mathcal{N}(0,1)$	follows a standard normal distribution

Acronyms

AIC	Akaike information criterion
AR	autoregressive
ARCH	autoregressive conditional heteroskedasticity
ARFIMA	autoregressive fractionally integrated moving average
ARMA	autoregressive moving average
BIC	Bayesian information criterion
CIR	cumulated impulse response
CLT	central limit theorem
CMT	continuous mapping theorem
CSS	conditional sum of squares
DCT	dominated convergence theorem
DFT	discrete Fourier transform
EXP	exponential model
fBm	fractional Brownian motion
FCLT	functional central limit theorem
FEXP	fractional EXP
FIN	fractionally integrated noise
GARCH	generalized autoregressive conditional heteroskedasticity
LLN	law of large numbers
LM	Lagrange multiplier
MA	moving average
MAC	memory and autocorrelation consistent
MDS	martingale difference sequence
ML	maximum likelihood
MSE	mean squared error
OLS	ordinary least squares
WLLN	weak law of large numbers

1

Introduction

1.1 Empirical Examples

Figure 1.1 displays 663 annual observations of minimal water levels of the Nile river. This historical data is from Beran (1994, Sect. 12.2) and ranges from the year 622 until 1284. The second panel contains the sample autocorrelations $\hat{\rho}(h)$ at lag $h \in \{1, 2, \ldots, 30\}$. The maximum value, $\hat{\rho}(1) = 0.57$, is not particularly large, but the autocorrelogram dies out only very slowly with $\hat{\rho}(30) = 0.15$ still being significantly positive. Such a slowly declining autocorrelogram is characteristic of what we will define as long memory or strong persistence. It reflects that the series exhibits a very persistent behavior in that we observe very long cyclical movements or (reversing) trends. Note, e.g. that from the year 737 until 805, there are only three data points above the sample average (=11.48), i.e. there are seven decades of data below the average. Then the series moves above the average for a couple of years, only to swing down below the sample mean for another 20 years from the year 826 on. Similarly, there is a long upward trend from 1060 on until about 1125, followed again by a long-lasting decline. Such irregular cycles or trends due to long-range dependence, or persistence, have first been discovered and discussed by Hurst, a British engineer who worked as hydrologist on the Nile river; see in particular Hurst (1951). Mandelbrot and Wallis (1968) coined the term *Joseph effect* for such a feature; see also Mandelbrot (1969). This alludes to the biblical seven years of great abundance followed by seven years of famine, only that *cycles* in Figure 1.1 do not have a period of seven years, not even a constant period.

Long memory in the sense of strong temporal dependence as it is obvious in Figure 1.1 has been reported in many fields of science. Hipel and McLeod (1994, Section 11.5) detected long memory in hydrological or meteorological series like annual average rainfall, temperature, and again river flow data; see also Montanari (2003) for a survey. A further technical area beyond geophysics with long memory time series is the field of data network traffic in computing; see Willinger et al. (2003).

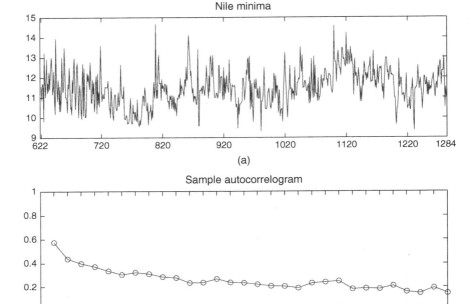

Figure 1.1 Annual minimal water levels of the Nile river.

The second data set that we look into is from political science. Let p_t denote the poll data on partisanship, i.e. the voting intention measured by monthly opinion polls in England. More precisely, p_t is the portion of people supporting the Labor Party. The sample ranges from September 1960 until October 1996 and has been analyzed by Byers et al. (1997).[1] Figure 1.2 contains the logit transformation of this poll data,

$$y_t = \ln\left(\frac{p_t}{1 - p_t}\right),$$

such that $y_t = 0$ for $p_t = 50\%$; here $\ln(x)$ stands for the natural logarithm of x. We observe long-lasting upswings followed by downswings amounting to a pseudocyclical pattern or reversing trends. This is well reflected and quantified by the sample autocorrelations in the lower panel, decreasing from $\hat{\rho}(1) = 0.9$ quite slowly to $\hat{\rho}(24) \approx 0.2$. Independently of Byers et al. (1997), Box-Steffensmeier and Smith (1996) detected long memory in US opinion poll data on partisanship. Long memory in political popularity has been confirmed

---

1 We downloaded the data from James Davidson's homepage on May 5, 2016. The link is http://people.exeter.ac.uk/jehd201/bdpdata.txt.

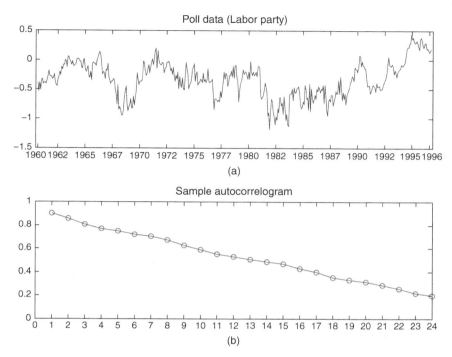

Figure 1.2 Monthly opinion poll in England, 1960–1996.

in a sequence of papers; see Byers et al. (2000, 2007), and Dolado et al. (2003); see also Byers et al. (2002) for theoretical underpinning of long memory in political popularity. Further evidence on long memory in political science has been presented by Box-Steffensmeier and Tomlinson (2000); see also the special issue of *Electoral Studies* edited by Lebo and Clarke (2000).

Since Granger and Joyeux (1980), the fractionally integrated autoregressive moving average (ARMA) model gained increasing popularity in economics. The empirical example in Granger and Joyeux (1980) was the monthly US index of consumer food prices. Granger (1980) had shown theoretically how the aggregation of a large number of individual series may result in an index that is fractionally integrated, which provided theoretical grounds for long memory as modeled by fractional integration in price indices. A more systematic analysis by Geweke and Porter-Hudak (1983) revealed long memory in different US price indices. These early papers triggered empirical research in long memory in inflation rates in independent work by Delgado and Robinson (1994) for Spain and by Hassler and Wolters (1995) and Baillie et al. (1996) for international evidence. Since then, there has been offered abundant evidence in favor of long memory in inflation rates; see, e.g. Franses and Ooms (1997),

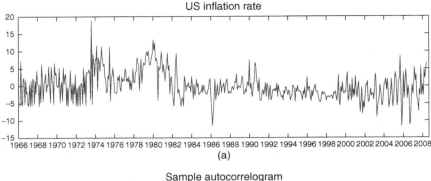

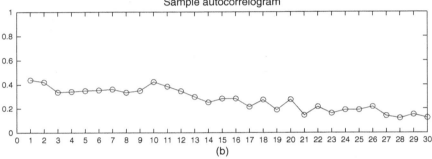

· **Figure 1.3** Monthly US inflation, 1966–2008.

Baum et al. (1999), Franses et al. (1999), Hsu (2005), Kumar and Okimoto (2007), Martins and Rodrigues (2014), and Hassler and Meller (2014), where the more recent research focused on breaks in persistence, i.e. in the order of fractional integration. For an early survey article on further applications in economics, see Baillie (1996).

Figure 1.3 gives a flavor of the memory in US inflation. The seasonally adjusted and demeaned data from January 1966 until June 2008 has been analyzed by Hassler and Meller (2014). The autocorrelations fall from $\hat{\rho}(1) = 0.44$ to a minimum of $0.12 = \min\{\hat{\rho}(h)\}$, $h = 1, 2, \ldots, 30$. Again, this slowly declining autocorrelogram mirrors the reversing trends in inflation, although Hassler and Meller (2014) suggested that the persistence may be superimposed by additional features like time-varying variance.

The fourth empirical example is from the field of finance. Figure 1.4 displays daily observations from January 4, 1993, until May 31, 2007. This sample of 3630 days consists of the logarithm of realized volatility of International Business Machines Corporation (IBM) returns computed from underlying five-minutes data; see Hassler et al. (2016) for details. Although the dynamics of the series

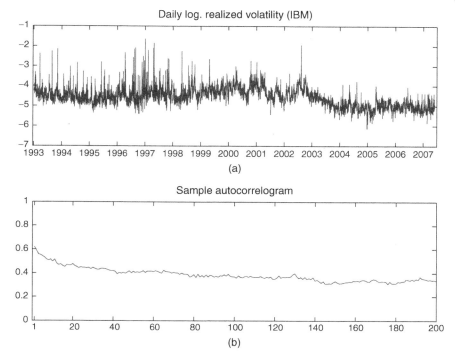

Figure 1.4 Daily realized volatility, 1993–2007.

is partly masked by extreme observations, one clearly may distinguish periods of weeks where the data tends to increase, followed by long time spans of decrease. The high degree of persistence becomes more obvious when looking at the sample autocorrelogram. Starting off with $\hat{\rho}(1) = 0.62$, the decline is extremely slow with $\hat{\rho}(400)$ still being well above 0.2. Long memory in realized volatility is sometimes considered to be a stylized fact since the papers by Andersen et al. (2001, 2003). Such a view is supported by the special issue in *Econometric Reviews* edited by Maasoumi and McAleer (2008).

Finally, with the last example we return to economics. Figure 1.5 shows 435 monthly observations from 1972 until 2008. The series is the logarithm of seasonally adjusted US unemployment rates (number of unemployed persons as a percentage of the civilian labor force); see Hassler and Wolters (2009) for details. The sample average of log-unemployment is 1.7926; compare the straight line in the upper panel of Figure 1.5. Here, the trending behavior is so strong that the sample average is crossed only eight times over the period of 35 years. The deviations from the average are very pronounced and very long relative to the sample size. In that sense the series from Figure 1.5 seems to be most persistent of all the five examples considered in this introduction.

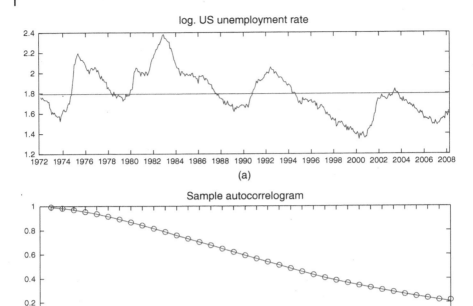

Figure 1.5 Monthly unemployment rate, 1972–2008.

This is also expressed by the sample autocorrelogram virtually beginning at one ($\hat{\rho}(1) = 0.992$) and $\hat{\rho}(h) > 0.2$ for $h \in \{1, 2, \ldots, 36\}$. What is more, the autocorrelations decline almost linearly in h, which is indicative of an $I(1)$ process or an $I(d)$ process with even $d > 1$; see Hassler (1997, Corollary 3) and Section 7.5. Hence, the log-unemployment data seems to be most persistent, or most strongly trending, among our empirical examples.

1.2 Overview

There are two natural approaches to long memory modeling by fractional integration. The first one takes the nonstationary $I(1)$ model as starting point, i.e. processes integrated of order 1. Such processes are often labeled as unit root processes in econometrics, where they play a major role within the cointegration framework; see, for instance, Johansen (1995), Lütkepohl (2005), or Pesaran (2015). The extension from the $I(1)$ model to the more general $I(d)$ model might be considered as a nearby approach from an econometric point of view. The second approach starts off with the classical stationary time series model, where the moving average coefficients from the Wold decomposition are assumed to be absolutely summable and to sum to a value different from 0. For this model, which may be called integrated of order 0, $I(0)$ (see Chapter 6), it holds true that the scaled sample average converges with the square root of the sample size to a nondegenerate normal distribution. This $I(0)$ model underlying the major body of time series books from Anderson (1971) over Brockwell and Davis (1991) and Hamilton (1994) to Fuller (1996) may be generalized to the stationary $I(d)$ process for $d < 1/2$. The latter can be further extended to the region of nonstationarity ($d \geq 1/2$). Here, we follow this second route starting with the $I(0)$ case. More precisely, the outline of the book is as follows.

A definition of stationarity of stochastic processes is given in the next chapter. Moreover, Chapter 2 contains a discussion of ergodicity that corrects expositions found in some books (see Example 2.2). Next, we show that a familiar sufficient condition for ergodicity in the mean (defined in Definition 2.3) is also necessary; see Proposition 2.2. Then we distinguish between (short and long) memory (Definition 2.4) and different degrees of persistence on statistical grounds: Short memory is separated from long memory to characterize under what circumstances the variance of the sample average is of order $1/T$, where T denotes the sample size; see Proposition 2.3. Persistence is defined (Definition 2.5) to characterize the absence or presence and strength of a trend component in a process; see also Eq. (4.22).

Chapter 3 focuses on moving average processes of infinite order, sometimes called linear processes. This is motivated by Wold's theorem in Section 3.2. We thus have a unified framework to embed the classical process of moderate persistence as well as processes with antipersistence or strong persistence, which may or may not display long memory at the same time. The discussion of memory vs. persistence is picked up again in Section 3.3. The discussion of Examples 3.2 through 3.5 shows that the series from Figures 1.1 to 1.5 display both long memory and strong persistence, which motivates the model of fractional integration in Chapter 6. Before leaving Chapter 3, we provide some interesting results on the summability of the classical ARMA process (Proposition 3.5) established with a sequence of technical lemmata.

Chapter 4 introduces to the frequency domain where much of the long memory analysis is settled. The frequency domain is not only useful for data analysis, but it also allows for a deeper theoretical study. For instance, the classical concept of invertibility can be recast following Bloomfield (1985) and Bondon and Palma (2007) in a way (Proposition 4.6) that extends the region of invertibility of fractionally integrated processes; see Proposition 6.2. Next, we introduce the so-called exponential model formulated in the frequency domain. This exponential model is typically not treated in time series books, although it is particularly convenient in the context of long memory as modeled by fractional integration. Similarly, time series books typically do not deal with so-called

Whittle estimation, which is a frequency domain approximation to maximum likelihood that we present in Section 4.6, thus laying the foundation for memory estimation in Chapters 8 and 9.

Chapter 5 opens the route to fractional integration. It is a short chapter on the fractional difference and integration operator, respectively. We provide four technical lemmata that will be used repeatedly in subsequent chapters. Chapter 6 defines the stationary fractionally integrated process (of type I), building on a precise definition of $I(0)$ processes; see Assumption 6.2. Conditions for (different degrees of) persistence follow under minimal restrictions from Lemma 5.4, while Proposition 6.1 translates this into the frequency domain. Corollary 6.1 and Proposition 6.3 reflect the persistence as (short or long) memory in the time domain. After a discussion of parametric fractionally integrated models in Section 6.2, two different types of nonstationarity are discussed in Section 6.3: First, type II fractionally integrated processes are only asymptotically stationary if $d < 1/2$. Second, the case $d \geq 1/2$ covers nonstationarity for both type I and type II processes. Proposition 6.6 shows that classical parametric models imply frequency domain assumptions often entertained in the literature. For the rest of the book, we assume the fractionally integrated models as introduced in Chapter 6.

Chapter 7 sets off with what seems to be the most general central limit theorem currently available for moving average processes. It is applied to the sample average of fractionally integrated processes, closing in particular the gap at $d = -1/2$ in the literature; see Corollary 7.1. Section 7.3 extends the central limit theorem to a functional central limit theory, where fractional Brownian motions show up in the limit. Two seemingly different representations of type II fractional Brownian motion are shown to be identical in Lemma 7.2. Finally, this chapter contains in Section 7.5 an exposition on the behavior of the sample autocorrelations under fractional integration.

The eighth chapter is dedicated to the estimation of all other parameters except for the mean, assuming a fully parametric model of fractional integration. Theorem 8.1 gives the general structure of the limiting covariance matrix of the asymptotic normal distribution for (different approximations to) maximum likelihood, while Corollary 8.1 focuses in particular on the integration parameter d. Approximations to maximum likelihood may be settled in the time domain (Proposition 8.2) or in the frequency domain (Whittle estimation, Proposition 8.3). In particular, we find that the nonstationarity-extended Whittle estimator (Proposition 8.4) overcomes all pitfalls of exact maximum likelihood, except for being parametric of course. Section 8.5 paves the way to semiparametric estimation in that it studies the log-periodogram regression in the presence of a so-called exponential model for the short memory component. While consistency is established in Proposition 8.5, we learn that the estimator is less efficient than corresponding estimators rooted in the maximum likelihood principle.

Chapter 9 begins with the already familiar log-periodogram regression, however, now in the presence of short memory, which is not parametrically modeled. The whole chapter is dedicated to procedures that are semiparametric in the sense that they are robust with respect to short memory. This comes at the price of reduced efficiency. Indeed, we obtain a slower rate of convergence compared with parametric estimators. Within the class of semiparametric estimators, there exist differences in efficiency, too, and the local Whittle estimator (Proposition 9.4 or 9.5), respectively its versions allowing for nonstationarity (Proposition 9.6 or 9.7), turn out to be superior.

Since semiparametric estimators are burdened with large variances, it is interesting to have powerful tests that allow to discriminate statistically, e.g. between short memory and long memory or between stationarity and nonstationarity. This issue is addressed in Chapter 10. The first test builds on a classical rescaled range analysis that can be traced back to Hurst (1951). It has been improved by the rescaled variance test that is designed to provide a better balance of power and size in finite samples. A different approach is adopted in Section 10.4 dedicated to Lagrange multiplier (LM) tests. In Section 10.6, the original LM test is recast in a convenient lag-augmented regression framework (Proposition 10.8), and it has the nice property of robustness against conditional and even unconditional heteroskedasticity (Proposition 10.9). At the same time it is asymptotically most powerful against local alternatives.

Long memory is a still rapidly growing field of applied and theoretical research. Therefore, we close the book with a collection of further topics in the final chapter.

All chapters contain a final section called "Technical Appendix: Proofs" (except for the last chapter "Further Topics"). There, we give the mathematical proofs of results provided and discussed in the main text. Some proofs just accomplish or spell out simple steps to adapt proofs from the literature to our context. Other proofs are truly original in that they establish new results that cannot be drawn from the literature. By separating the proofs from the propositions in the main text, we hope to improve the readability of the book. Finally, it should be stressed that the book is not fully self-contained. While in some propositions we spell out all required assumptions, there are many cases where we refer to the literature. For brevity and convenience one finds in the latter case formulations like "…satisfying Assumption 6.3 […] and some further restrictions by Robinson (1995b);" such that the reader is expected to read up details from the provided reference, namely, Robinson (1995b), for this example from Proposition 9.4.

2

Stationary Processes

To begin with, we briefly recap the concept of stationary stochastic processes. Then we discuss ergodicity and how this property is related to consistent estimation of the expected value of a stationary process. Finally, we present precise definitions of (short and long) memory and of (moderate, strong, or anti-) persistence.

2.1 Stochastic Processes

We begin with a couple of definitions, introducing some notational conventions at the same time. Most books on time series analysis define a stochastic process as a sequence of random variables on a common probability space. While this is correct, we find it also helpful to define a stochastic process as a more abstract object. Let us briefly review the concept of a random element; see Billingsley (1968, p. 22). It builds on a probability space, which is the triple $(\Omega, \mathcal{F}, P)$: For a given set Ω, $\mathcal{F}$ is a set of subsets meeting the requirements of a σ-field, and P is a measure mapping from $\mathcal{F}$ to the interval $[0, 1]$. A random element X is a mapping into a metric space $\mathbb{S}$, where measures are defined on the class $\mathcal{S}$ of Borel sets, which is the smallest σ-field containing all open sets of $\mathbb{S}$. Consider the inverse image $X^{-1}(S)$ of $S \in \mathcal{S}$:

$$X^{-1}(S) = \{\omega \mid X(\omega) \in S\} \subseteq \Omega.$$

X is required to be measurable, or $\mathcal{F}$-measurable, in that the inverse image is contained in $\mathcal{F}$ for any $S \in \mathcal{S}$: $X^{-1}(S) \in \mathcal{F}$. The distribution of X is the probability measure $P_x = P(X^{-1})$ given by

$$P_x(S) = P(X^{-1}(S)).$$

If $\mathbb{S}$ equals the real line, $\mathbb{S} = \mathbb{R}$, then X is called a (real) random variable, and $\mathcal{S} = \mathcal{B}$ is the Borel set of the real line. If $\mathbb{S} = \mathbb{R}^n$ is the n-dimensional Euclidean space and $\mathcal{B}^n$ is the Cartesian product,

$$\mathcal{B}^n = \mathcal{B} \times \mathcal{B} \times \cdots \times \mathcal{B},$$

Time Series Analysis with Long Memory in View, First Edition. Uwe Hassler.
© 2019 John Wiley & Sons, Inc. Published 2019 by John Wiley & Sons, Inc.

then X is a random vector. $\mathbb{S}$ may also be a functional space, e.g. the space of continuous function on $[0, 1]$ (see Billingsley (1968, Chapter 2)), such that X denotes a random function. Here, we focus on $\mathbb{S} = \mathbb{R}^n$ with $n \to \infty$, which is the space consisting of all infinite sequences of real numbers. Kolmogorov's theorem guarantees that P_x exists uniquely and consistently for $n \to \infty$; see Billingsley (1968, p. 228), Breiman (1992, Cor. 2.19), or Davidson (1994, Theorem. 12.4).

To become precise, we define a univariate real-valued discrete-time process **x** for an index set $\mathbb{T}$ of time indices, where *discrete time* means that the index set contains integers only, $\mathbb{T} \subseteq \mathbb{N}$ or $\mathbb{T} \subseteq \mathbb{Z}$. For finite $\mathbb{T} = \{1, \ldots, n\}$, the stochastic process **x** is a mapping from the underlying probability space $(\Omega, \mathcal{F}, P)$ to $(\mathbb{R}^n, \mathcal{B}^n, P_x)$; for infinite, countable $\mathbb{T} = \mathbb{N}$ or $\mathbb{T} = \mathbb{Z}$, we let $n \to \infty$:

$$\mathbf{x} : \Omega \to \mathbb{R}^n \quad \text{or} \quad \mathbf{x} : \Omega \to \mathbb{R}^\infty.$$

Alternatively, one may consider the sequence of univariate random variables $\{x_t\}_{t \in \mathbb{T}}$ mapping from Ω for each $t \in \mathbb{T}$:

$$x_t : \mathbb{T} \times \Omega \to \mathbb{R}$$
$$(t; \, \omega) \mapsto x_t(\omega).$$

By Breiman (1992, Proposition 2.13), this constitutes a jointly measurable mapping from $(\Omega, \mathcal{F})$ to $(\mathbb{R}^\infty, \mathcal{B}^\infty)$, and $\{x_t\}_{t \in \mathbb{T}}$ is tantamount to **x**.

A process is said to be strictly stationary if the distribution of any n-vector $(x_{t_1}, \ldots, x_{t_n})'$ is, for any indices $t_1 < t_2 < \cdots < t_n$, invariant over time. This means that a shift from $t_1 < t_2 < \cdots < t_n$ to $t_1 + h < t_2 + h < \cdots < t_n + h$ leaves the joint distribution unaltered. A stochastic process is called covariance stationary if the expected value and the variance of x_t are constant over time and if the covariance between x_t and x_{t+h} (called autocovariance) depends on the time lag h only, where we assume these moments to exist. Then we write

1. $E(x_t) = \mu_x = \mu$ for $t \in \mathbb{T}$,
2. $\text{Cov}(x_t, x_{t+h}) = E[(x_t - E(x_t))\,(x_{t+h} - E(x_{t+h}))] = \gamma_x(h) = \gamma(h)$ for all $t, t + h \in \mathbb{T}$,

such that $\gamma_x(0)$ is short for the variance $\text{Var}(x_t)$. Sometimes we assume implicitly that $h \geq 0$ since $\gamma_x(h) = \gamma_x(-h)$. Often we will suppress the index x and simply write μ and $\gamma(h)$ if there is no risk of confusion. Clearly, strict stationarity implies covariance stationarity as long as we assume $E(x_t^2) < \infty$. In the sequel, by *stationarity*, we mean covariance stationarity unless stated otherwise, and we will always assume a finite variance, $E(x_t^2) < \infty$, implying $\gamma_x(h) < \infty$.

Definition 2.1 (*Stationarity*) *A stochastic process $\{x_t\}_{t \in \mathbb{T}}$ with $E(x_t^2) < \infty$ is called (covariance) stationary if it holds that $E(x_t) = \mu$ and $\text{Cov}(x_t, x_{t+h}) =$*

$\gamma(h)$ *for all* $t, t + h \in \mathbb{T}$. *It is called strictly stationary if the joint distribution of* $(x_{t_1}, \ldots, x_{t_n})'$ *equals that of* $(x_{t_1+h}, \ldots, x_{t_n+h})'$ *for all n and h and* $t_1 < t_2 < \cdots < t_n$.

In general, the property of stationarity will depend on the choice of the index set. This is illustrated for the example of an autoregressive process of order 1 in Hassler (2016, p. 57). Under stationarity one may define the autocorrelations:

$$\rho_x(h) = \frac{\gamma_x(h)}{\gamma_x(0)}, \quad h \in \mathbb{Z}.$$

The autocorrelations (or temporal correlation coefficients) at lag h reflect the linear dependence between random variables being h time periods apart. Time series analysis is all about the detection and modeling of temporal dependence and in particular temporal correlation (or serial correlation). Nevertheless, stochastic processes as models for time series typically build on a sequence of innovations that is free of serial correlation, here denoted by $\{\varepsilon_t\}$:

$$\text{Cov}(\varepsilon_t, \varepsilon_{t+h}) = 0, \quad h \neq 0. \tag{2.1}$$

Further, we always assume that the sequence of innovations or shocks is not systematic in that the expected value is 0:

$$\text{E}(\varepsilon_t) = 0.$$

The variance may be time dependent, typically denoted by $\sigma_t^2 = \text{Var}(\varepsilon_t)$.

Now, we introduce three sets of assumptions on $\{\varepsilon_t\}$ that we will work with in the sequel. First, if a sequence of innovations is (covariance) stationary, $\sigma_t^2 = \sigma^2$, then we call $\{\varepsilon_t\}$ white noise, abbreviated as

$$\{\varepsilon_t\} \sim \text{WN}(0, \sigma^2).$$

Second, $\{\varepsilon_t\}$ is called a martingale difference sequence (MDS) if the *conditional* expectation (given its own past) is 0:[1]

$$\text{E}(\varepsilon_t | \varepsilon_{t-1}, \varepsilon_{t-2}, \ldots) = 0. \tag{2.2}$$

A consequence of this defining property is that MDS are free of serial correlation, such that Eq. (2.1) holds; see, e.g. Davidson (1994, Theorem 15.3). Hence, a stationary MDS ($\sigma_t^2 = \sigma^2$) meets the definition of white noise. This does not rule out temporal dependence as it is, e.g. captured by the class of autoregressive conditional heteroskedasticity (ARCH) models pioneered by Engle (1982). Third, we may assume that the innovations of the covariance stationary white noise process are not only uncorrelated but even independent over time; if we further assume an identical distribution over time, then $\{\varepsilon_t\}$ is called a pure

---

1 On top we assume integrability in that $\text{E}(|\varepsilon_t|) < \infty$. The conditional expectation is understood as conditioned on the σ-field generated by the past $\varepsilon_{t-1}, \varepsilon_{t-2}, \ldots$; see, e.g. Breiman (1992, Chapter 4) or Davidson (1994, Chapter 10) for details.

random process, or also an iid process (independent identically distributed), for short:

$$\{\varepsilon_t\} \sim \text{iid}(0, \sigma^2).$$

Note that the joint distribution of an iid vector $(\varepsilon_{t_1}, \ldots, \varepsilon_{t_n})'$ is given by the product of the identical distributions, which hence is constant by assumption. Therefore, the pure random process is not only covariance stationary but also strictly stationary.

Clearly, every iid process satisfies the MDS condition Eq. (2.2), while the converse does not hold true. And every stationary MDS is white noise, while again the converse does not hold. The last claim is illustrated by an example.

Example 2.1 *White noise, not MDS* Define the process

$$\varepsilon_t = \eta_t + \eta_{t-1}^2 - 1,$$

where $\{\eta_t\}$ is a standard normal iid sequence: $\eta_t \sim \text{ii}\mathcal{N}(0, 1)$. It is straightforward to verify that

$$\text{E}(\varepsilon_t) = 0, \quad \text{Var}(\varepsilon_t) = 3, \quad \text{and} \quad \text{Cov}(\varepsilon_t, \varepsilon_{t-j}) = 0, \ j \neq 0.$$

Hence, the process is white noise. Next, we apply the so-called law of iterated expectations (see, e.g. Breiman (1992, Proposition 4.20) or Davidson (1994, Theorem 10.26)),

$$\text{E}\left(\varepsilon_t \varepsilon_{t-1}^2\right) = \text{E}\left(\text{E}\left(\varepsilon_t \varepsilon_{t-1}^2 \mid \varepsilon_{t-1}\right)\right),$$

such that

$$\text{E}\left(\varepsilon_t \varepsilon_{t-1}^2\right) = \text{E}\left(\varepsilon_{t-1}^2 \text{E}\left(\varepsilon_t \mid \varepsilon_{t-1}\right)\right);$$

if $\{\varepsilon_t\}$ was an MDS, this last term would be zero because $\text{E}(\varepsilon_t \mid \varepsilon_{t-1}) = 0$; see, e.g. Hassler (2016, Proposition 2.2). It is elementary, however, to see that $\text{E}(\varepsilon_t \varepsilon_{t-1}^2) = 2$, such that this white noise does not meet the MDS condition.

2.2 Ergodicity

Time series analysis is fundamentally different from other fields of statistics. Why is that? Imagine you wish to forecast today (at time T) the interest rate tomorrow (time $T + 1$). The interest rate tomorrow (say, x_{T+1}) is a random variable from today's perspective with unknown expectation μ_{T+1} and variance. How can we estimate tomorrow's expectation? Let us assume that the mean and variance tomorrow are the same as today and yesterday, which is implied by stationarity: $\mu_{T+1} = \mu$. Further, assume for simplicity that the interest rate is reported with limited precision and can take on only N possible values r_i with probabilities p_i:

$$P(x_{T+1} = r_i) = p_i.$$

The set of possible outcomes (in %) may be, e.g. with $N = 20$,

$$R = \{r_1, r_2, r_3, \ldots, r_{20}\} = \{1.1, 1.2, 1.3, \ldots, 3.0\}.$$

The unknown mean or expectation is then defined as

$$\mu = E(x_{T+1}) = \sum_{i=1}^{N} r_i p_i.$$

Usually, we estimate means (and variances) by averaging over (squared) observations. With time series we have two problems. First and trivially, future observations are not available at time T. Second and more delicately, even tomorrow, at time $T + 1$, we will observe only *one* realization out of the set R of all possible outcomes. So, polemically, how can we possibly estimate today tomorrow's μ with one observation only that we don't even know!

The only thing that we can do is to estimate the population mean μ (which is constant by the stationarity assumption) by the time average over T variables $x_1, \ldots, x_T$, where T denotes the sample size for the rest of the book:

$$\bar{x} = \frac{1}{T} \sum_{t=1}^{T} x_t.$$

Under what circumstances will $\bar{x}$ converge to μ? The issue is related to *ergodicity* in the analysis of stochastic processes. Ergodicity implies that the sample average is consistent for the population mean in that $\bar{x}$ converges to μ:

$$\bar{x} \to \mu \quad \text{as } T \to \infty. \tag{2.3}$$

Such results are also labeled *law of large numbers* (LLN), where we distinguish different versions depending on the type of convergence in Eq. (2.3). We recommend Pötscher and Prucha (2001) and White (2001, Chapter 2) for a concise review of modes of convergence. One says $\bar{x}$ converges to μ almost surely (or with probability 1) if the following holds. Let $C \subseteq \Omega$ be such that the average converges in the classical sense for every element in C:

$$\frac{1}{T} \sum_{t=1}^{T} x_t(\omega) \to \mu \quad \text{for all } \omega \in C.$$

If $P(C) = 1$, then convergence is said to be almost sure. The shorthand notation is

$$\bar{x} \overset{a.s.}{\to} \mu \quad \text{as } T \to \infty.$$

Such a result has been called strong LLN, because it implies the weaker convergence in probability (weak LLN),

$$\bar{x} \overset{p}{\to} \mu \quad \text{as } T \to \infty,$$

which signifies

$$\lim_{T \to \infty} P\left(|\bar{x} - \mu| > \epsilon\right) = 0$$

for arbitrary $\epsilon > 0$. A third mode is convergence in mean square (or quadratic mean), where the mean squared error, $E((\bar{x} - \mu)^2)$, vanishes asymptotically:

$$E\left((\bar{x} - \mu)^2\right) \to 0 \quad \text{as } T \to \infty. \tag{2.4}$$

Symbolically, we write

$$\bar{x} \overset{2}{\to} \mu \quad \text{as } T \to \infty$$

for mean square convergence and note that it implies convergence in probability, too. Note, however, that mean square convergence does not imply and is not implied by almost sure convergence.

When speaking of an LLN, we typically understand μ to be the expectation of a stationary sequence. More generally, however, the limit of convergence (almost sure, in probability or in mean square) does not have to be a constant. The above definitions equally hold when replacing μ by some random variable. In fact, this is exactly the case in the following theorem. Before we turn to consistency and ergodicity in greater detail, we pin down what is sometimes called the ergodic theorem.

Theorem 2.1 (Ergodic Theorem) *Let $\{x_t\}$ be a strictly stationary process with* $E(|x_t|) < \infty$. *Then*

$$\bar{x} \overset{a.s.}{\to} \mathcal{E} \quad \text{as } T \to \infty,$$

where $\mathcal{E}$ is a random variable with $E(\mathcal{E}) = \mu = E(x_t)$, which is defined in Doob (1953, Theorem 2.1, p. 465); see also Stout (1974, Theorem 3.5.7) or Breiman (1992, Theorem 6.28).

Proof: Doob (1953), Stout (1974), or Breiman (1992); see also Davidson (1994, p. 198, 199). □

In general, the limit $\mathcal{E}$ is a random variable defined as conditional expectation. We do not need details here except for $E(\mathcal{E}) = \mu$. The shortcoming of the ergodic theorem lies in the fact that $\mathcal{E}$ does not have to equal $\mu = E(x_t)$. Ergodicity is a delicate concept that assumes more than $\mathcal{E} = \mu$, although this is one implication given in Theorem 2.2. The general concept of ergodicity requires some excursion into the probability theory behind a stochastic process

$\mathbf{x} = \{x_t\}_{t\in\mathbb{T}}, \mathbf{x} : \ \Omega \to \mathbb{R}^\infty$, and its so-called coordinate representation denoted by $\hat{\mathbf{x}} = \{\hat{x}_t\}_{t\in\mathbb{T}}$. The latter does not map from $(\Omega, \ \mathcal{F})$ to $(\mathbb{R}^\infty, \ \mathcal{B}^\infty)$, but rather from $(\mathbb{R}^\infty, \ \mathcal{B}^\infty)$ to $(\mathbb{R}^\infty, \ \mathcal{B}^\infty)$. It does so by projecting the tth coordinate of $\mathbf{x}$ from $\mathbb{R}^\infty$ to $\mathbb{R}$ for all $t \in \mathbb{T}$: $\hat{x}_t(\mathbf{x}) = x_t$. Admittedly, the difference between $\hat{\mathbf{x}}$ and $\mathbf{x}$ is only subtle, and not surprisingly, the coordinate representation $\hat{\mathbf{x}}$ has the same distribution P_x as $\mathbf{x}$; see Breiman (1992, p. 22). To illustrate the difference between $\hat{\mathbf{x}}$ and $\mathbf{x}$, the following example adopted from Breiman (1992) may help. Let $\mathbf{x}$ on $(\Omega, \ \mathcal{F}, \ \mathrm{P})$ be the process where a fair coin is tossed independently, and heads or tails are mapped to zero or one, respectively; further, let $\mathbf{x}^*$ on $(\Omega^*, \ \mathcal{F}^*, \ \mathrm{P}^*)$ be the process where a fair die is cast independently, and even or odd outcomes are mapped to zero or one, respectively. With obvious definitions of $(\Omega, \ \mathcal{F}, \ \mathrm{P})$ and $(\Omega^*, \ \mathcal{F}^*, \ \mathrm{P}^*)$, we have two different processes. However, if $\mathbf{x}$ and $\mathbf{x}^*$ are performed in separate rooms behind closed doors and if only the numbers of $\mathbf{x}$ and $\mathbf{x}^*$ are reported, one cannot distinguish between the two processes. They are observationally equivalent, or in other words, they have the same coordinate representation. From a distributional point of view, one may hence equate a stochastic process $\mathbf{x}$ with the coordinate representation process $\hat{\mathbf{x}}$.

Next, we define a shift transformation τ from $(\mathbb{R}^\infty, \ \mathcal{B}^\infty)$ onto $(\mathbb{R}^\infty, \ \mathcal{B}^\infty)$. As the name suggests, τ maps $(\ldots, x_1, x_2, \ldots)$ to $(\ldots, x_2, x_3, \ldots)$, and this mapping is measurable according to Breiman (1992, Proposition 6.11). It provides an alternative way to write the coordinate representation of a stationary process,

$$\hat{x}_t(\mathbf{x}) = \hat{x}_{t-1}(\tau \ \mathbf{x}) = \hat{x}_{t-2}(\tau^2 \ \mathbf{x}),$$

where τ^t denotes the iterated mapping, e.g. τ^2 maps $(\ldots, x_1, x_2, \ldots)$ to $(\ldots, x_3, x_4, \ldots)$. Consequently,

$$\hat{x}_t(\mathbf{x}) = \hat{x}_1(\tau^{t-1} \ \mathbf{x}).$$

Moreover, we define for $(\Omega, \ \mathcal{F})$ underlying the stationary process $\mathbf{x}$ an invariant event. $I \in \mathcal{F}$ is called invariant if there exists a subset B of $\mathbb{R}^\infty$, $B \in \mathcal{B}^\infty$, such that any shift of the process stays in B, i.e.

$$I = \{\omega| \ \tau^t \ \mathbf{x}(\omega) \in B\} \quad \text{for all } t \in \mathbb{T}.$$

Trivial examples of invariant events are Ω with $B = \mathbb{R}^\infty$ and the empty set. If all invariant events have either probability zero or one, then the strictly stationary process $\mathbf{x}$ is called ergodic (see Breiman (1992, Definition 6.30)), while White (2001, Definition 3.33) used a different but equivalent definition (see also Davidson (1994, Theorem 13.13)). Note that Doob (1953, p. 457) called an ergodic process *metrically transitive*. Whenever we maintain ergodicity in the sense of Definition 2.2, we assume (implicitly) strict stationarity.

Definition 2.2 (*Ergodicity*)   *A strictly stationary process* $\mathbf{x} = \{x_t\}_{t\in\mathbb{T}}$ *with* $\mathbb{T} = \mathbb{N}$ *or* $\mathbb{T} = \mathbb{Z}$ *is called ergodic if every invariant event has probability zero or one.*

There is an intuitive understanding how ergodicity ensures convergence of $\bar{x}$ to $\mu = E(x_t)$. Let $\mathbf{x}(\Omega)$ denote the range of $\mathbf{x}$, the sample space of $\{x_t\}$. Assume that there exists an invariant event I with $0 < P(I) < 1$ and image B, and let $\omega \in I$. Consequently, $\tau \, \mathbf{x}(\omega) \in B$ (at least almost surely; see Breiman (1992, Proposition 6.15)), and any iteration of τ does not leave B with probability one. Consequently, the coordinate process $\{\hat{x}_t(\mathbf{x})\}$ with $\hat{x}_t(\mathbf{x}) = \hat{x}_{t-1}(\tau \, \mathbf{x})$ will never leave the subspace given by the image $\mathbf{x}(I)$ and will never visit certain regions of $\mathbf{x}(\Omega)$. Since $\{\hat{x}_t(\mathbf{x})\}$ does not have access to the total information contained in $\mathbf{x}(\Omega)$, it is clear that the sample average over the coordinate process may not converge to the population mean $\mu = E(x_t)$. Since $\mathbf{x}$ and the coordinate process have the same distribution, the same holds true for $\bar{x}(\omega) = T^{-1} \sum_{t=1}^{T} x_t(\omega)$. And this is exactly what is ruled out under ergodicity where no invariant event exists other than the trivial ones, Ω, and the empty set, such that the (coordinate) process will eventually visit all regions of the sample space $\mathbf{x}(\Omega)$. The following theorem becomes more precise than this intuition.

Theorem 2.2 (*Ergodicity*)   *Let $\{x_t\}$ be a strictly stationary, ergodic process in the sense of Definition 2.2 with $E(|x_t|) < \infty$.*

(a) Then

$$\bar{x} \overset{a.s.}{\to} E(x_t) \quad as \ T \to \infty.$$

(b) If f is a measurable mapping, $\mathbb{R}^{k+\ell+1} \to \mathbb{R}$, then the process $\{f(x_{t-k}, \ldots, x_t, \ldots, x_{t+\ell})\}$ is strictly stationary and ergodic, too.

Proof: Doob (1953, Theorem 2.1, p. 465) for (a) and Doob (1953, p. 458) for (b); see also Stout (1974, Theorem 3.5.8) or Breiman (1992, Proposition 6.31).   □

Note that Breiman (1992) and Stout (1974) considered only $\mathbb{T} = \mathbb{N}$, while Doob (1953, p. 465) argued that one may replace $\mathbb{N}$ by $\mathbb{Z}$ and further that

$$\frac{1}{2T+1} \sum_{t=-T}^{T} x_t$$

has the same limit as $\bar{x} = T^{-1} \sum_{t=1}^{T} x_t$.

Theorem 2.2 has interesting implications. First, it guarantees for stationary, ergodic processes that the sample average is consistent for the population mean, where convergence is with probability 1 (almost surely). Second, it implies that the same holds true for any appropriate transformation of the process. For $k = \ell = 0$, e.g.

$$\frac{1}{T} \sum_{t=1}^{T} f(x_t) \overset{a.s.}{\to} E(f(x_t)) \quad as \ T \to \infty,$$

for all measurable f with $E(|f(x_t)|) < \infty$. With $f(x) = x^m$, this guarantees the consistent estimation of higher moments, provided they exist. Third, Theorem 2.2 implies consistent autocovariance estimation by choosing $k = 0$ and $f(x_t, x_{t+\ell}) = x_t x_{t+\ell}$:

$$\frac{1}{T} \sum_{t=1}^{T-\ell} x_t x_{t+\ell} \overset{a.s.}{\to} E(x_t x_{t+\ell}) \quad \text{as } T \to \infty.$$

To the best of our knowledge, the properties covered by Theorem 2.2 are not equivalent to Definition 2.2, but simply an implication. Still, this theorem provides a simple and understandable characterization of what ergodicity means. Hence, we state for the sake of simplification and clarity the following not fully rigorous definition.

Characterization of Ergodicity: *A strictly stationary, ergodic process* $\{x_t\}$ *is characterized by*

$$\frac{1}{T} \sum_{t=1}^{T} f(x_{t-k}, \dots, x_t, \dots, x_{t+\ell}) \overset{a.s.}{\to} E(f(x_{t-k}, \dots, x_t, \dots, x_{t+\ell}))$$

as $T \to \infty$ *for all f with* $E(|f(x_{t-k}, \dots, x_t, \dots, x_{t+\ell})|) < \infty$ *from Theorem* 2.2.

There is a special case of important processes that satisfy Definition 2.2, namely, independent and identically distributed sequences.

Proposition 2.1 *Let* $\{\varepsilon_t\}$ *be iid. Then this process is ergodic in the sense of Definition 2.2.*

Proof: Doob (1953, Theorem 1.2, p. 460) or Breiman (1992, Cor. 6.33). □

From that if follows of course that

$$\frac{1}{T} \sum_{t=1}^{T} \varepsilon_t \overset{a.s.}{\to} 0 \quad \text{and} \quad \frac{1}{T} \sum_{t=1}^{T} \varepsilon_t^2 \overset{a.s.}{\to} \sigma^2, \tag{2.5}$$

if $\{\varepsilon_t\}$ is iid, because then $\{\varepsilon_t^2\}$ is iid as well.

It is worth noting that k and ℓ in Theorem 2.2 may equal infinity. Hence, if we have an infinite sequence $\{c_j\}$ such that $\sum_{j=0}^{\infty} c_j \varepsilon_{t-j}$ constructed from an iid process $\{\varepsilon_t\}$ is strictly stationary, then the infinite sum $\sum_{j=0}^{\infty} c_j \varepsilon_{t-j}$ is ergodic, too; see Proposition 3.3.

Another important class of strictly stationary and ergodic processes was characterized by Nelson (1990). Consider the so-called GARCH(1,1) model, which is a special case of more general GARCH processes suggested by Bollerslev (1986) as a generalization of ARCH processes by Engle (1982). Let $\{\eta_t\}$ be iid with unit variance, and define

$$\varepsilon_t = \sigma_t \eta_t \quad \text{with} \quad \sigma_t^2 = \omega + \alpha \varepsilon_{t-1}^2 + \beta \sigma_{t-1}^2,$$

where $\omega > 0$. Then $\{\varepsilon_t\}$ is called a GARCH(1,1) process, and it is well known that it forms an MDS. Nelson (1990, Theorem 2) further showed that the condition

$$E(\ln(\beta + \alpha\eta_t^2)) < 0$$

is sufficient for strict stationarity and ergodicity of this process.[2]

Some econometric books claim that ergodicity is a form of asymptotic independence (see Hayashi (2000, p. 101)) and similarly in Davidson and MacKinnon (1993, Definition 4.11). The following example shows that this is not correct.

Example 2.2 *Ergodic, not asymptotically independent* Consider the alternating process given by $x_t = (-1)^t C$, where C is a random variable taking on the values -1 and 1 with equal probability. This produces a sequence of alternating sign where the starting value is drawn by tossing a fair coin:

$$P(C = 1) = P(C = -1) = \frac{1}{2}$$

with

$$E(C) = 0 \quad \text{and} \quad \text{Var}(C) = 1.$$

Clearly, $E(x_t) = 0$ and $\text{Var}(x_t) = 1$, such that

$$|E(x_t x_{t+\ell+n})| = E(C^2) = 1$$

differs for all n and for $n \to \infty$ from

$$|E(x_t)||E(x_{t+\ell+n})| = 0.$$

Hence, the process is not asymptotically independent. However, adopting the arguments by Davidson (1994, Example 13.15), one establishes that this strictly stationary sequence is ergodic in the sense of Definition 2.2.

In addition to the mathematical difficulties with the general concept of ergodicity, a drawback of Theorem 2.2 is the assumption of strict stationarity, which is hard to verify in practice, except for the cases like iid or GARCH(1,1) discussed above. Still, consistency in some sense of Eq. (2.3) is a minimum requirement for practical time series analysis, where one typically observes only *one* realization of a stochastic process. Many time series books hence follow the tradition of considering convergence of the arithmetic mean in mean square without caring about the more involved ergodicity. Following Hamilton (1994, p. 47) or Fuller (1996, p. 308), we call this *ergodic for the mean*.

---

2 This even continues to hold for $\alpha + \beta = 1$, which is the special case of a so-called integrated GARCH process, where no constant variance exists; see Engle and Bollerslev (1986).

Definition 2.3 (*Ergodic for the mean*)   *A covariance stationary process with* $E(|x_t|) < \infty$ *is called ergodic for the mean when the sample average converges to* $\mu = E(x_t)$ *in mean square; see Eq. (2.4).*

Hamilton (1994, Proposition 7.5), e.g. proved Eq. (2.4) assuming an absolutely summable autocovariance sequence. Brockwell and Davis (1991, Theorem 7.1.1) and Fuller (1996, Cor. 6.1.1.1) established a weaker sufficient condition: If the autocovariances converge to 0, then Eq. (2.4) holds:

$$[\gamma(h) \xrightarrow{h \to \infty} 0] \;\Rightarrow\; [\bar{x} \xrightarrow{2} \mu]. \tag{2.6}$$

An even weaker condition is given in the next proposition.[3] It only requires that the average over the autocovariances converges to 0. This sufficient condition cannot be weakened, since it is also necessary. Hence, a necessary and sufficient condition for convergence in the mean square is average asymptotic uncorrelatedness; see Eq. (2.7). From now on, *stationary* stands again for covariance stationary.

Proposition 2.2   *Let* $\{x_t\}$ *be a stationary process with expectation* μ *and autocovariance sequence* $\{\gamma(h)\}$. *The sample average converges in mean square to* μ, *i.e. Eq. (2.4) holds true, if and only if*

$$\frac{1}{H} \sum_{h=1}^{H} \gamma(h) \;\to\; 0 \tag{2.7}$$

as $H \to \infty$.

Proof: See Appendix to this chapter.

We want to close this section by highlighting some relations between stationarity and ergodicity for the mean to shed additional light on the two concepts and their relations:

1. Strict stationarity and ergodicity for the mean do not imply ergodicity in the sense of Definition 2.2; a counterexample is given in Hassler (2017, Example 1).
2. Strict stationarity and ergodicity in the sense of Definition 2.2 do not imply asymptotic independence; see Example 2.2.
3. Covariance stationarity and ergodicity for the mean do not imply asymptotic uncorrelatedness; a counterexample is given in Hassler (2017, Example 2).
4. (Covariance) Stationarity does not imply ergodicity (for the mean); see Hassler (2017, Example 3) for a counterexample.

---

3 It follows from Lemma 3.2 (*a*) that Eq. (2.7) is implied by $\gamma(h) \to 0$.

5. By Davidson (1994, Cor. 13.14), strict stationarity and ergodicity in the sense of Definition 2.2 imply Eq. (2.7) and hence ergodicity for the mean by Proposition 2.2.

We will come back to the issue of ergodicity (for the mean) in Proposition 3.3 (and 3.2).

2.3 Memory and Persistence

Next, we want to address a stronger result than convergence of the sample mean. It is on the variance of the sample mean under the more restrictive assumptions that the autocovariances are absolutely summable. We label such processes as *short memory*.

Definition 2.4 (*Short and long memory*)   *A stationary process $\{x_t\}$ is said to display long memory if the sequence of autocovariances dies out so slowly that it is not absolutely summable:*

$$\sum_{h=0}^{H} |\gamma(h)| \to \infty \quad as \ H \to \infty. \tag{2.8}$$

Otherwise it is said to have short memory:

$$\sum_{h=0}^{\infty} |\gamma(h)| < \infty. \tag{2.9}$$

Our definition of long memory coincides with the one by Giraitis et al. (2012, Definition 3.1.2). While long memory is defined in terms of *absolute* autocovariances, it will turn out in Proposition 2.3 that it makes sense to consider the summation over $\{\gamma(h)\}$, too. For convenience, we define the long-run variance ω^2 of a stationary process as

$$0 \le \omega^2 := \sum_{h=-\infty}^{\infty} \gamma(h), \tag{2.10}$$

provided this quantity exists finitely. Note that $0 \le \omega^2$ is not an assumption but follows from Brockwell and Davis (1991, Theorem 1.5.1) or Fuller (1996, Theorem 1.4.1). The long-run variance is necessarily finite under short memory; in Section 3.1 we will further elaborate on summability. The special case where $\omega^2 = 0$ will be treated in greater detail in Section 3.3.

Our distinction between long and short memory is motivated by the following result.

Proposition 2.3 *Let $\{x_t\}$ be a stationary process with short memory in the sense of Definition 2.4. It then holds that*

$$\lim_{T \to \infty} T \ Var(\bar{x}) = \omega^2$$

with ω^2 from Eq. (2.10).

Proof: Fuller (1996, Cor. 6.1.1.2) or Brockwell and Davis (1991, Theorem 7.1.1).

□

Proposition 2.3 suggests a further distinction on top of short or long memory. To that end, we approximate the variance of the sample mean of $x_1, \ldots, x_T$ as follows:

$$Var(\bar{x}) \approx \frac{\omega^2}{T}.$$

We distinguish three cases: (a) If $0 < \omega^2 < \infty$, this means that $Var(\bar{x})$ is of order $1/T$, i.e. this rate $1/T$ characterizes how quickly the sample mean converges. (b) If $\omega^2 = 0$, then $Var(\bar{x})$ is of a smaller order than $1/T$, i.e. the sample mean converges at a faster rate; this case arises if the data oscillate more closely around the true population mean μ than in the first case. (c) The long-run variance may not exist finitely, which means that $Var(\bar{x})$ vanishes to 0 more slowly than $1/T$. In this last case the sample mean converges more slowly than in the two previous cases. Those three cases characterize three different degrees of what we call persistence. In case (c) the process is driven by what one may call local but reversing trends: It has a constant mean, but this is hard to estimate ($Var(\bar{x})$ converges slowly) because there are long spells where $\{x_t\}$ deviates from its expectation μ. In case (a) the process fluctuates around its expected value only moderately and returns to μ quite regularly, such that $Var(\bar{x})$ converges at the conventional rate $1/T$. In case (b), which we call *antipersistence*, the process crosses its expectation μ more often, oscillating around μ, such that the estimation of μ is most precise. Those three types of behavior can be observed in Figure 2.1.[4] We define the following terms to characterize such features; see also the definition of antipersistence in Giraitis et al. (2012, Definition 3.1.2).

Definition 2.5 (*Persistence*) *A stationary process $\{x_t\}$ is said to be*

(a) *moderately persistent if the long-run variance from Eq. (2.10) exists and $0 < \omega^2 < \infty$,*
(b) *antipersistent if the long-run variance from Eq. (2.10) exists and $\omega^2 = 0$,*

---

4 The processes are simulated as (a) AR(1) with parameter $\phi = 0.5$, as (b) fractionally integrated noise of order $d = -0.4$, and as (c) fractionally integrated noise of order $d = 0.45$; further details on these models will be given in the subsequent chapters. Throughout, the innovational variance is $Var(\varepsilon_t) = 1$.

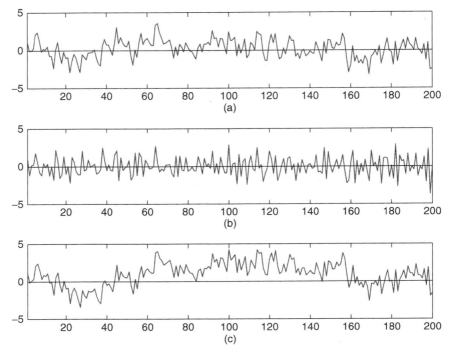

Figure 2.1 Processes with (top to bottom) (a) moderate persistence, (b) antipersistence, and (c) strong persistence.

(c) *strongly persistent if the long-run variance does not exist and* $\sum_{h=-H}^{H} \gamma_x(h)$
diverges with H:

$$\sum_{h=-H}^{H} \gamma_x(h) \to \infty \quad \text{as } H \to \infty.$$

Antipersistence is ruled out by assumption in conventional time series analysis: $\omega^2 > 0$. At the same time, traditional textbooks assume short memory processes. A short memory process with $\omega^2 > 0$ is a requirement of what is often called *integrated of order* 0, $I(0)$, in econometrics; for a general definition of integration, see Chapter 6.

Clearly, persistence and memory in the sense of Definitions 2.4 and 2.5 are related but not equivalent. While memory is defined in absolute terms, $|\gamma(h)|$, the long-run variance is not. Examples of different combinations of memory and persistence will be discussed in Section 3.3, where an additional interpretation of ω^2 will be given. Note that Giraitis et al. (2012, Definition 3.1.2) define antipersistence only for the case of short memory. In Section 3.3, however, we will see an example of an antipersistent process displaying long memory in the sense of our definitions.

Note that certain processes do not fit into Definition 2.5. Consider again Example 2.2 with $\gamma(h) = (-1)^h$. In this case, $\sum_{h=-H}^{H} \gamma(h)$ does not converge, i.e. ω^2 does not exist although $\sum_{h=-H}^{H} \gamma(h)$ does not diverge in the sense of strong persistence. But this is a rather special example.

Finally, it should be noted that the long-run variance is a traditional measure of persistence in the moderate case where $0 < \omega^2 < \infty$. In particular, Cochrane (1988) or Cogley and Sargent (2005) considered the variance ratio of long-run variance and variance,

$$\text{VR} := \frac{\omega^2}{\gamma(0)},$$

which has more recently been advocated again as measure of persistence by Hassler (2014). In the case of a white noise process (no persistence), it holds that $\text{VR} = 1$. By persistence, we understand the feature that subsequent observations form clusters: Observations above average tend to be followed by more data above average, and the other way around one expects clusters of data below average. The larger the persistence measure $\text{VR} > 1$, the stronger is such a tendency, which depends not only on the autocorrelation coefficient at lag 1 but also on higher order lags. In the case of negative persistence ($\text{VR} < 1$), observations above average tend to be followed by values below average, thus generating a kind of zigzag behavior of the time series. Antipersistence is an extreme form of negative persistence ($\text{VR} = 0$).

2.4 Technical Appendix: Proofs

Proof of Proposition 2.2

The result is given in Woodward et al. (2017, Theorem 1.2) with a reference to Yaglom (1962). Although a recent proof is contained in Hassler (2017), we present a slightly shorter version here. It will rely on an auxiliary result that may be useful in other cases, too. According to Knopp (1990) it is attributed to Leopold Kronecker. Further details on summability and infinite series are discussed in the introductory section of the next chapter.

Lemma 2.1 (*Kronecker*)   *Assume that the real-valued sequence $\{a_n\}$, $n \in \mathbb{N}$, is summable, i.e. that $\sum_{n=1}^{N} a_n$ converges as $N \to \infty$. Further, let $\{b_n\}$ be positive and monotonically increasing with $1/b_N \to 0$. It then holds that*

$$\frac{1}{b_N} \sum_{n=1}^{N} a_n b_n \ \to \ 0$$

as $N \to \infty$.

Proof: Knopp (1990, Theorem 3, p. 129). □

An obvious implication of this lemma is that

$$\frac{1}{N} \sum_{n=1}^{N} a_n n \to 0 \quad \text{as } N \to \infty. \tag{2.11}$$

Since $E(\overline{x}) = \mu$, the mean squared error equals the variance, and

$$\text{Var}(\overline{x}) = \frac{1}{T^2} \sum_{t=1}^{T} \sum_{s=1}^{T} \text{Cov}(x_t, x_s) = \frac{1}{T^2} \sum_{h=-T+1}^{T-1} (T - |h|)\gamma(h)$$

$$= \frac{1}{T} \sum_{h=-T+1}^{T-1} \left(1 - \frac{|h|}{T}\right) \gamma(h). \tag{2.12}$$

Hence, $\text{Var}(\overline{x}) \to 0$ if and only if

$$\sum_{h=1}^{T} \left(1 - \frac{h}{T}\right) \frac{\gamma(h)}{T} \to 0. \tag{2.13}$$

Consequently, convergence in mean square, Eq. (2.4), is equivalent to Eq. (2.13).
First, we show that Eq. (2.7) is sufficient. Note that Eq. (2.7) implies by Eq. (2.11) that

$$\sum_{h=1}^{T} \frac{h}{T} \frac{\gamma(h)}{T} \to 0,$$

which immediately implies Eq. (2.13) as required. Second, we show that Eq. (2.7) is necessary. From Hassler (2017, Eq. (8)) we learn that under covariance stationarity

$$\frac{1}{T} \sum_{h=1}^{T} \gamma(h) \to c, \tag{2.14}$$

where c is some constant (defined in terms of the so-called spectral distribution function). Again, by Eq. (2.11) this implies

$$\sum_{h=1}^{T} \left(1 - \frac{h}{T}\right) \frac{\gamma(h)}{T} \to c - 0 = c.$$

By Eq. (2.13) we conclude that $c = 0$, which implies Eq. (2.7) by Eq. (2.14) as required. Hence, mean square convergence implies via Eq. (2.13) that Eq. (2.7) is necessary for mean square convergence, and the proof is complete. □

3

Moving Averages and Linear Processes

By assumption, the fundamental time series model underlying this book is the moving average (MA) process of infinite order. To set the stage for this, we first review some useful results on infinite series. Further, we define invertibility as the property that an MA process can be expressed as an autoregression (AR) of infinite order. In a separate section we discuss examples that illustrate that (long) memory and (strong) persistence are related but not necessarily equivalent. Section 3.4 turns to the well-known autoregressive moving average (ARMA) processes; some interesting properties are added in Proposition 3.5.

3.1 Infinite Series and Summability

Any stationary process $\{x_t\}$ with expectation 0 can be decomposed into two random components (see Theorem 3.2):

$$x_t = \sum_{j=0}^{\infty} c_j \varepsilon_{t-j} + \delta_t, \quad E(x_t) = 0,$$

where $\{c_j\}$ is a deterministic sequence. The first component is an infinite sum of weighted past white noise shocks. The second one, δ_t, will be discussed in Section 3.2. To avoid problems with ergodicity, it will be suggested to assume $\delta_t = 0$, and this assumption is maintained for the next chapters. Hence, $\sum c_j \varepsilon_{t-j}$ is the major building block for the long memory models introduced later. Therefore, we study processes of this type in detail in this chapter. For a solid discussion we require a sound understanding of summability and convergence of infinite series. This is what the following exposition is about.

It is clear what we mean by an infinite sum of real numbers $\{c_j\}_{j \in \mathbb{N}_0}$, where $\mathbb{N}_0 = \{0, 1, 2, \ldots\}$. The infinite series $\sum_{j=0}^{\infty} c_j$ is understood as limit of the partial sum series $\{C_N\}_{N \in \mathbb{N}_0}$ as $N \to \infty$:

$$C_N = \sum_{j=0}^{N} c_j \quad \text{with} \quad \lim_{N \to \infty} \sum_{j=0}^{N} c_j = C,$$

Time Series Analysis with Long Memory in View, First Edition. Uwe Hassler.
© 2019 John Wiley & Sons, Inc. Published 2019 by John Wiley & Sons, Inc.

where C is short for the limit $C = \sum_{j=0}^{\infty} c_j$ provided it exists. Such a limit is defined in the usual way: For any $\epsilon > 0$ it holds that

$$C_N \in [C - \epsilon, C + \epsilon]$$

for all integers N larger than a number depending on ϵ, i.e. for all $N \geq N(\epsilon)$.

Let us now define a stochastic series in terms of white noise $\{\varepsilon_t\} \sim$ WN$(0, \sigma^2)$:

$$x_{t,N} = \sum_{j=0}^{N} c_j \varepsilon_{t-j} .$$

Note that $\{x_{t,N}\}$ is double indexed. Can one define a sequence $\{x_t\}$ at any point in time t for $N \to \infty$? The answer is yes, and given more explicitly in the following proposition, where $\{x_t\}$ is defined as a mean square limit.

Proposition 3.1 *Let $\{c_j^2\}_{j \in \mathbb{N}_0}$ be summable, and $\{\varepsilon_t\} \sim$ WN$(0, \sigma^2)$. Then there exists a sequence $\{x_t\}_{t \in \mathbb{Z}}$ with*

$$\lim_{N \to \infty} \mathrm{E}\left[\left(x_t - \sum_{j=0}^{N} c_j \varepsilon_{t-j} \right)^2 \right] = 0.$$

This sequence is stationary with autocovariances

$$\gamma_x(h) = \sigma^2 \sum_{j=0}^{\infty} c_j \, c_{j+h}, \quad h = 0, 1, \ldots ,$$

and $\mathrm{E}(x_t) = 0.$

Proof: This proposition is a special case of Fuller (1996, Theorem 2.2.3); see Theorem 3.1. $\qquad \square$

The stationary sequence $\{x_t\}$ characterized in Proposition 3.1 will be denoted in the following as

$$x_t = \sum_{j=0}^{\infty} c_j \, \varepsilon_{t-j}, \quad \sum_{j=0}^{\infty} c_j^2 < \infty, \quad c_0 = 1, \tag{3.1}$$

where $c_0 = 1$ is a normalization to identify σ^2. Since x_t is given as a weighted average of $\varepsilon_t, \varepsilon_{t-1}, \varepsilon_{t-2}$, and so on, the process is often called a moving average of infinite order, MA(∞). The stochastic and statistical properties of such processes will depend on the convergence properties of the cumulated MA coefficients $\{c_j\}$. Before dedicating the next section to a discussion of MA(∞) processes, we therefore collect some technical results on summability of real sequences next.

Lemma 3.1 *Let $\{c_j\}_{j\in\mathbb{N}_0}$ be a sequence of real numbers. Then the following holds:*

(a) $\sum_{j=0}^{\infty} c_j < \infty \Rightarrow \lim_{j\to\infty} c_j = 0 \Rightarrow \lim_{N\to\infty} \frac{1}{N}\sum_{j=1}^{N} c_j = 0,$

(b) $\sum_{j=0}^{\infty} |c_j| < \infty \Rightarrow \begin{cases} \sum_{j=0}^{\infty} c_j < \infty \\ \\ \sum_{j=0}^{\infty} c_j^2 < \infty \end{cases},$

(c) $\sum_{j=0}^{\infty} g^j < \infty \iff |g| < 1,$

(d) $\sum_{j=1}^{\infty} \frac{1}{j^p} < \infty \iff p > 1,$

(e) $\sum_{j=1}^{\infty} \frac{1}{j^2} = \frac{\pi^2}{6},$

(f) $\sum_{j=1}^{\infty} \frac{(-1)^{j+1}}{j} = \ln(2),$

(g) $\sum_{j=1}^{\infty} \frac{j}{(j+1)!} = 1.$

Discussion and Proof

(a) A necessary condition for summability is that $\{c_j\}$ is a null sequence, $c_j \to 0$; see, e.g. Knopp (1990, Chapter 15). Clearly, it is not sufficient, since, e.g. $1/j \to 0$ for $j \to \infty$, while $1/j$ is not summable; see (d). Further, a necessary condition for $\{c_j\}$ to be a null sequence is that the average converges to 0; see, e.g. Knopp (1990, Theorem 1, p. 72) and Lemma 3.2.

(b) If $\{|c_j|\}$ is summable, then $\{c_j\}$ is said to be absolutely summable. If $\{c_j^2\}$ is summable, then $\{c_j\}$ is said to be square summable. Hence, absolute summability implies summability and square summability. The converse does not hold: (f) together with (d) shows that summability is not sufficient for absolute summability, and (d) shows with $p = 1$ that a square summable sequence is not necessarily (absolutely) summable. For these and further results, see also Fuller (1996, Section 2.2).

(c) This is the well-known result for the geometric series following from

$$\sum_{j=0}^{N} g^j = \frac{1-g^{N+1}}{1-g}, \quad g \neq 1.$$

(d) This is a well-known result on generalized harmonic series, also called p series. For a proof, see, e.g. Hassler (2016, Problem 5.1) or Knopp (1990, p. 115).

(e) This is a well-known example of a convergent series due to Leonhard Euler; see also Gradshteyn and Ryzhik (2000, 0.233(3)).

(f) Note that the following Taylor expansion of the natural logarithm is uniformly convergent on $(-1, 1]$:

$$\ln(1 + x) = \sum_{j=1}^{\infty} \frac{(-1)^{j+1}}{j} \, x^j.$$

For $x = 1$ one has the required result.

(g) This last result is an example of so-called telescoping sums or series, where the sum over any N terms is reduced to the sum of two simple terms:

$$\sum_{j=1}^{N} \frac{j}{(j+1)!} = \sum_{j=1}^{N} \left(\frac{1}{j!} - \frac{1}{(j+1)!} \right) = \frac{1}{1} - \frac{1}{(N+1)!}.$$

Then, letting $N \to \infty$ gives the required result.

Hence, the proof of Lemma 3.1 is complete. □

The second implication in Lemma 3.1 (a) can be generalized. To that end, we define a sequence $\{c_j\}$ to be Cesàro summable if the average

$$A_N = \frac{1}{N} \sum_{j=1}^{N} c_j$$

converges with $N \to \infty$; see, e.g. Fuller (1996, p. 129). If $c_j \to c$, then $A_N \to c$ as well, which is a well-known result attributed to Cauchy in a footnote by Knopp (1990, p. 72). The converse, however, does not hold true. Further, if $\{c_j\}$ is bounded, this is not sufficient for Cesàro summability, as we will show by a counterexample. We summarize these findings in the following lemma.

Lemma 3.2   *Let $\{c_j\}$ be a real sequence, $j \in \mathbb{N}$, and $A_N = \frac{1}{N} \sum_{j=1}^{N} c_j$.*

(a) If $\lim_{j \to \infty} c_j = c < \infty$, then $\lim_{N \to \infty} A_N = c$, while the converse is not true.

(b) Let now $|c_j| \le c < \infty$ for all j. This does not imply that A_N converges.

Proof: See Appendix to this chapter.

We isolate two results on the harmonic series, which corresponds to $p = 1$ in Lemma 3.1 (d).

Lemma 3.3 *For a fixed natural number h, it holds as $N \to \infty$ that*

(a) $\displaystyle\sum_{j=1}^{N} \frac{1}{j(h+j)} \to \frac{1}{h} \sum_{j=1}^{h} \frac{1}{j}$,

(b) $\displaystyle\sum_{j=1}^{N} \frac{1}{j} - \ln(N) \to \gamma$,

where $\gamma \approx 0.5772$ is Euler's constant.

Proof: See Appendix to this chapter.

Lemma 3.3 (a) is again an example of so-called telescoping series. In the proof, the partial sum $\sum_{j=1}^{N} \frac{1}{j(h+j)}$ will be reduced for arbitrary large N to the sum over only h simple terms, which is the telescoping property. And then we will let $N \to \infty$. The number γ showing up in Lemma 3.3 (b) is sometimes called Euler–Mascheroni constant, too. It has been evaluated up to many digits, but still nowadays it is unknown whether it is a rational number or not.

We close this section with a generalization of Proposition 3.1. The motivation is that many processes or time series are defined by linearly filtering an input sequence, say, $\{x_t\}$. The special case of Proposition 3.1 consists of filtering a white noise sequence $\{\varepsilon_t\}$. Now we consider more generally

$$y_t = \sum_{j=0}^{\infty} \psi_j x_{t-j},$$

and ask: Under what conditions is $\{y_t\}$ defined (and in what sense)? If the input $\{x_t\}$ of the filter is given according to Eq. (3.1), does the output $\{y_t\}$ possess a representation like in Eq. (3.1), too, and what are its properties? Answers to these questions are given in the following theorem.

Theorem 3.1 (*Filtering MA(∞) processes*) Let

$$x_t = \sum_{j=0}^{\infty} c_j \varepsilon_{t-j}, \quad \sum_{j=0}^{\infty} |c_j| < \infty, \quad c_0 = 1,$$

$\{\varepsilon_t\} \sim \mathrm{WN}(0, \sigma^2)$, *and* $\{\psi_j^2\}_{j \in \mathbb{N}_0}$ *is summable. Then there exists a sequence* $\{y_t\}_{t \in \mathbb{Z}}$ *with*

$$\lim_{N \to \infty} \mathrm{E}\left[\left(y_t - \sum_{j=0}^{N} \psi_j x_{t-j} \right)^2 \right] = 0$$

with $\mathrm{E}(y_t) = 0$. *Further, the sequence* $\{b_j^2\}_{j \in \mathbb{N}_0}$ *is summable, where*

$$b_j := \sum_{k=0}^{j} \psi_k c_{j-k}, \tag{3.2}$$

and $\{y_t\}$ is stationary with autocovariances

$$\gamma_y(h) = \sigma^2 \sum_{j=0}^{\infty} b_j\, b_{j+h}, \quad h = 0, 1, \ldots,$$

and

$$\lim_{N\to\infty} E\left[\left(y_t - \sum_{j=0}^{N} b_j \varepsilon_{t-j}\right)^2\right] = 0.$$

The same continues to hold when $\{c_j\}$ is only square summable while $\{\psi_j\}$ is absolutely summable.

Proof: Fuller (1996, Theorem 2.2.3), and the statement of the last sentence is from Fuller (1996, Corollary 2.2.3). □

The process $\{y_t\}$ characterized in Theorem 3.1 will be written as follows in the subsequent chapters:

$$y_t = \sum_{j=0}^{\infty} \psi_j x_{t-j} = \sum_{j=0}^{\infty} \psi_j \left(\sum_{k=0}^{\infty} c_k \varepsilon_{t-j-k}\right) = \sum_{j=0}^{\infty} b_j \varepsilon_{t-j},$$

where $\{b_j\}$ is given by so-called convolution in Eq. (3.2).

3.2 Wold Decomposition and Invertibility

Any stationary process $\{x_t\}$ with $E(x_t) = 0$ can be decomposed into two components. The first one is the so-called moving average process of infinite order, MA(∞), and the second one, $\{\delta_t\}$, is called *singular* or *deterministic*:

$$x_t = \sum_{j=0}^{\infty} c_j \varepsilon_{t-j} + \delta_t, \quad E(\delta_t) = 0. \tag{3.3}$$

Further aspects are given in Theorem 3.2, although we do not present mathematical details on $\{\delta_t\}$, which require Hilbert space arguments. The label *deterministic* may be misleading in the beginning, because $\{\delta_t\}$ is a random sequence; however, δ_t at time t is perfectly predictable from its past. We begin with an intuitive discussion of singular, or deterministic, processes; for details see also Anderson (1971, p. 420), Brockwell and Davis (1991, p. 187), or Fuller (1996, p. 81).

Let $\hat{\delta}_t$ be the best linear predictor of δ_t from its own past with some sequence $\{p_k\}$:

$$\hat{\delta}_t = \sum_{k=1}^{\infty} p_k \delta_{t-k}.$$

This linear predictor is assumed best in that the mean squared prediction error, $E[(\delta_t - \widehat{\delta}_t)^2]$, is minimized. It then holds for a singular, or deterministic, process by definition that

$$E[(\delta_t - \widehat{\delta}_t)^2] = 0 \quad \text{for all } t.$$

In that sense such a process is perfectly predictable from its past.

All advanced textbooks on time series analysis contain instructive proofs of the Wold decomposition due to Wold (1938). We particularly recommend in alphabetical order Anderson (1971, Theorem 7.6.7), Brockwell and Davis (1991, Theorem 5.7.1), Fuller (1996, Theorem 2.10.2), or Hannan (1970, Theorem 2, p. 137). Our formulation follows most closely (Anderson, 1971).

Theorem 3.2 (*Wold decomposition*) *Let $\{x_t\}$ be a stationary process with* $E(x_t) = 0$. *Then, x_t can be decomposed as in Eq. (3.3) with $\sum_{j=0}^{\infty} c_j \varepsilon_{t-j}$ satisfying*

$$\sum_{j=0}^{\infty} c_j^2 < \infty \quad \text{and} \quad \{\varepsilon_t\} \sim \text{WN}(0, \sigma^2).$$

The white noise sequence $\{\varepsilon_t\}$ and the square summable coefficients $\{c_j\}$ are unique. Further, the white noise innovations are uncorrelated with the deterministic process $\{\delta_t\}$: $E(\varepsilon_t \delta_s) = 0$ for all s and t.

Proof: Anderson (1971, Theorem 7.6.7). □

Before we move on, we wish to consider an example taken from Hassler (2017) that illustrates what may happen in the presence of a deterministic process.

Example 3.1 *Deterministic process: lack of ergodicity* Let

$$x_t = C + \varepsilon_t, \quad \varepsilon_t \sim \text{WN}(0, \sigma^2),$$

where the white noise sequence is independent of the coin-tossing variable C from Example 2.2:

$$P(C = c) = \frac{1}{2} \quad \text{for } c \in \{-1, 1\}.$$

With the notation of Theorem 3.2, we have that $\delta_t = C$ is deterministic: δ_t is perfectly predictable from δ_{t-1}. Consequently,

$$E(x_t) = 0 \quad \text{and} \quad \text{Var}(x_t) = 1 + \sigma^2,$$

and ($h \neq 0$)

$$\gamma_x(h) = E[(C + \varepsilon_t)(C + \varepsilon_{t+h})] = 1.$$

Hence, the process is (covariance) stationary, but

$$\frac{1}{H} \sum_{h=1}^{H} \gamma_x(h) = 1,$$

which violates the sufficient and necessary condition Eq. (2.7). Consequently, the sample mean does not converge to the expectation in mean square. This can be verified directly, too:

$$\frac{1}{T} \sum_{t=1}^{T} x_t = C + \bar{\varepsilon} \xrightarrow{2} C + 0 \neq 0 = E(x_t).$$

Clearly, this non-ergodicity stems from the fact that the random variable C does not vary over time, rather taking on the values -1 or 1 different from its mean 0 for all time, which makes it perfectly predictable. Imagine now that a coin is tossed at each point in time: C_t with $x_t = C_t + \varepsilon_t$. The sequence of variables C_t varies around the true expectation $E(C_t) = 0$, and the process $\{C_t + \varepsilon_t\}$ would be ergodic (for the mean); but of course C_t would no longer be perfectly predictable, i.e. no longer deterministic.

We hence learn from the example that $\delta_t = C$ causes the lack of ergodicity. In practice, it is impossible to verify whether or not a given times series is generated from an ergodic process. Therefore, we will assume for the rest of the book that we deal with stationary processes $\{x_t\}$ without deterministic component δ_t in order to rule out the complication encountered in Example 3.1. This is tantamount to assuming a Wold decomposition with $\delta_t = 0$ for all t. This is a common (although mostly implicit) assumption in time series analysis. Under stationarity one assumes $\{x_t\}$ given in Eq. (3.1) above. We call a stationary process $\{x_t\}$ from Eq. (3.3) with $\delta_t = 0$ an infinite moving average process, or MA(∞); see Eq. (3.1).

The absence of a deterministic component δ_t in the sense of Theorem 3.2 does of course not rule out expected values different from 0. In Theorem 3.2 we maintain $E(x_t) = 0$. With a constant number μ, this allows to define the process

$$x_t = \mu + \sum_{j=0}^{\infty} c_j \, \varepsilon_{t-j}$$

with expected value $\mu = E(x_t)$.

Ruling out the deterministic component from the Wold decomposition by assumption is well motivated by the following proposition that guarantees consistent estimation of the mean (ergodic for the mean in the sense of Definition 2.3).

Proposition 3.2 *Let $\{x_t\}$ be a stationary process with autocovariance sequence $\{\gamma(h)\}$ and*

$$x_t = \mu + \sum_{j=0}^{\infty} c_j \varepsilon_{t-j}$$

for some constant μ, where $\sum_{j=0}^{\infty} c_j \varepsilon_{t-j}$ satisfies the assumptions of Theorem 3.2. It then holds that

$$\bar{x} = \frac{1}{T} \sum_{t=1}^{T} x_t \xrightarrow{2} \mu$$

as $T \to \infty$.

Proof: See Appendix to this chapter.

Some authors call the MA(∞) component from Theorem 3.2,

$$\sum_{j=0}^{\infty} c_j \varepsilon_{t-j},$$

a linear process. This is justified by the fact that the past innovations ε_{t-j} enter x_t linearly. Other authors call $\sum_{j=0}^{\infty} c_j \varepsilon_{t-j}$ a linear process only if the innovations are iid. Prominent examples for the first convention are in alphabetical order: Brockwell and Davis (1991, p. 404), Giraitis et al. (2012, Definition 3.2.1), Phillips and Solo (1992, Eq. (13)), and Priestley (1981, p. 141). The second convention is adopted by, e.g. Beran et al. (2013, p. 43), Davidson (2000, p. 91), Hannan (1970, p. 209, 210), and Taniguchi and Kakizawa (2000, Eq. (2.2.1) or (3.1.1)). Here, we will stick to the second tradition and call $\sum_{j=0}^{\infty} c_j \varepsilon_{t-j}$ a linear process only if $\{\varepsilon_t\} \sim$ iid$(0, \sigma^2)$, because (see Davidson, 2000, p. 91) "The key feature is that the serial dependence of the process depends wholly on the coefficient sequence [...]. Other forms of serial dependence are called nonlinear."

It is worth highlighting the case of iid innovations, because all linear processes are ergodic in the sense of Definition 2.2, such that the sample average converges almost surely and so do the covariance estimators; see Theorem 2.2. The corresponding result is given in Proposition 3.3, which we adopt from Hannan (1970).

Proposition 3.3 *Let $\{x_t\}$ be a linear process with expected value μ and autocovariances $\{\gamma(h)\}$,*

$$x_t = \mu + \sum_{j=0}^{\infty} c_j \varepsilon_{t-j}, \quad \sum_{j=0}^{\infty} c_j^2 < \infty,$$

where $\{\varepsilon_t\}$ is iid$(0, \sigma^2)$. It then holds that this process is strictly stationary and ergodic, such that

$$\bar{x} \overset{a.s.}{\to} \mu,$$

and

$$\hat{\gamma}(h) := \frac{1}{T} \sum_{t=1}^{T-h} (x_t - \bar{x})(x_{t+h} - \bar{x}) \overset{a.s.}{\to} \gamma(h), \quad h = 0, 1, \dots,$$

as $T \to \infty$.

Proof: Theorem 2 on page 203 together with Theorem 3 on page 204 in Hannan (1970). □

Distributional results for the sample mean are postponed until Chapter 7.

Under certain conditions, an MA(∞) process has an autoregressive representation of infinite order, abbreviated as AR(∞). Brockwell and Davis (1991, Definition 3.1.4) called a process invertible if an autoregressive representation exists with absolutely summable coefficients. Such a definition was tailored for ARMA processes discussed in Section 3.4. For more general processes like MA(∞), we adopt the more general definition by Bondon and Palma (2007, Eq. (6)) (see also Palma (2007, p. 7)), which was pioneered by Bloomfield (1985).[1]

Definition 3.1 (*Invertibility*) *Let $\{x_t\}$ be a stationary MA process,*

$$x_t = \sum_{j=0}^{\infty} c_j \varepsilon_{t-j}, \quad \sum_{j=0}^{\infty} c_j^2 < \infty, \quad c_0 = 1,$$

with $\{\varepsilon_t\} \sim WN(0, \sigma^2)$, $t \in \mathbb{Z}$. The process is said to be invertible if there exists a sequence $\{a_j\}_{j \in \mathbb{N}_0}$ such that

$$E\left[\left(\varepsilon_t - \sum_{j=0}^{N} a_j x_{t-j}\right)^2\right] \to 0$$

as $N \to \infty$.

For invertible processes we write in short

$$\sum_{j=0}^{\infty} a_j x_{t-j} = \varepsilon_t,$$

---

1 Yet another approach to invertibility has been put forward by Granger and Andersen (1978), relying on forecastability.

or

$$x_t = - \sum_{j=1}^{\infty} a_j x_{t-j} + \varepsilon_t.$$

Although $\{x_t\}$ alone is called invertible, it must be noted that invertibility is a property of the relationship between the sequence $\{x_t\}$ and $\{\varepsilon_t\}$; see Brockwell and Davis (1991, p. 86). More importantly, note that no summability condition is imposed on $\{a_j\}$ in Definition 3.1. In fact, we know from Bondon and Palma (2007) that not even square summability is required; see also Proposition 6.2. In the next chapter, we will adopt a sufficient condition for invertibility formulated in the frequency domain by Bondon and Palma (2007, Theorem 3); see also Palma (2007, Lemma 3.1) and Proposition 4.6.

3.3 Persistence versus Memory

In this section, we assume that $\{x_t\}_{t \in \mathbb{Z}}$ is given by the MA(∞) process from Eq. (3.1),

$$x_t = \sum_{j=0}^{\infty} c_j \varepsilon_{t-j}, \quad \sum_{j=0}^{\infty} c_j^2 < \infty, \quad c_0 = 1,$$

where the innovations are white noise with finite variance:

$$\{\varepsilon_t\} \sim \text{WN}(0, \sigma^2), \quad 0 < \sigma^2 < \infty.$$

The MA coefficients are often called impulse response coefficients and measure the effect of a shock in period $t - j$ on x_t:

$$\frac{\partial x_t}{\partial \varepsilon_{t-j}} = c_j.$$

By assumption, $c_j \to 0$ as $j \to \infty$, and hence the effect of past shocks is not permanent. The speed at which the impulse responses vanish to 0 characterizes the persistence of the process (or the persistence of shocks driving the process). For a review of alternative measures of persistence, see, e.g. Paya et al. (2007). Sometimes, one is interested in the cumulated effect and computes

$$\text{CIR}(N) = \sum_{j=0}^{N} c_j.$$

The accumulated responses up to N periods back measure the total effect if there occurred a unit shock in each past period, including the present period at time t. For $N \to \infty$ one obtains the so-called long-run effect, often called total multiplier in economics. This measure is defined as

$$\text{CIR} := \lim_{N \to \infty} \text{CIR}(N), \tag{3.4}$$

provided that this quantity exists. The CIR has been advocated by Andrews and Chen (1994) as being superior to alternative measures of persistence. We will now show that the CIR contains the same information as the long-run variance ω^2 from Eq. (2.10). Further, the following proposition says that absolute summability of the MA impulse responses is sufficient for short memory.

Proposition 3.4 *Let the stationary MA process $\{x_t\}$ from Eq. (3.1) have summable MA coefficients:*

$$x_t = \sum_{j=0}^{\infty} c_j \, \varepsilon_{t-j}, \quad \sum_{j=0}^{\infty} c_j < \infty, \quad \varepsilon_t \sim \mathrm{WN}(0, \sigma^2).$$

(a) It then holds

$$\omega^2 = \sigma^2 \mathrm{CIR}^2 = \sigma^2 \left(\sum_{j=0}^{\infty} c_j \right)^2 \tag{3.5}$$

for the long-run variance $\omega^2 = \sum_{h=-\infty}^{\infty} \gamma(h)$ from Eq. (2.10).
(b) Further, if $\{c_j\}$ is absolutely summable, then $\{\gamma(h)\}$ is absolutely summable, too, where the autocovariances $\{\gamma(h)\}$ are given in Proposition 3.1.

Proof: See, e.g. Problems 3.3 and 14.1 in Hassler (2016). □

It has sometimes been claimed in the literature that long memory and strong persistence are equivalent. While this may hold under specific model assumptions, it is not generally true. With some examples we discuss four cases of different combinations of memory and persistence.

Example 3.2 *Short memory under Antipersistence* It is very simple to construct an antipersistent process. First, consider

$$x_t = \varepsilon_t - \varepsilon_{t-1}, \quad \text{i.e. } c_0 = -c_1 = 1, \quad c_2 = c_3 = \cdots = 0.$$

Clearly,

$$\mathrm{CIR} = c_0 + c_1 = 0,$$

and $\{|c_j|\}$ is summable. By Proposition 3.4, $\{x_t\}$ has short memory and is antipersistent in the sense of Definitions 2.4 and 2.5. In Figure 3.1 we have simulated such an antipersistent process. The upper graph simply displays a sequence of (standard normal) pseudorandom numbers, which is a simulation of an iid process; the second graph from the top shows the differences thereof. Clearly, the second series has a larger variance than the first one: $\mathrm{Var}(\varepsilon_t - \varepsilon_{t-1}) = 2\,\mathrm{Var}(\varepsilon_t)$. The dynamic structure becomes more obvious

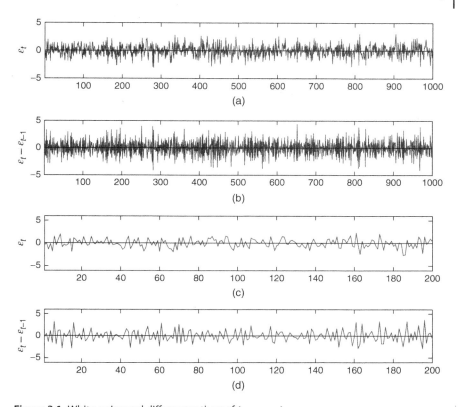

Figure 3.1 White noise and differences thereof ($\varepsilon_t - \varepsilon_{t-1}$).

when we look only at a subsample, e.g. the first 200 observation; see the lower graphs. Note that few subsequent observations of $\{\varepsilon_t\}$ may be negative (or positive) by coincidence, while the realizations of $\{\varepsilon_t - \varepsilon_{t-1}\}$ change signs regularly. Generally, the last plot is largely a zigzag curve where changes in the sign are rather the rule than the exception. This reflects the negative autocorrelation of $\{\varepsilon_t - \varepsilon_{t-1}\}$ at lag 1. The iid sequence on the other hand does not reveal such a structure.

Second, consider $\{x_t\}$ from Eq. (3.1) with

$$c_0 = 1, \quad c_j = -\left(\frac{1}{2}\right)^j, \quad j \geq 1.$$

By Lemma 3.1 (c) we have

$$\sum_{j=1}^{\infty} c_j = -\sum_{j=1}^{\infty} \left(\frac{1}{2}\right)^j = -\sum_{j=0}^{\infty} \left(\frac{1}{2}\right)^j + 1$$
$$= -2 + 1 = -1.$$

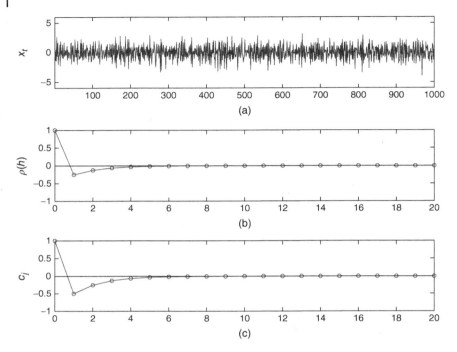

Figure 3.2 Antipersistent process with $c_j = -(1/2)^j$.

Consequently,

$$\text{CIR} = 1 - 1 = 0,$$

while at the same time $\{|c_j|\}_{j \in \mathbb{N}}$ is summable. Again, Proposition 3.4 implies short memory and antipersistence. Clearly, the sequence of impulse responses is negative; see the last graph in Figure 3.2. Similarly, the autocorrelogram is below 0. It converges to 0 very fast, such that the simulated series (first graph) looks similar to the differenced noise ($\{\varepsilon_t - \varepsilon_{t-1}\}$ in Figure 3.1).[2]

Third and very similarly, by Lemma 3.1 (g) we have with

$$c_0 = 1, \quad c_j = -\frac{j}{(j+1)!}, \quad j \geq 1,$$

---

2 Note that all simulated data in this section build on one and the same iid sequence $\{\varepsilon_t\}$ with unit variance shown in the upper panel of Figure 3.1. All MA(∞) series are simulated as

$$x_t = \sum_{j=0}^{1000+t-1} c_j \varepsilon_{t-j},$$

which approximates the infinite moving average filter.

that

$$\sum_{j=1}^{\infty} c_j = -1 \ .$$

Consequently, CIR= $1 - 1 = 0$, while again $\{|c_j|\}$ is summable. The impulse response sequence shown in Figure 3.3 parallels a lot the one from the previous example. Consequently, Figures 3.2 and 3.3 look very much alike.

Example 3.3 *Long memory under antipersistence* We now construct the extreme example of an antipersistent process displaying long memory. Define $\{c_j\}$ from Eq. (3.1) as

$$c_0 = 1, \quad c_j = \frac{(-1)^j}{j \ln(2)}, \quad j \geq 1.$$

By Lemma 3.1 (f) we have

$$\sum_{j=1}^{\infty} c_j = -\frac{1}{\ln(2)} \sum_{j=1}^{\infty} \frac{(-1)^{j+1}}{j} = -1.$$

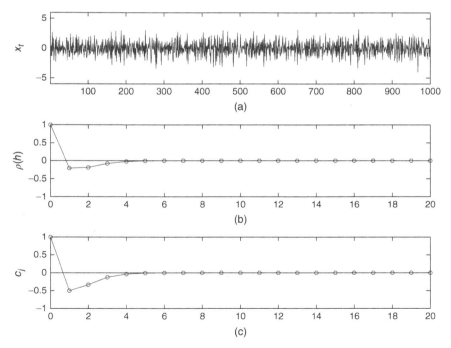

Figure 3.3 Antipersistent process with $c_j = -j/(j+1)!$.

Hence, trivially CIR= $1 - 1 = 0$. For the autocovariances we obtain by Proposition 3.1 that

$$
\begin{aligned}
\frac{\gamma_x(h)}{\sigma^2} &= c_0 c_h + c_1 c_{h+1} + \cdots + c_j c_{h+j} + \cdots \\
&= \frac{(-1)^h}{h \ln(2)} + \frac{(-1)^{h+2}}{(h+1) \ln^2(2)} + \cdots + \frac{(-1)^{h+2j}}{j(h+j) \ln^2(2)} + \cdots \\
&= \frac{(-1)^h}{h \ln(2)} + \frac{(-1)^h}{\ln^2(2)} \sum_{j=1}^{\infty} \frac{1}{j(h+j)} \\
&= \frac{(-1)^h}{h \ln(2)} + \frac{(-1)^h}{h \ln^2(2)} \sum_{k=1}^{h} \frac{1}{k} \\
&= \frac{(-1)^h}{h \ln(2)} + \frac{(-1)^h}{h \ln^2(2)} (\ln(h) + \gamma_h)
\end{aligned}
$$

with $\gamma_h \to \gamma$ as $h \to \infty$, where for the last two equalities Lemma 3.3 (a) and (b) were used. Hence,

$$
\frac{h}{\ln(h)} |\gamma_x(h)| \to \frac{\sigma^2}{\ln^2(2)} (0 + 1 + 0) = \frac{\sigma^2}{\ln^2(2)},
$$

such that $|\gamma_x(h)|$ behaves like $\sigma^2 \ln(h)/(h \ln^2(2))$ with large h. Since, $1/h$ is not summable, this implies that $\ln(h)/h$ is not summable. Hence, the process displays indeed long memory notwithstanding its antipersistence.

Figure 3.4 illustrates this antipersistent long memory process. The sequence of impulse responses has alternating signs; see the last graph in Figure 3.4. Analogously, the autocorrelogram oscillates around 0. Although the absolute values of the autocorrelogram are not summable, the autocorrelations *annihilate* in that the impulse responses sum up to 0. Hence, the simulated data (first graph) resemble more the previous antipersistent cases than the long memory examples below.

Example 3.4 ***Long memory under moderate persistence*** The previous example can be modified to allow for long memory and moderate persistence. We assume

$$
c_0 = 1, \quad c_j = \frac{(-1)^{j+1}}{j}, \quad j \geq 1,
$$

with

$$
\text{CIR} = 1 + \sum_{j=1}^{\infty} \frac{(-1)^{j+1}}{j} = 1 + \ln 2
$$

by Lemma 3.1 (f), such that

$$
0 < \omega_x^2 = \sigma^2 \text{CIR}^2 < \infty.
$$

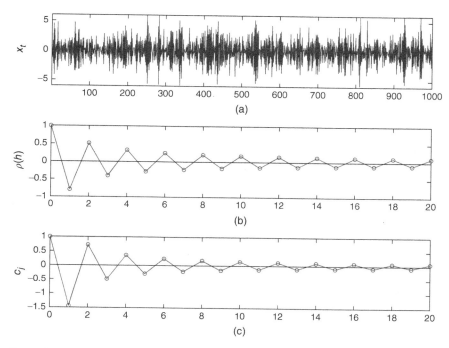

Figure 3.4 Long memory under antipersistence with $c_j = (-1)^j/(j \ln 2)$.

At the same time one obtains just as in Example 3.3:

$$(-1)^h \frac{h}{\ln(h)} \gamma_x(h) \to \sigma^2, \quad h \to \infty,$$

such that $\{|\gamma_x(h)|\}$ is not summable. Again, the sequence of impulse responses has alternating signs, and the autocorrelogram oscillates around 0, while the absolute values of the autocorrelogram are not summable, and the impulse responses sum up to a constant different from 0. Comparing Figures 3.5 and 3.4, we see that this long memory model under moderate persistence differs from the long memory process under antipersistence in that it (vaguely) displays spells of reversing trends.

Example 3.5 *Long memory under strong persistence* Now, we move to the extreme case and discuss processes with strong persistence displaying long memory. First, we construct a process where the autocovariances decay at rate $\ln(h)/h$, while the $\{c_j\}$ are not summable at the same time. To that end, we choose

$$c_j = \frac{1}{j+1}, \quad j \geq 0,$$

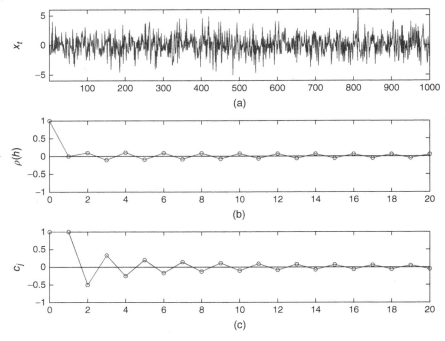

Figure 3.5 Long memory under moderate persistence, $c_j = (-1)^{j+1}/j$.

such that $\sum_{j=0}^{N} c_j$ diverges at rate $\ln(N)$ by Lemma 3.3 (b). The autocovariances satisfy by Proposition 3.1

$$\frac{\gamma_x(h)}{\sigma^2} = c_0 c_h + c_1 c_{h+1} + \cdots + c_j c_{h+j} + \cdots$$

$$= \frac{1}{h+1} + \frac{1}{2(h+2)} + \cdots + \frac{1}{(j+1)(h+j+1)} + \cdots$$

$$= \sum_{j=1}^{\infty} \frac{1}{j(h+j)},$$

such that by Lemma 3.3 (a)

$$\frac{\gamma_x(h)}{\sigma^2} = \sum_{j=1}^{\infty} \frac{1}{j(h+j)} = \frac{1}{h} \sum_{k=1}^{h} \frac{1}{k}.$$

This proves that the autocovariances converge to 0 with $\ln(h)/h$ and are hence not summable. Figure 3.6 illustrates this strongly persistent long memory process. The sequence of harmonically decaying impulse responses is reflected

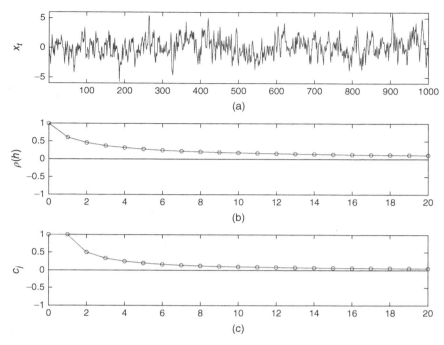

Figure 3.6 Long memory under strong persistence. $c_j = 1/(j+1)$.

in the slowly converging autocorrelations; see the graphs in Figure 3.6. The simulated series displays long time spans of growth or decrease; such reversing trends are what we typically understand by strong persistence of a stationary process.

The second example of long memory and strong persistence is even more extreme. We now assume

$$c_0 = 1, \quad c_j = \frac{1}{j^{0.8}}, \quad j \geq 1.$$

Clearly, the impulse responses are not summable, but square summable; see Lemma 3.1 (d). Similar impulse responses show up in Eq. (5.26) resulting from so-called fractionally integrated processes; see Chapter 6.[3] Comparing Figures 3.7 with 3.6, we observe the same features but even more pronounced:

---

3 Figure 3.7 corresponds to what we will call an $I(d)$ process with $d = 0.2$; see Definition 6.1. The autocorrelations at lag h were approximated as follows:

$$\rho(h) = \frac{\sum_{j=0}^{\infty} c_j c_{j+h}}{\sum_{j=0}^{\infty} c_j^2} = \frac{\sum_{j=0}^{N} c_j c_{j+h}}{\sum_{j=0}^{N} c_j^2}, \quad N = 10^6.$$

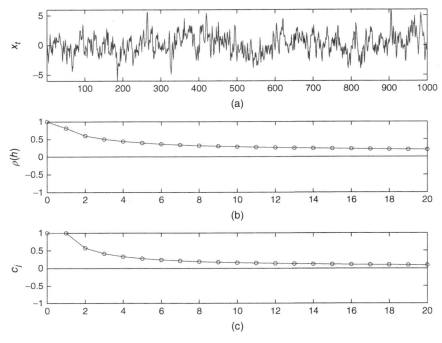

Figure 3.7 Long memory process under strong persistence with $c_j = 1/j^{0.8}$.

long spells of reversing trends in the simulated data and very slow convergence of autocorrelograms and impulse response sequences.

The trajectories of the empirical examples given in the introductory chapter rather look like the one shown in Figure 3.6 or even in Figure 3.7, where we observe reversing trends as distinct periods of upswings and downswings. Similarly, the sample autocorrelograms in Chapter 1 are all positive and slowly declining, mimicked by the ones in Figure 3.6 or 3.7. For that reason, we will focus in Chapter 6 on fractionally integrated processes, where long memory only occurs under strong persistence. Statistically, we have seen that it makes sense to distinguish between memory and persistence: Short or long memory determines whether the variance of the sample means vanishes with $1/T$ or not; persistence allows to further discriminate between a vanishing (to 0) long-run variance, a positive and finite or an unbounded long-run variance, implying different statistical properties. From a subject-matter point of view, however, Examples 3.2 through 3.5 suggest that in applied sciences typically long memory goes hand in hand with strong persistence.

3.4   Autoregressive Moving Average Processes

The most widely used time series model is the autoregressive moving average model,

$$x_t = \phi_1\, x_{t-1} + \cdots + \phi_p\, x_{t-p} + \varepsilon_t + \theta_1\, \varepsilon_{t-1} + \cdots + \theta_q\, \varepsilon_{t-q}, \quad t \in \mathbb{Z},$$

where $\phi_p \neq 0$ and $\theta_q \neq 0$. Such processes are denoted by ARMA(p, q). More compactly we write them in terms of the lag operator L with $L^k x_t = x_{t-k}, k \in \mathbb{Z}$, which may also shift into the future for $k < 0$. The time interpretation is by no means crucial, and we may equally apply L to other sequences, for instance, $L^k c_j = c_{j-k}$. For finite orders p and q, we now define

$$\Phi(L)\, x_t = \Theta(L)\, \varepsilon_t, \quad t \in \mathbb{Z}, \tag{3.6}$$

where the polynomials

$$\Phi(L) = 1 - \phi_1\, L - \cdots - \phi_p\, L^p \quad \text{and} \quad \Theta(L) = 1 + \theta_1\, L + \cdots + \theta_q\, L^q$$

are assumed to have no common roots. The possibly complex-valued roots or zeros of $\Phi(L)$ are the solutions $z_1, \ldots, z_p$ of

$$\Phi(z) = 0 \quad \text{with} \quad \Phi(z) = \left(1 - \frac{z}{z_1}\right)\left(1 - \frac{z}{z_2}\right)\cdots\left(1 - \frac{z}{z_p}\right),$$

where $z_j \neq 0$. Similarly, $\Theta(z) = 0$ provides q roots $\zeta_1, \ldots, \zeta_q$ with

$$\Theta(z) = \left(1 - \frac{z}{\zeta_1}\right)\left(1 - \frac{z}{\zeta_2}\right)\cdots\left(1 - \frac{z}{\zeta_q}\right).$$

The assumption of no common roots is tantamount to $\Phi(L)$ and $\Theta(L)$ having no common factor, which would cancel. Further, one typically works under the assumption that both polynomials have roots or zeros only outside of the unit circle:

$$\Phi(z) = 0 \Rightarrow |z| > 1, \quad z \in \mathbb{C}, \tag{3.7}$$

$$\Theta(z) = 0 \Rightarrow |z| > 1, \quad z \in \mathbb{C}. \tag{3.8}$$

The meaning of those two conditions will be discussed in the sequel. Technically, these conditions are equivalent to the existence of a one-sided absolutely summable expansion of the inverse polynomials Φ^{-1} and Θ^{-1}, respectively; see also Hassler (2016, Proposition 3.3). For instance, we can write under Eq. (3.7)

$$\Phi^{-1}(L) = \frac{1}{\Phi(L)} = \sum_{j=0}^{\infty} b_j L^j, \quad \sum_{j=0}^{\infty} |b_j| < \infty.$$

Consequently, one may define by convolution of $\Theta(L)$ and $\sum_{j=0}^{\infty} b_j L^j$

$$C(L) = \sum_{j=0}^{\infty} c_j L^j := \frac{\Theta(L)}{\Phi(L)} = \Theta(L) \sum_{j=0}^{\infty} b_j L^j,$$

which gives an MA(∞) representation of the ARMA process under condition Eq. (3.7):

$$x_t = \frac{\Theta(L)}{\Phi(L)} \varepsilon_t = \sum_{j=0}^{\infty} c_j \varepsilon_{t-j}, \quad \sum_{j=0}^{\infty} |c_j| < \infty. \tag{3.9}$$

The ARMA process with the index set $\mathbb{T} = \mathbb{Z}$ has an absolutely summable MA(∞) representation in terms of past innovations if and only if the so-called stability condition Eq. (3.7) holds (see, e.g. Brockwell and Davis (1991, Theorem 3.1.1.)); consequently, the process is stationary. Note that the case of autoregressive roots inside the unit circle, $\Phi(z) = 0 \Rightarrow |z| < 1$, allows for stationary solutions, too. However, they do not have the MA representation in terms of *past* innovations considered here. Rather, their MA representation is given in terms of *future* innovations (see Brockwell and Davis (1991, Theorem 3.1.3)): $x_t = \sum_{j=0}^{\infty} c_j \varepsilon_{t+j}$. Ruling out this case amounts to assuming what Brockwell and Davis (1991) call a *causal* process. The special case where $\Phi(z) = 0 \Rightarrow |z| = 1$ is often called the (autoregressive) unit root case. For $z = 1$ this means that $\Phi(L)$ contains the difference $1 - L$ as factor: The process $\{x_t\}$ is then not stationary but has to be differenced in order to obtain stationarity. Sometimes, processes meeting Eq. (3.7) are also called stable.

Let us now return to the MA(∞) representation of the ARMA process in Eq. (3.9). By definition of $C(L)$, it holds that

$$1 + \theta_1 L + \cdots + \theta_q L^q = (1 - \phi_1 L - \cdots - \phi_p L^p)(c_0 + c_1 L + c_2 L^2 + \cdots).$$

From this one may explicitly obtain the sequence $\{c_j\}$ by comparison of coefficients:

$$L^0 : 1 = c_0,$$
$$L^1 : \theta_1 = c_1 - \phi_1 c_0 = c_1 - \phi_1 : \quad c_1 = \phi_1 + \theta_1,$$
$$L^2 : \theta_2 = c_2 - \phi_1 c_1 - \phi_2 c_0 : \quad c_2 = \phi_2 + \phi_1(\phi_1 + \theta_1) + \theta_2,$$
$$\vdots$$
$$L^j : 0 = c_j - \phi_1 c_{j-1} - \cdots - \phi_p c_{j-p}, \quad j > q.$$

The last equation is written more compactly as

$$\Phi(L)c_j = 0, \quad j > q, \tag{3.10}$$

which is a homogeneous linear difference equation of order p with constant coefficients.

Under the so-called invertibility condition Eq. (3.8), we can treat $\Theta(L)$ in exactly the same way as just demonstrated for $\Phi(L)$, such that we obtain a one-sided absolutely summable infinite polynomial

$$A(L) = \sum_{j=0}^{\infty} a_j L^j := \frac{\Phi(L)}{\Theta(L)}, \quad \sum_{j=0}^{\infty} |a_j| < \infty, \tag{3.11}$$

with $a_0 = 1$. An ARMA process with (3.8) is invertible in the sense of Definition 3.1, and it has the absolutely summable AR(∞) representation:

$$A(L)\, x_t = \varepsilon_t \quad \text{or} \quad x_t = -\sum_{j=1}^{\infty} a_j x_{t-j} + \varepsilon_t.$$

The sequence $\{a_j\}$ of autoregressive coefficients satisfies again a homogeneous linear difference equation with constant coefficients, this time of order q:

$$\Theta(L)a_j = 0, \quad j > q. \tag{3.12}$$

Moreover, the autocovariances of a stationary ARMA process satisfy Eq. (3.10), too (see, for instance, Brockwell and Davis (1991, p. 93)):

$$\Phi(L)\gamma_x(h) = 0, \quad h \geq \max(p, q + 1), \tag{3.13}$$

where L operates on h: $L^k \gamma_x(h) = \gamma_x(h - k)$.

Now, we are equipped to prove the following proposition. It states that the autocovariance structure and MA(∞) dynamics of stationary ARMA processes are bounded geometrically and that the AR(∞) dynamics of invertible ARMA processes are geometrically bounded, too.

Proposition 3.5 *Let $\{x_t\}$ be a stationary ARMA process (Eq. (3.6) under Eq. (3.7)) with MA(∞) representation (3.9). It then holds*

(a) for the autocovariances that there exist constants γ_0 and γ_1 with

$$0 < \gamma_1 < 1 \quad \text{and} \quad \gamma_0 > 0$$

such that for all $h \in \mathbb{N}$

$$|\gamma_x(h)| \leq \gamma_0 \gamma_1^h,$$

(b) and for the impulse responses that there exist constants g_0 and g_1 with

$$0 < g_1 < 1 \quad \text{and} \quad g_0 > 0$$

such that for all $j \in \mathbb{N}$

$$|c_j| \leq g_0 g_1^j,$$

(c) and that further the impulse response coefficients are s-summable for any s > 0 in that

$$\sum_{j=0}^{\infty} j^s |c_j| < \infty.$$

(d) If the process is invertible (i.e. under Eq. (3.8)), it holds for the AR(∞) coefficients $\{a_j\}$ from Eq. (3.11), first, that there exist constants α_0 and α_1 with

$$0 < \alpha_1 < 1 \quad and \quad \alpha_0 > 0$$

such that for all $j \in \mathbb{N}$

$$|a_j| \le \alpha_0 \alpha_1^j,$$

and, second, that the AR coefficients are s-summable for any s > 0,

$$\sum_{j=0}^{\infty} j^s |a_j| < \infty,$$

and, third, that $\sum_{j=0}^{\infty} a_j \ne 0$.

Proof: See Appendix to this chapter.

Note that invertibility in the sense of Eq. (3.8) implies that $\Theta(1) \ne 0$, which in turn implies $\sum_{j=0}^{\infty} c_j \ne 0$ for the MA(∞) representation given in Eq. (3.9). Invertibility, however, is not assumed in Proposition 3.5 (a) through (c).

To prove Proposition 3.5, we establish two technical lemmata, which are also interesting for the sake of their own. The first one says that the product of polynomial growth (t^s) and geometric decay (g^t with $|g| < 1$) is bounded geometrically at all points in time. The proof does actually not require s to be integer.

Lemma 3.4   *For $0 < g < 1$ and $s > 0$, there exist g_0 and g_1 with*

$$g_0 > 0 \quad and \quad 0 < g_1 < 1,$$

such that

$$t^s g^t \le g_0 g_1^t$$

for all $t > 0$.

Proof: See Appendix to this chapter.

Next we need to bound the solution of a stable difference equation like in, e.g. Eq. (3.10). To that end, we define a homogeneous linear difference equation of order k with constant (real-valued) coefficients $d_1, \dots, d_k$:

$$D(L)y_t = 0, \quad t \in \mathbb{N}, \quad D(L) = 1 + d_1 L + \dots + d_k L^k. \tag{3.14}$$

We will now characterize all sequences solving Eq. (3.14), which will be called the general homogeneous solution: $\{y_t^h\}_{t \in \mathbb{N}}$. To do so we define $z_1, \ldots, z_\ell, \ell \leq k$, as all distinct zeros or roots of $D(z)$ with multiplicity $r_1, \ldots, r_\ell$ each. Note that z_i may be complex valued, and

$$
D(z) = \prod_{i=1}^{\ell} \left(1 - \frac{z}{z_i} \right)^{r_i}, \quad z \in \mathbb{C}, \quad z_i \neq 0.
$$

Then there exist coefficients $c_{i,m}$ such that the general homogeneous solution becomes (Brockwell and Davis, 1991, Eq. (3.6.7))

$$
y_t^h = \sum_{i=1}^{\ell} (c_{i,0} + c_{i,1}t + \cdots + c_{i,\rho_i}t^{\rho_i}) \left(\frac{1}{z_i} \right)^t \quad \text{where} \quad \rho_i := r_i - 1.
$$

Equation (3.14) is called stable if

$$
D(z) = 0 \Rightarrow |z| > 1, \quad z \in \mathbb{C}, \tag{3.15}
$$

because under this condition the solution converges to a constant value (equal to 0):

$$
y_t^h \to 0 \quad \text{as} \quad t \to \infty.
$$

More precisely, the second lemma states that this convergence is at a geometric rate.

Lemma 3.5 *Let $\{y_t^h\}$, $t \in \mathbb{N}$, be the general solution of Eq. (3.14) under (3.15). Then there exist constants g_0 and g_1 with*

$$
0 < g_1 < 1 \quad \text{and} \quad g_0 > 0
$$

such that

$$
|y_t^h| \leq g_0 g_1^t
$$

for all $t \in \mathbb{N}$.

Proof: See Appendix to this chapter.

3.5 Technical Appendix: Proofs

Proof of Lemma 3.2

(a) The implication of Cesàro summability is established, e.g. in Knopp (1990, Theorem 2, p. 72). The converse does not hold because of the following obvious counterexample: $c_j = (-1)^j$ with $A_N \to 0$.

(b) While A_N is obviously bounded by assumption,

$$|A_N| \le \frac{1}{N} \sum_{j=1}^{N} |c_j| \le c,$$

this is not sufficient for convergence. Consider the dilated sequence of 1 and -1,

$$c_1 = 1, \quad c_2 = c_3 = -1, \quad c_4 = c_5 = c_6 = c_7 = 1, \dots,$$

i.e.

$$c_{2^k} = \cdots = c_{2^{k+1}-1} = (-1)^k,$$

and consider the Cesàro sum or average A_N. In particular, define $N_k = 2^k - 1$, such that

$$\sum_{j=1}^{N_k} c_j = 1 - 2 + (-2)^2 + \cdots + (-2)^{k-1} = \frac{1 - (-2)^k}{3}.$$

Consequently,

$$A_{N_k} = \frac{1 - (-2)^k}{3(2^k - 1)},$$

which clearly does not converge as $N_k \to \infty$ ($k \to \infty$). Hence, the proof is complete. $\qquad\square$

Proof of Lemma 3.3

(a) We simply rearrange terms for $h < N$:

$$\sum_{j=1}^{N} \frac{1}{j(h+j)} = \sum_{j=1}^{N} \left[\frac{h+j}{hj(h+j)} - \frac{j}{hj(h+j)} \right]$$

$$= \frac{1}{h} \sum_{j=1}^{N} \left[\frac{1}{j} - \frac{1}{h+j} \right]$$

$$= \frac{1}{h} \sum_{j=1}^{h} \left[\frac{1}{j} - \frac{1}{N+j} \right].$$

Obviously, $\sum_{j=1}^{h} \frac{1}{N+j} \to 0$ as N gets large. Hence,

$$\sum_{j=1}^{\infty} \frac{1}{j(h+j)} = \frac{1}{h} \left[1 + \frac{1}{2} + \cdots + \frac{1}{h} \right],$$

which proves the claim.

(b) We define

$$\gamma_N := \sum_{j=1}^{N} \frac{1}{j} - \ln(N)$$

$$= \sum_{j=1}^{N} \frac{1}{j} - \int_{1}^{N} \frac{1}{x} \, dx$$

$$= \frac{1}{N} + \sum_{j=1}^{N-1} \int_{j}^{j+1} \left(\frac{1}{j} - \frac{1}{x} \right) \, dx,$$

where $\frac{1}{j} - \frac{1}{x} > 0$ for $x \in (j, j+1]$. Consequently, $\gamma_N > 0$. Further,

$$\gamma_N < \frac{1}{N} + \sum_{j=1}^{N-1} \int_{j}^{j+1} \left(\frac{1}{j} - \frac{1}{j+1} \right) \, dx,$$

$$= \frac{1}{N} + \sum_{j=1}^{N-1} \left(\frac{1}{j} - \frac{1}{j+1} \right) = 1.$$

Hence, $0 < \gamma_N < 1$, and γ_N is decreasing monotonically with N:

$$\gamma_{N+1} - \gamma_N = \frac{1}{N+1} - \int_{N}^{N+1} \frac{1}{x} \, dx$$

$$< \frac{1}{N+1} - \int_{N}^{N+1} \frac{1}{N+1} \, dx = 0.$$

Therefore, $\{\gamma_N\}$ converges. We denote the limit by γ. The numerical approximation of this constant (Euler's constant with $\gamma = 0.5772\ldots$) is known from the literature; see also Knopp (1990, p. 225) or Gradshteyn and Ryzhik (2000, Eq. (0.131))). $\qquad \square$

Proof of Proposition 3.2

The proof draws on results that will be provided later in Chapter 4. In particular, Proposition 4.3 maintains the same assumptions as Proposition 3.2. We hence know from Eq. (4.17) that the autocovariances equal the sequence of Fourier coefficients of the integrable spectrum f. By Fuller (1996, Lemma 3.1.1.A) this implies that they form a null sequence, although they do not have to be summable; this result corresponds to the famous Riemann–Lebesgue lemma. From Lemma 3.2 (a) we conclude that the average of the autocovariances must form a null sequence, too. This amounts just to the (necessary and) sufficient condition Eq. (2.7) from Proposition 2.2. Hence, the proof is complete. $\qquad \square$

Proof of Lemma 3.4

Given g and s we may choose some number a with

$$0 < a < -\frac{1}{s}\ln g,$$

and

$$g_1 := \exp(as + \ln g) = e^{as}g < 1.$$

Since $\ln(at) \le at - 1$, we have

$$\ln t \le -1 - \ln a + at$$

or

$$t \le e^{-1-\ln a}e^{at},$$

resulting in

$$t^s \le e^{-s}a^{-s}e^{ast}.$$

Upon multiplication with g^t, one obtains

$$t^s g^t \le e^{-s}a^{-s}(e^{as}g)^t.$$

With $g_0 := e^{-s}a^{-s}$, this proves Lemma 3.4. $\qquad\qquad\square$

Proof of Lemma 3.5

By the triangle inequality one has

$$|y_t^h| \le \sum_{i=1}^{\ell}(|c_{i,0}| + |c_{i,1}|t + \cdots + |c_{i,p_i}|t^{p_i})\frac{1}{|z_i|^t}.$$

Since $|z_i| > 1$, it holds by Lemma 3.4

$$\frac{|c_{i,s}|t^s}{|z_i|^t} \le g_{0,i,s}g_{1,i,s}^t,$$

where with obvious notation $g_{0,i,s}$ and $g_{1,i,s}$ are defined as in the proof of Lemma 3.4. Consequently,

$$|y_t^h| \le \sum_{i=1}^{\ell} r_i \max_{0 \le s \le p_i}\{g_{0,i,s}\}\left(\max_{0 \le s \le p_i}\{g_{1,i,s}\}\right)^t$$

$$\le \ell \max_{1 \le i \le \ell}\left\{r_i \max_s\{g_{0,i,s}\}\right\}\left(\max_{1 \le i \le \ell}\left\{\max_s\{g_{1,i,s}\}\right\}\right)^t,$$

which defines g_0 and g_1 from Lemma 3.5. Hence, the proof is complete. $\qquad\square$

Proof of Proposition 3.5

Since the autocovariances satisfy the stable difference equation (3.13), Lemma 3.5 yields the result in (a). The same argument holds true for the MA coefficients in (b). Since the MA(∞) coefficients are geometrically bounded, $j^s|c_j|$ is geometrically bounded by Lemma 3.4 and hence summable as required for (c). Finally, the AR(∞) coefficients behave under invertibility as the MA(∞) coefficients under stability; consequently, the first two statements from (d) are established in the same way as the claims in (b) and (c). Since $\Phi(1) \neq 0$ under stationarity, we have $A(1) \neq 0$, which is the last claim in (d). Hence, the proof is complete. $\square$

4

Frequency Domain Analysis

Sometimes it is convenient to map the serial correlation of time series or stochastic processes from the time domain to the so-called frequency domain. We define the periodogram that expresses how strongly contribute the cycles with certain periods to the sample variance of the data. Then we define the spectrum as theoretical counterpart for stationary processes. This allows for the exponential (EXP) time series model settled in the frequency domain. Asymptotic properties of the periodogram are given under increasingly general assumptions. Finally, we introduce so-called Whittle estimation as a frequency domain approximation to maximum likelihood (ML) estimation. Whittle estimation is particularly convenient when estimating EXP models; further, it will turn out to be especially fruitful when estimating long memory.

4.1 Decomposition into Cycles

In this section, we work with a given time series of length T, $x_1, \ldots, x_T$. At the end, we will have arrived at a definition and some understanding of the periodogram, which is an empirical measure of how strongly certain cycles with a given frequency contribute to the variation of the data at hand. We will be able to decompose the sample variance,

$$\hat{\gamma}_x(0) := \frac{1}{T} \sum_{t=1}^{T} (x_t - \bar{x})^2, \tag{4.1}$$

accordingly. Since the periodogram as defined below does not require stochastic assumptions, we do not consider a stationary process $\{x_t\}$ as in the previous chapters, but simply a sample $\{x_t\}_{t=1,\ldots,T}$ without caring whether the data is generated from a stationary process or not. Stochastic assumptions are only required from Section 4.3 on.

The periodic cycles considered are trigonometric functions like

$$a \cos(\lambda t) + b \sin(\lambda t) = a \cos(\lambda(t + P)) + b \sin(\lambda(t + P)),$$

Time Series Analysis with Long Memory in View, First Edition. Uwe Hassler.
© 2019 John Wiley & Sons, Inc. Published 2019 by John Wiley & Sons, Inc.

where P stands for the period. The period P and the frequency λ are inversely related. By the properties of the sine and cosine, we have

$$\lambda P = 2\pi \quad \text{or} \quad P = \frac{2\pi}{\lambda}$$

as long as $\lambda \neq 0$. In particular, we will focus on the so-called harmonic frequencies, also called Fourier frequencies

$$\lambda_1 = 2\pi/T, \quad \lambda_2 = 2\lambda_1, \dots, \lambda_j = j\lambda_1 = 2\pi j/T.$$

The first harmonic frequency, λ_1, is also called the fundamental with period $P_1 = 2\pi/\lambda_1 = T$. Clearly, a cycle of a longer period (i.e. smaller frequency) cannot be observed from a sample of length T. Similarly, frequencies larger than π are not considered since they correspond to periods $2\pi/\lambda$ shorter than 2, which is not observable in discrete time with $t = 1, 2, \dots$. Hence, the set of harmonic frequencies typically consists of

$$\lambda_j = \frac{2\pi j}{T}, \quad j = 1, 2, \dots, M = \left\lfloor \frac{T-1}{2} \right\rfloor, \tag{4.2}$$

where $\lfloor x \rfloor$ denotes the largest integer smaller or equal to x. In the case of an odd or even sample size, respectively, we hence have

$$M = \begin{cases} \dfrac{T-1}{2}, & \text{if } T \text{ is odd} \\[2mm] \dfrac{T}{2} - 1, & \text{if } T \text{ is even} \end{cases}.$$

The first result that we obtain says that any given time series can be decomposed into the sum of (weighted) trigonometric functions evaluated at $\lambda_j t$.

Proposition 4.1 *Let* $\{x_t\}, t = 1, \dots, T$, *be a sequence of data. It then holds that*

$$x_t = a_0 + \sum_{j=1}^{M} (a_j \cos(\lambda_j t) + b_j \sin(\lambda_j t)) + b_{M+1}(-1)^t \quad \text{if } T = 2M + 2,$$

and

$$x_t = a_0 + \sum_{j=1}^{M} (a_j \cos(\lambda_j t) + b_j \sin(\lambda_j t)) \quad \text{if } T = 2M + 1,$$

where

$$a_0 = \frac{1}{T} \sum_{t=1}^{T} x_t = \bar{x},$$

$$a_j = \frac{2}{T} \sum_{t=1}^{T} x_t \cos(\lambda_j t), \quad j = 1, \dots, \left\lfloor \frac{T-1}{2} \right\rfloor = M,$$

$$b_j = \frac{2}{T} \sum_{t=1}^{T} x_t \sin(\lambda_j t), \quad j = 1, \dots, \left\lfloor \frac{T-1}{2} \right\rfloor = M,$$

$$b_{M+1} = \frac{1}{T} \sum_{t=1}^{T} x_t (-1)^t \quad \text{if } T = 2M + 2.$$

Proof: See Appendix to this chapter.

The properties in Proposition 4.1 heavily rely on the fact that the trigonometric functions evaluated at $\lambda_j t$ have a couple of classical orthogonality properties that are worth being remembered. They are collected in the following lemma proven, for example, by Fuller (1996, Theorem 3.1.1).

Lemma 4.1 *Let* $\lambda_j \in (0, \pi)$ *be a harmonic frequency from Eq. (4.2).*

(a) It then holds

$$\sum_{t=1}^{T} \cos(\lambda_j t) = \sum_{t=1}^{T} \sin(\lambda_j t) = 0,$$

$$\sum_{t=1}^{T} (\cos(\lambda_j t))^2 = \sum_{t=1}^{T} (\sin(\lambda_j t))^2 = \frac{T}{2},$$

$$\sum_{t=1}^{T} \cos(\lambda_j t) \, \sin(\lambda_j t) = 0.$$

(b) Further, for harmonic frequencies $\lambda_k \in (0, \pi)$ *with* $j \neq k$:

$$\sum_{t=1}^{T} \cos(\lambda_j t) \, \cos(\lambda_k t) = \sum_{t=1}^{T} \cos(\lambda_j t) \, \sin(\lambda_k t)$$

$$= \sum_{t=1}^{T} \sin(\lambda_j t) \, \sin(\lambda_k t)$$

$$= 0.$$

Proof: Fuller (1996, Theorem 3.1.1). □

Not only x_t itself can be decomposed into the sum of cycles with multiples of the Fourier frequencies λ_j. A related result holds for the sample variance.

For the following proposition we assume an odd sample size for simplicity.

Proposition 4.2 *Let $\{x_t\}$, $t = 1, \ldots, T = 2M + 1$, be a sequence of data. It then holds for the sample variance from Eq. (4.1) that*

$$\hat{\gamma}_x(0) = \frac{2}{T^2} \sum_{j=1}^{M} \left[\left(\sum_{t=1}^{T} x_t \cos(\lambda_j t) \right)^2 + \left(\sum_{t=1}^{T} x_t \sin(\lambda_j t) \right)^2 \right].$$

Proof: See Appendix to this chapter.

We now define the following entity called periodogram of $\{x_t\}_{t=1,\ldots,T}$ at the Fourier frequencies:

$$I_x(\lambda_j) = \frac{1}{2\pi T} \left[\left(\sum_{t=1}^{T} x_t \cos(\lambda_j t) \right)^2 + \left(\sum_{t=1}^{T} x_t \sin(\lambda_j t) \right)^2 \right]. \tag{4.3}$$

For $T = 2M + 1$ with M from Eq. (4.2), we have hence in Proposition 4.2

$$\hat{\gamma}_x(0) = \frac{4\pi}{T} \sum_{j=1}^{M} I_x(\lambda_j). \tag{4.4}$$

In that sense the periodogram allows to decompose the sample variance into M cycles or frequencies.

We now follow Brockwell and Davis (1991) or Fuller (1996) and define $I_x(\cdot)$ as a piecewise constant function on $[0, \pi]$ coinciding with $I_x(\lambda_j)$:

$$I_x(\lambda) = \begin{cases} 0, & 0 \leq \lambda \leq \dfrac{\pi}{T} \\[2mm] I_x(\lambda_1), & \dfrac{\pi}{T} < \lambda \leq \lambda_1 + \dfrac{\pi}{T} \\[2mm] I_x(\lambda_2), & \lambda_2 - \dfrac{\pi}{T} < \lambda \leq \lambda_2 + \dfrac{\pi}{T} \\[2mm] \vdots & \vdots \\[2mm] I_x(\lambda_M), & \lambda_M - \dfrac{\pi}{T} < \lambda \leq \lambda_M + \dfrac{\pi}{T} \end{cases}.$$

Obviously, $I_x(\lambda)$, $\lambda \in [0, \pi]$, is a nonnegative step function. The area under it is given by the following integral:

$$\int_0^{\pi} I_x(\lambda) \, d\lambda = \sum_{j=1}^{M} I_x(\lambda_j) \frac{2\pi}{T}.$$

Consequently, from Proposition 4.2 for $T = 2M + 1$

$$\hat{\gamma}_x(0) = 2 \int_0^{\pi} I_x(\lambda) \, d\lambda. \tag{4.5}$$

Hence, the periodogram at frequency $\lambda \in \left[\lambda_j \pm \frac{\pi}{T}\right]$ tells how strongly cycles of frequencies neighboring λ_j (or of period T/j) contribute to the sample variance of the data.

If $\{x_t\}$ is generated from a stationary, ergodic process, then we expect

$$\hat{\gamma}_x(0) \xrightarrow{p} \gamma_x(0) \quad \text{as} \quad T \to \infty.$$

How does the right-hand side of Eq. (4.5) behave in such a situation? This is what we turn to in Section 4.3. In Section 4.5 we will be concerned again with the periodogram and learn more about its statistical properties when assuming a stationary moving average $MA(\infty)$ process behind the data.

Before moving on, we want to briefly mention four further representations with properties of the periodogram.

First, in Hamilton (1994, Proposition 6.2 and p. 163) we find a definition of the periodogram that is equivalent to Eq. (4.3), although t is replaced by $(t - 1)$ in the trigonometric terms:

$$I_x(\lambda_j) = \frac{1}{2\pi T}\left[\left(\sum_{t=1}^{T} x_t \cos(\lambda_j(t - 1))\right)^2 + \left(\sum_{t=1}^{T} x_t \sin(\lambda_j(t - 1))\right)^2\right].$$

Second, by Lemma 4.1 we have

$$\sum_{t=1}^{T} \cos(\lambda_j t) = \sum_{t=1}^{T} \sin(\lambda_j t) = 0,$$

such that a constant term drops out of the periodogram. Demeaning does not affect the periodogram, i.e. replacing x_t in Eq. (4.3) by $x_t - \bar{x}$ leaves the periodogram unaltered,

$$I_x(\lambda_j) = I_{x-\bar{x}}(\lambda_j), \tag{4.6}$$

where $I_{x-\bar{x}}(\lambda_j)$ is defined as

$$\frac{1}{2\pi T}\left[\left(\sum_{t=1}^{T}(x_t - \bar{x})\cos(\lambda_j t)\right)^2 + \left(\sum_{t=1}^{T}(x_t - \bar{x})\sin(\lambda_j t)\right)^2\right].$$

Third, the periodogram is often defined as the square of the so-called discrete Fourier transform (DFT), which is the following complex-valued entity $w_x(\lambda_j)$[1*]:

$$w_x(\lambda_j) = \frac{1}{\sqrt{2\pi T}} \sum_{t=1}^{T} x_t e^{i\lambda_j t} \quad \text{with} \quad I_x(\lambda_j) = |w_x(\lambda_j)|^2. \tag{4.7}$$

1* The next section contains a short digression on complex numbers.

Fourth, we give an equivalent definition of the periodogram in terms of the sample autocovariances,

$$\hat{\gamma}_x(h) = \frac{1}{T} \sum_{t=1}^{T-h} (x_t - \bar{x})(x_{t+h} - \bar{x}), \quad h = 0, 1, \ldots, T-1. \tag{4.8}$$

Fuller (1996, p. 313, 314) argues that $\hat{\gamma}_x(h)$ is superior to the alternative estimator:

$$\frac{1}{T-h} \sum_{t=1}^{T-h} (x_t - \bar{x})(x_{t+h} - \bar{x}) = \frac{T}{T-h} \hat{\gamma}_x(h),$$

in that $\hat{\gamma}_x(h)$ guarantees positive definiteness of the sample autocovariance matrix. From Brockwell and Davis (1991, Proposition 10.1.2), we have

$$I_x(\lambda_j) = \frac{1}{2\pi} \left(\hat{\gamma}_x(0) + 2 \sum_{h=1}^{T-1} \hat{\gamma}_x(h) \cos(\lambda_j h) \right), \tag{4.9}$$

see also Hamilton (1994, Eq. (6.2.5)) or Fuller (1996, Result 7.1.1). Equation (4.9) shows how the time domain information, i.e. the sequence of all autocovariances, is mapped into the frequency domain. A corresponding result will be obtained for theoretical population terms in Section 4.3.

4.2 Complex Numbers and Transfer Functions

Before turning to the spectrum, this section contains a short technical digression. In particular, it will be helpful to be familiar with complex numbers $z \in \mathbb{C}$. For

$$i^2 = -1 \quad \text{or} \quad i = \sqrt{-1}$$

we define

$$z = a + i b, \quad a, b \in \mathbb{R}$$

with complex conjugate

$$\bar{z} = a - i b$$

and absolute value

$$|z| = \sqrt{z\bar{z}} = \sqrt{a^2 - iab + iab - i^2b} = \sqrt{a^2 + b^2},$$

such that

$$|z|^2 = a^2 + b^2. \tag{4.10}$$

We will also regularly apply Euler's formula:

$$e^{ix} = \cos x + i \sin x, \quad x \in \mathbb{R}, \tag{4.11}$$

with complex conjugate

$$\overline{e^{ix}} = e^{-ix} = \cos x - i \sin x.$$

For a stationary process we assume that it consists of a purely stochastic $MA(\infty)$ component and a constant $\mu = E(x_t)$, which we could set equal to 0 without loss of generality; see Brockwell and Davis (1991, p. 166):

$$x_t = \mu + \sum_{j=0}^{\infty} c_j \varepsilon_{t-j}, \quad c_0 = 1, \quad \sum_{j=0}^{\infty} c_j^2 < \infty, \quad \varepsilon_t \sim WN(0, \sigma^2). \tag{4.12}$$

The spectrum will be defined in terms of the MA coefficients or impulse responses $\{c_j\}$. To that end, we now define the so-called transfer function of the polynomial $C(L) = \sum_{k=0}^{\infty} c_k L^k$ as

$$C(e^{-i\lambda}) = \sum_{k=0}^{\infty} c_k e^{-i\lambda k}.$$

The transfer function squared to the power 2 is called power transfer function and abbreviated as follows:

$$T_C(\lambda) := |C(e^{-i\lambda})|^2 = \left| \sum_{k=0}^{\infty} c_k e^{-i\lambda k} \right|^2. \tag{4.13}$$

By (4.11) and (4.10) it holds that

$$T_C(\lambda) = \left(\sum_{k=0}^{\infty} c_k \cos(\lambda k) \right)^2 + \left(\sum_{k=0}^{\infty} c_k \sin(\lambda k) \right)^2. \tag{4.14}$$

An alternative representation building on

$$e^{i\lambda} + e^{-i\lambda} = 2 \cos \lambda$$

is

$$T_C(\lambda) = \sum_{k=0}^{\infty} c_k^2 + 2 \sum_{j=1}^{\infty} \sum_{k=0}^{\infty} c_k c_{k+j} \cos(\lambda j). \tag{4.15}$$

These are the building blocks to define the spectrum of a stationary process.

4.3 The Spectrum

We now carry the ideas from the first section for a sample of size T to the theoretical population, that is, to the realm of stochastic processes. We assume that $\{x_t\}_{t \in \mathbb{T}}$ is a stationary stochastic process with autocovariances $\gamma_x(h)$. In analogy to (4.5) we want to determine a function $f_x(\lambda)$, called spectrum, such that

$$\gamma_x(0) = 2 \int_0^\pi f_x(\lambda)\, d\lambda \tag{4.16}$$

with the previous interpretation that $f_x(\lambda)$ measures how strongly the frequency λ or period P contributes to the variance of the process. For $\lambda \to 0$ it holds that $P = 2\pi/\lambda \to \infty$. This may be interpreted the following way: A cycle with infinite period is no longer cyclic; it is rather a trend. Hence, $f_x(0)$ at the origin measures how strongly a trending or persistent behavior affects the variance of the process. Examples of spectra given below will support this intuition.

The following proposition combines results from Brockwell and Davis (1991, Corollary 4.3.1, Theorem 5.7.2, and Theorem 5.8.1).

Proposition 4.3   *Let $\{x_t\}$ be a stationary process from Eq. (3.1) or (4.12),*

$$x_t = \mu + \sum_{j=0}^\infty c_j\, \varepsilon_{t-j}, \qquad \sum_{j=0}^\infty c_j^2 < \infty, \qquad \{\varepsilon_t\} \sim \mathrm{WN}(0, \sigma^2).$$

Then the autocovariances can be written as

$$\gamma_x(h) = \int_{-\pi}^\pi \cos(\lambda h) f_x(\lambda)\, d\lambda, \tag{4.17}$$

where the so-called spectrum f_x is Lebesgue integrable on $(-\pi, \pi]$ and given by

$$f_x(\lambda) := T_C(\lambda)\, \frac{\sigma^2}{2\pi}, \tag{4.18}$$

with T_C defined in (4.13), and $\sigma^2 = \mathrm{Var}(\varepsilon_t)$. Further, it holds (Kolmogorov's formula) that

$$\sigma^2 = 2\pi \exp\left[\frac{1}{2\pi} \int_{-\pi}^\pi \ln f_x(\lambda)\, d\lambda\right] \tag{4.19}$$

as long as $\sigma > 0$.

Proof: From Brockwell and Davis (1991), see Appendix to this chapter.

We add a couple of comments on the spectrum and Proposition 4.3. First, the spectrum is symmetric as can be seen from Eq. (4.15) in that

$$T_C(-\lambda) = T_C(\lambda).$$

Therefore, we often consider $f_x(\lambda)$ only on $[0, \pi]$:

$$\gamma_x(h) = 2 \int_0^\pi \cos(\lambda h) f_x(\lambda)\, d\lambda.$$

For $h = 0$ we obtain

$$\gamma_x(0) = 2 \int_0^{\pi} f_x(\lambda) \, d\lambda,$$

which is exactly Eq. (4.16). Second, by Proposition 4.3 $f_x(\lambda)$ is always integrable, although it is not necessarily continuous. The spectrum may be unbounded for some frequency $\lambda_0 \in [0, \pi]$. This case (for $\lambda_0 = 0$) will characterize fractional integration as the leading case of long memory in Chapter 6. Strictly speaking, we would have to write in such a case:

$$f_x(\lambda) \quad \text{for } \lambda \in [0, \pi] \setminus \{\lambda_0\}.$$

Notwithstanding such a potential pole, or singularity, we will write for brevity $f_x(\lambda)$ with $\lambda \in [0, \pi]$, even if $f_x(\lambda_0)$ does not exist. Third, many authors, including Brockwell and Davis (1991), call f_x *spectral density*. Clearly, $f_x(\lambda) \geq 0$ by definition in Eq. (4.18), and

$$\int_{-\pi}^{\pi} \frac{f_x(\lambda)}{\gamma_x(0)} \, d\lambda = 1.$$

This is why we reserve the label *spectral density* for the normalized spectrum $f_x/\gamma_x(0)$. Fourth, the spectrum is periodic due to the periodicity of sine and cosine: $f_x(\lambda) = f_x(\lambda + 2\pi)$. We collect these more elementary properties in the following proposition.

Proposition 4.4 *For the spectrum from Proposition 4.3, it holds that*

$$f_x(-\lambda) = f_x(\lambda),$$

$$f_x(\lambda) = f_x(\lambda + 2\pi),$$

$$f_x(\lambda) \geq 0.$$

Further, $f_x(\lambda)$ is continuous on $[0, \pi]$ if $\{x_t\}$ has short memory (in the sense of Definition 2.4).

Proof: The first three statements are obvious, and the last statement on continuity is given, e.g. in Fuller (1996, Theorem 3.1.9). □

We want to highlight Kolmogorov's formula (4.19). We learn that $\sigma > 0$ if and only if $f_x(\lambda) > 0$ almost everywhere. Further, with $f_x(\lambda) = T_C(\lambda) \frac{\sigma^2}{2\pi}$ it is elementary to show that

$$\sigma^2 = \sigma^2 \exp\left[\frac{1}{2\pi} \int_{-\pi}^{\pi} \ln T_C(\lambda) \, d\lambda\right],$$

such that because of symmetry

$$\int_0^\pi \ln T_C(\lambda)\, d\lambda = 0. \tag{4.20}$$

This identity will be used in Section 4.6.

To better understand the link between autocovariances and spectrum, we suggest a little detour on what is called Fourier analysis in mathematics. Formally, the sequence $\{\gamma_x(h)\}$ given in Eq. (4.17) forms the sequence of so-called Fourier coefficients of the periodic function f_x. As corresponding Fourier sequence one defines

$$f_n(\lambda) := \frac{1}{2\pi} \sum_{h=-n}^{n} \gamma_x(h) e^{-i\lambda h}, \quad n \in \mathbb{N}.$$

Two questions naturally arise. Does $f_n(\lambda)$ converge as $n \to \infty$? And if so, does the limit coincide with $f_x(\lambda)$? We adopt the answers from Fuller (1996, Theorem 3.1.7).[2]

Proposition 4.5 *Let $f_x(\lambda)$ of $\{x_t\}$ from Proposition 4.3 be continuous at λ, and derivatives exist from the left and from the right. It then holds that*

$$\frac{1}{2\pi} \sum_{h=-n}^{n} \gamma_x(h) e^{-i\lambda h} \to f_x(\lambda)$$

as $n \to \infty$.

Proof: Fuller (1996, Theorem 3.1.7). □

By $\gamma_x(h) = \gamma_x(-h)$ and $e^{-i\lambda h} + e^{i\lambda h} = 2\cos(\lambda h)$, we get from Proposition 4.5 that

$$2\pi f_x(\lambda) = \gamma_x(0) + 2 \sum_{h=1}^{\infty} \gamma_x(h) \cos(\lambda h) \tag{4.21}$$

This is the population counterpart to the sample result Eq. (4.9) for the periodogram. Hence, we observe that one may move back and forth from the time domain to the frequency domain. The time domain information in terms of autocovariances is mapped into the frequency domain by Eq. (4.21); and vice versa, frequency domain information contained in $f_x(\lambda)$ is translated into serial correlation in the time domain by Eq. (4.17).

We want to highlight one particularly important link from time to frequency domain. Consider the spectrum at the origin ($\lambda = 0$). From Eq. (4.18) as well as

---

2 This proposition guarantees only pointwise convergence for each λ. Under the stronger assumption that f_x is differentiable with square integrable derivative, Fuller (1996, Theorem 3.1.8) establishes uniform convergence.

Eq. (4.21), we observe that $f_x(0)$ is proportional to the long-run variance used to define persistence in Definition 2.5 and to the accumulated impulse responses; see Proposition 3.4:

$$f_x(0) = \frac{\omega^2}{2\pi} = \text{CIR}^2 \frac{\sigma^2}{2\pi}. \tag{4.22}$$

Hence, the spectrum at the origin characterizes the persistence of a process. Antipersistence amounts to $f_x(0) = 0$, and moderate persistence has been defined as

$$0 < f_x(0) < \infty,$$

while in case of strong persistence, the spectrum does not take on a finite value at the origin.

We return to the existence of AR(∞) representations of MA processes, which was called invertibility:

$$\sum_{j=0}^{\infty} a_j x_{t-j} = \varepsilon_t.$$

The infinite sum on the left-hand side is understood to be convergent in mean square; see Definition 3.1. We now adopt a sufficient condition for invertibility.

Proposition 4.6 *Let $\{x_t\}$ be a stationary MA(∞) process with bounded spectrum f_x on $[0, \pi]$, and $1/f_x$ is integrable. Then $\{x_t\}$ is invertible in the sense of Definition 3.1.*

Proof: Bloomfield (1985, Theorem 2). □

Under short memory in the sense of Definition 2.4, $f_x(\lambda)$ is continuous on $[0, \pi]$ by Proposition 4.4. Consequently, as long as $f_x(\lambda)$ is bounded away from 0 everywhere, then the assumptions of Proposition 4.6 are automatically met by short memory processes. The special case of ARMA processes will be discussed in the next section. The condition of a bounded spectrum f_x in Proposition 4.6 does not allow for fractional integration considered from Chapter 6 on. Therefore, we will provide a generalization in Proposition 6.2 to cover situations where the spectrum is not bounded (at the origin), but has a singularity instead.

In Theorem 3.1 we saw the effect that linear filtering has on the autocovariance structure of a process. An analogous result is now given in the frequency domain. We consider

$$y_t = \sum_{j=0}^{\infty} \psi_j x_{t-j}, \quad \sum_{j=0}^{\infty} |\psi_j| < \infty.$$

The following proposition relates the spectra of $\{x_t\}$ and $\{y_t\}$ through the power transfer function of the filter $\Psi(L) = \sum_{j=0}^{\infty} \psi_j L^j$. This result for absolutely summable filters is adopted from Brockwell and Davis (1991, Theorem 4.4.1).

Proposition 4.7 *Let $\{x_t\}$ be a stationary process with spectrum $f_x(\lambda)$, and assume the filter*

$$\Psi(L) = \sum_{j=0}^{\infty} \psi_j L^j$$

to be absolutely summable, $\sum_{j=0}^{\infty} |\psi_j| < \infty$. Consider $\{y_t\}$ as

$$y_t = \Psi(L)x_t.$$

Then $\{y_t\}$ has a spectrum given by

$$f_y(\lambda) = T_\Psi(\lambda)f_x(\lambda), \quad \lambda \in [0, \pi],$$

where the power transfer function $T_\Psi(\lambda)$ of $\Psi(L)$ is defined in (4.13).

Proof: Brockwell and Davis (1991, Theorem 4.4.1). □

This result will be used to obtain spectra of certain parametric models.

4.4 Parametric Spectra

Proposition 4.7 allows to determine theoretical spectra for many processes characterized by a given parameterization. First, let us consider a white noise sequence $\{\varepsilon_t\}$. By the definition given in Eq. (4.18), it obviously holds that

$$f_\varepsilon(\lambda) = \frac{\sigma^2}{2\pi}.$$

This means that all frequencies contribute equally strongly to the variance of a white noise process. For a stationary MA process $\{x_t\}$ with

$$x_t = \sum_{j=0}^{\infty} c_j \varepsilon_{t-j} = C(L)\varepsilon_t$$

Proposition 4.7 yields

$$f_x(\lambda) = T_C(\lambda)f_\varepsilon(\lambda) = T_C(\lambda)\frac{\sigma^2}{2\pi}.$$

In light of Eq. (4.15), this is particularly simple to evaluate for MA processes of finite order. For the MA(1) process

$$x_t = \varepsilon_t + \theta\,\varepsilon_{t-1}$$

one immediately gets

$$f_{\text{MA1}}(\lambda) = (1 + \theta^2 + 2\theta\cos(\lambda))\frac{\sigma^2}{2\pi}.$$

For stationary ARMA processes defined in Eq. (3.6),$\Phi(L)x_t = \Theta(L)\epsilon_t$, it holds due to the stability assumption Eq. (3.7) that

$$T_\Phi(\lambda) \neq 0 \quad \text{on } [0, \pi].$$

Consequently, it holds for stationary ARMA processes $\{x_t\}$ that

$$f_{ARMA}(\lambda) = \frac{T_\Theta(\lambda)}{T_\Phi(\lambda)} \frac{\sigma^2}{2\pi}, \tag{4.23}$$

where $T_\Theta(\lambda)$ and $T_\Phi(\lambda)$ are the power transfer functions of the MA and AR polynomials, respectively. By Eq. (3.8) it holds under invertibility that $\Theta(e^{-i\lambda}) \neq 0$, such that the spectrum is positive everywhere: $T_\Theta(\lambda) > 0$. Hence, ARMA processes meeting Eqs. (3.7) and (3.8) are invertible in the sense of Definition 3.1 by Proposition 4.6. Because of Eq. (4.22) the value at the origin is of particular interest,

$$f_{ARMA}(0) = \frac{\Theta^2(1)}{\Phi^2(1)} \frac{\sigma^2}{2\pi}. \tag{4.24}$$

The special case of an AR(1) process,

$$x_t = \phi\, x_{t-1} + \varepsilon_t,$$

reduces to

$$f_{AR1}(\lambda) = \frac{1}{1 + \phi^2 - 2\phi\cos(\lambda)} \frac{\sigma^2}{2\pi}.$$

Consider the limiting case of $\phi \to 1$ where the AR(1) process becomes nonstationary. For this nonstationary case, the spectrum is actually not defined, but we may take f_{ps} from Eq. (4.25) as a definition of a pseudo-spectrum interpreted as the limit of the expected periodogram; see, e.g. Solo (1992):

$$f_{ps}(\lambda) = \frac{1}{2 - 2\cos(\lambda)} \frac{\sigma^2}{2\pi}, \quad \lambda \neq 0. \tag{4.25}$$

Note that the pseudo-spectrum has a pole at the origin, diverging to infinity as $\lambda \to 0$ (see Eq. (5.6) below):

$$f_{ps}(\lambda) \sim \frac{1}{\lambda^2} \frac{\sigma^2}{2\pi}, \quad \lambda \to 0.$$

This reflects the trending behavior of the nonstationary AR(1) process with $\phi = 1$; see also the above discussion around Eq. (4.22).

A discussion and graphical illustration of many more concrete ARMA spectra is given, e.g. in Hassler (2016, Chapter 4).

Proposition 4.7 allows to rewrite a parametric model given in the time domain in spectral terms. One may proceed the other way around, too, and define a parametric model directly in the frequency domain. One such model was proposed by Bloomfield (1973) and called *exponential model*. We maintain the spectral assumption:

$$f_x(\lambda; \psi) = \frac{\sigma^2}{2\pi} \exp\left(\sum_{\ell=1}^{k} \psi_\ell \cos(\ell \lambda)\right), \tag{4.26}$$

given the parameters $\psi_1, \ldots, \psi_k$. With $\psi_0 = \ln(\sigma^2/2\pi)$, one often writes

$$f_x(\lambda; \psi) = \exp\left(\sum_{\ell=0}^{k} \psi_\ell \cos(\ell \lambda)\right).$$

We abbreviate this model as EXP(k); it will turn out to be particularly useful when modeling fractional integration. If we stick to the assumption of a stationary MA(∞) process with spectrum from Eq. (4.26), we obtain from Eq. (4.13)

$$\left|\sum_{j=0}^{\infty} c_j e^{-i\lambda j}\right|^2 = \exp\left(\sum_{\ell=1}^{k} \psi_\ell \cos(\lambda \ell)\right),$$

i.e.

$$\sum_{j=0}^{\infty} c_j e^{-i\lambda j} = \exp\left(\frac{1}{2} \sum_{\ell=1}^{k} \psi_\ell e^{-i\lambda \ell}\right).$$

A Taylor expansion of the right-hand side hence provides the MA coefficients $\{c_j\}$; see Pourahmadi (1983). From Hurvich (2002, Eq. (4)) we adopt the following recursive relation:

$$c_j = \frac{1}{2j} \sum_{n=1}^{j} n\psi_n c_{j-n}, \quad j \geq 1, \quad c_0 = 1. \tag{4.27}$$

From that one may see that the MA coefficients decay exponentially fast to 0. Consider, for instance, the EXP(1) model with

$$c_j = \frac{1}{2j} \psi_1 c_{j-1},$$

such that

$$c_j = \frac{1}{j!}\left(\frac{\psi_1}{2}\right)^j; \tag{4.28}$$

clearly, the impulse responses vanish even faster than geometrically irrespective of the value of ψ_1.

In Figure 4.1 we provide a comparison of the simplest AR and EXP models, namely, models of order 1. We choose $\psi_1 = \psi = 1.6$, such that $c_1 = 0.8$ for the EXP(1) model; see Eq. (4.28). The AR coefficient $\phi_1 = \phi = 0.8$ is picked such that $c_1 = 0.8$ for the AR(1) model, too. Although the first impulse response coefficients are equal, the sequence $\{c_j\}$ dies out much faster for the EXP(1) model

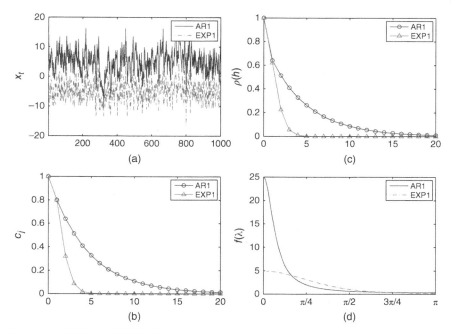

Figure 4.1 AR(1) and EXP(1) with $\phi = 0.8$ and $\psi = 1.6$.

than in the AR(1) case. This less persistent behavior is clearly reflected in the frequency domain, $f_x(\lambda)$, where the EXP(1) spectrum is much less peaked at the origin.[3] Consequently, the autocorrelations $\rho(h)$ are close to 0 from $h = 3$ on. Further, we display simulated series of $T = 1000$ observations (adding 5 and -5 to make them distinguishable): Not surprisingly, the AR(1) series $\{x_t\}$ is more persistent than the EXP(1) series in that the AR(1) case displays repeated, longer periods of trending behavior.

Finally, we want to stress what stationary ARMA processes and EXP models have in common: Both have spectra that are twice differentiable and that have a first derivative f_x' equal to 0 at the origin.

Proposition 4.8   *Let the stationary process $\{x_t\}$ follow either an ARMA model or an EXP model. Then the spectrum is twice differentiable with first derivative $f_x'(\lambda)$ and $f_x'(0) = 0$, and with bounded second derivative, $f_x''(\lambda) < \infty$, $\lambda \in [0, \pi]$.*

Proof: See Appendix to this chapter.

---

3 For the graphs we have set $\sigma^2 = 2\pi$.

More trivially, the ARMA and EXP spectra are bounded and bounded away from 0 at the origin (as long as the ARMA process is invertible), which directly follows from Eqs. (4.24) and (4.26).

4.5 (Asymptotic) Properties of the Periodogram

As a synthesis of the previous sections, we now assume a sample $\{x_1, \ldots, x_T\}$ generated from a stationary linear process $\{x_t\}$. In such a setup we study the stochastic properties of the periodogram.

Before presenting the first result, we remind the reader of the following definition of an $\mathcal{EXP}$ distribution. The random variable X follows an exponential distribution with parameter $\theta > 0$,

$$X \sim \mathcal{EXP}(\theta) \quad \text{with} \quad \mathrm{E}(X) = \frac{1}{\theta}, \quad \mathrm{Var}(X) = \frac{1}{\theta^2},$$

if the density is given by

$$\phi_x(x) = \begin{cases} \theta e^{-\theta x}, & x \geq 0 \\ 0, & \text{else} \end{cases}.$$

Let us now start with the motivating exposition where $\{x_t\}$ is normal white noise. We will establish the following distributional result at the harmonic frequencies. The exponential distribution is often rephrased in terms of a chi-squared distribution with 2 degrees of freedom, $\chi^2(2)$, for short.

Proposition 4.9 *Assume a sample $\{\varepsilon_1, \ldots, \varepsilon_T\}$ from normal white noise, i.e.*

$$\varepsilon_t \sim \mathcal{N}(0, \sigma^2).$$

It then holds for the periodogram $I_\varepsilon(\lambda_j)$ at the harmonic frequencies $\lambda_1, \ldots, \lambda_M$ from Eq. (4.2) that

$$I_\varepsilon(\lambda_j) \sim \mathcal{EXP}\left(\frac{2\pi}{\sigma^2}\right),$$

or equivalently

$$2\frac{I_\varepsilon(\lambda_j)}{\sigma^2/2\pi} \sim \chi^2(2),$$

and $I_\varepsilon(\lambda_1)$ through $I_\varepsilon(\lambda_M)$ are independent.

Proof: See Appendix to this chapter.

Due to the very restrictive assumptions, we have an exact finite sample result in Proposition 4.9, which will be carried to the case of more realistic processes asymptotically further down. Remember that in the case of white noise, the theoretical spectrum is

$$f_\varepsilon(\lambda_j) = \frac{\sigma^2}{2\pi}.$$

Not surprisingly, the distribution of the periodogram as sample entity depends on the spectrum as population counterpart. The equivalence of the stated $\mathcal{EXP}$ distribution to the χ^2 distribution is given in Corollary 4.1. This corollary is an application of the following lemma on variable transformation, which is fundamental in the probability literature; see e.g. Mood et al. (1974).

Lemma 4.2 *Let X be a continuous random variable with density ϕ_x on the domain D_x:*

$$\phi_x(x) \begin{cases} > 0, & x \in D_x \\ = 0, & x \notin D_x \end{cases}.$$

Let $Y = g(X)$ be a transformation where g is invertible with differentiable g^{-1}:

$$g : D_x \to D_y, \quad g^{-1} : D_y \to D_x.$$

Denote the density of Y by ϕ_y with

$$\phi_y(y) \begin{cases} > 0, & y \in D_y \\ = 0, & y \notin D_y \end{cases}.$$

It then holds that

$$\phi_y(y) = \left| \frac{dg^{-1}(y)}{dy} \right| \phi_x(g^{-1}(y)), \quad y \in D_y.$$

Proof: Mood et al. (1974, p. 200). □

This lemma yields the following corollary, on which the equivalence of the distributional statements in Proposition 4.9 rests.

Corollary 4.1 *Let $X \sim \chi^2(2)$, and $Y = c\, X$ with $c > 0$. It then holds that*

$$Y \sim \mathcal{EXP}\left(\frac{1}{2c}\right).$$

Proof: See Appendix to this chapter.

Of course, this corollary can also be read the other way around:

$$Y \sim \mathcal{EXP}\left(\frac{1}{2c}\right) \quad \text{implies} \quad \frac{Y}{c} \sim \chi^2(2).$$

Now, we move on to processes with serial correlation. The result from Proposition 4.9 carries over asymptotically, however, only for a finite number p of harmonic frequencies. From Giraitis et al. (2012, Theorem 5.3.1) we take the following limiting result; see also Brillinger (1975, Theorem 5.2.6). Here, "$\xrightarrow{D}$" stands for the usual convergence in distribution as $T \to \infty$.

Proposition 4.10 *Assume a sample* $\{x_1, \ldots, x_T\}$ *from an absolutely summable, linear process, i.e.*

$$x_t = \sum_{j=0}^{\infty} c_j \, \varepsilon_{t-j}, \qquad \sum_{j=0}^{\infty} |c_j| < \infty, \quad \varepsilon_t \sim iid(0, \sigma^2),$$

with positive spectrum, $f_x(\lambda) > 0$, $\lambda \in [0, \pi]$. *Consider harmonic frequencies* λ_{j_i} *where the indices may vary with* T, $j_i = j_i(T)$, $i = 1, \ldots, p$, *with*

$$1 \le j_1 < \cdots < j_p < \frac{T}{2}.$$

Define

$$v_{j_i} = 2 \, \frac{I_x(\lambda_{j_i})}{f_x(\lambda_{j_i})}.$$

It then holds that[4]

$$v_{j_i} \xrightarrow{D} \chi^2(2), \quad i = 1, \ldots, p,$$

and v_{j_1} *through* v_{j_p} *are asymptotically independent.*

Proof: Giraitis et al. (2012, Theorem 5.3.1). □

Under a slightly different set of assumptions, Brillinger (1975, Theorem 5.2.6) provided the same finding, cf. also the underlying result on discrete Fourier transforms in Brillinger (1975, Theorem 4.4.1). Also, Fuller (1996, Theorem 7.1.2) or Brockwell and Davis (1991, Theorem 10.3.2) have similar results, although not allowing for the potential time dependence of the indices, $1 \le j_1 < \cdots < j_p < T/2$. The asymptotic independence is remarkable, since the distance between any two harmonic frequencies vanishes for fixed j and k,

$$|\lambda_j - \lambda_k| = \frac{2\pi}{T} \, |j - k| \to 0,$$

---

4 The notation $v_{j_i} \xrightarrow{D} \chi^2(2)$ is short for "v_{j_i} converges in distribution to a random variable that follows a $\chi^2(2)$ distribution." A similar notation will be used for the rest of the book.

as $T \to \infty$. Nevertheless, the periodogram entities $I_x(\lambda_j)$ and $I_x(\lambda_k)$ are asymptotically independent. A special case arises when all harmonic frequencies are in the vicinity of some *fixed* frequency λ_0, such that

$$\lambda_{j_i} \overset{T\to\infty}{\to} \lambda_0, \quad i = 1, \dots, p. \tag{4.29}$$

In that case, we obtain that the finite collection of p periodograms becomes iid asymptotically with

$$I_x(\lambda_{j_i}) \overset{D}{\to} \frac{f_x(\lambda_0)}{2} \chi^2(2),$$

or

$$I_x(\lambda_{j_i}) \overset{D}{\to} \mathcal{EXP}\left(\frac{1}{f_x(\lambda_0)}\right)$$

according to Corollary 4.1. In particular for $\lambda_0 = 0$, the result from Proposition 4.10 had been provided by Hannan (1973b, Theorem 3) under the condition Eq. (4.29).

Proposition 4.10 has been extended to the case where the collection of p periodogram values is allowed to grow with the sample size T but more slowly. To that end, Phillips (2007, Theorem 3.2) had to strengthen the assumptions with respect to higher moments of the innovations and with respect to summability. Further, $\lambda_1, \dots, \lambda_p$ are now again the first Fourier frequencies next to the origin.

Proposition 4.11 *Assume a sample $\{x_1, \dots, x_T\}$ meeting the assumptions from Proposition 4.10 where in addition*

$$E(|\varepsilon_t|^q) < \infty \quad \text{for some } q > 2,$$

and $\sum_{j=0}^{\infty} j|c_j| < \infty$. At the first p harmonic frequencies, it holds asymptotically that $\{I_x(\lambda_j)\}$ forms an iid sequence with

$$I_x(\lambda_j) \overset{D}{\to} \mathcal{EXP}\left(\frac{1}{f_x(0)}\right), \quad j = 1, \dots, p,$$

as long as

$$\frac{p}{T^{1/2-1/q}} \to 0 \quad \text{as } T \to \infty. \tag{4.30}$$

Proof: Follows from the limiting complex normal distribution obtained by Phillips (2007, Theorem 3.2) for the DFT, see Eq. (4.7). Details are given in the Appendix to this chapter. □

Note that the assumption of 1-summability in Proposition 4.11,

$$\sum_{j=0}^{\infty} j|c_j| < \infty,$$

is met by all stationary ARMA processes; see Proposition 3.5. Proposition 4.11 works under (4.29) with $\lambda_0 = 0$. With respect to $\lambda_0 > 0$, Phillips (2007, p. 109) suggested that the analogous result holds: "Theorem 3.2 is given for collections of dft's in the vicinity in the origin [...]. However, with some modifications to the proof of Theorem 3.2, a related result can be proved for asymptotically infinite collections of dft's in the vicinity of an arbitrary fixed frequency." If Phillips (2007, Theorem 3.2) holds for $\lambda_0 > 0$, then Proposition 4.11 holds for $\lambda_0 > 0$, too.

From Proposition 4.10 we learn under Eq. (4.29) because of the mean and variance of a χ^2 distribution that

$$E(I_x(\lambda_0)) \to f_x(\lambda_0), \quad T \to \infty,$$

$$\text{Var}(I_x(\lambda_0)) \to f_x^2(\lambda_0) \neq 0, \quad T \to \infty.$$

Hence, for any fixed frequency, the periodogram is asymptotically unbiased for the theoretical spectrum; however, the variance of $I_x(\lambda_0)$ does not converge to 0, such that the estimation of $f_x(\lambda_0)$ by $I_x(\lambda_0)$ is not consistent. Consistent spectral estimation of $f_x(\lambda_0)$ for some fixed λ_0 can be achieved by smoothing $I_x(\lambda)$ over a neighborhood of λ_0; details are included in, e.g. Anderson (1971), Brockwell and Davis (1991), or Fuller (1996), but we do not need them here.

4.6 Whittle Estimation

We close this chapter by heuristically introducing a frequency domain approximation to ML estimation. It is not treated in standard textbooks but has turned out to be particularly fruitful when analyzing long memory models. The general proposal is due to Whittle (1953a) and has been extended for multivariate time series by Whittle (1953b). In fact, it builds on a frequency domain approximation of the Gaussian time domain likelihood. We circumvent the mathematical derivation and rather build on intuition using Proposition 4.11. Although this is not rigorous, we now treat[5] $I(\lambda_1), \ldots, I(\lambda_M)$ at all harmonic frequencies as if $I(\lambda_j)$ followed $\mathcal{EXP}(1/f(\lambda_j))$ asymptotically and as if $I(\lambda_j)$ and $I(\lambda_k)$ were asymptotically independent for $j \neq k$, i.e. we ignore condition Eq. (4.30). If this was generally true (e.g. under Proposition 4.9), then the likelihood function would asymptotically become

$$\mathcal{L}(\psi, \sigma; I(\lambda_1), \ldots, I(\lambda_M)) = \prod_{j=1}^{M} \frac{1}{f(\lambda_j; \psi, \sigma)} \exp\left(\frac{-I(\lambda_j)}{f(\lambda_j; \psi, \sigma)}\right).$$

---

5 Note that we drop the index x on the periodogram.

The spectrum f depends on a parameter vector ψ and on σ. In case of an ARMA(p,q) process from Eq. (3.6), we have

$$\psi = (\phi_1, \ldots, \phi_p, \theta_1, \ldots, \theta_q)'.$$

But the parametric model could equally well be an EXP(k) process as defined in Eq. (4.26). Generally, we assume an MA(∞) process like in Proposition 4.3, such that

$$f(\lambda_j; \psi, \sigma) = T_C(\lambda_j; \psi) \frac{\sigma^2}{2\pi}$$

with power transfer function T_C. Consequently, the log-likelihood in the frequency domain approximately becomes, upon dividing by M, see Whittle (1953b, Theorem 6),

$$\ell_W(\psi, \sigma) := M^{-1} \ln(\mathcal{L}(\psi, \sigma; I(\lambda_1), \ldots, I(\lambda_M))) \tag{4.31}$$

$$= -\frac{1}{M} \sum_{j=1}^{M} \ln(f(\lambda_j; \psi, \sigma)) - \frac{1}{M} \sum_{j=1}^{M} \frac{I(\lambda_j)}{f(\lambda_j; \psi, \sigma)}.$$

With $M = \lfloor (T-1)/2 \rfloor$ we may approximate

$$\frac{1}{M} \sum_{j=1}^{M} \ln(f(\lambda_j; \psi, \sigma)) \approx \frac{1}{\pi} \int_0^{\pi} \ln(f(\lambda; \psi, \sigma)) d\lambda$$

$$= \frac{1}{2\pi} \int_{-\pi}^{\pi} \ln(f(\lambda; \psi, \sigma)) d\lambda = \ln \frac{\sigma^2}{2\pi},$$

where the latter equality follows from Eq. (4.20). The ML principle hence requires to maximize

$$\ell_W(\psi, \sigma) \approx -\ln \sigma^2 + \ln 2\pi - \frac{1}{M} \frac{2\pi}{\sigma^2} \sum_{j=1}^{M} \frac{I(\lambda_j)}{T_C(\lambda_j; \psi)}.$$

Concentrating out σ^2 yields as Whittle estimator for the innovation variance[6]

$$\hat{\sigma}_W^2 := \frac{2\pi}{M} \sum_{j=1}^{M} \frac{I(\lambda_j)}{T_C(\lambda_j; \hat{\psi}_W)}.$$

Plugging into $\ell_W(\psi, \sigma)$ yields approximately

$$-\ln \frac{2\pi}{M} \sum_{j=1}^{M} \frac{I(\lambda_j)}{T_C(\lambda_j; \psi)} + \ln 2\pi - 1,$$

such that maximization provides as Whittle estimator

$$\hat{\psi}_W := \arg \min Q(\psi)$$

---

6 Consider the example $x_t = \varepsilon_t \sim \text{WN}(0, \sigma^2)$, i.e. $T_C = 1$. Consequently, $\hat{\sigma}_W^2 = \frac{2\pi}{M} \sum_{j=1}^{M} I(\lambda_j)$. Not surprisingly this corresponds to $\hat{\gamma}_x(0)$ from Eq. (4.4).

with

$$Q(\psi) := \frac{2\pi}{M} \sum_{j=1}^{M} \frac{I(\lambda_j)}{T_C(\lambda_j; \psi)}, \quad M = \left\lfloor \frac{T-1}{2} \right\rfloor. \tag{4.32}$$

Note that many authors define Q over the full range $(0, 2\pi)$:

$$Q(\psi) = \frac{2\pi}{T} \sum_{j=1}^{T-1} \frac{I(\lambda_j)}{T_C(\lambda_j; \psi)}.$$

Both definitions are equivalent except for the effect when T is odd or even. Generally, the objective function Q from Eq. (4.32) is computationally very simple.

Before we obtain properties of Whittle estimation, we have to fix some assumptions.

Assumption 4.1 (*Parametric model*) *Let $\{x_t\}$ be given by*

$$x_t = \sum_{j=0}^{\infty} c_j \varepsilon_{t-j}, \quad \varepsilon_t \sim \mathrm{WN}(0, \sigma^2), \quad \sum_{j=0}^{\infty} c_j^2 < \infty,$$

where $c_0 = 1$ and $c_j = c_{j,\psi}$ depend on a k-dimensional vector ψ contained in Ψ, $\psi \in \Psi$.[7] Further, we denote the true parameter vector by ψ_0, and it is assumed to be an interior point of Ψ. The spectrum becomes

$$f(\lambda) = f(\lambda; \psi) = \frac{\sigma^2}{2\pi} T_C(\lambda; \psi),$$

where the power transfer function T_C according to Eq. (4.13) depends on $\{c_j\} = \{c_{j,\psi}\}$. The spectrum is assumed to fulfill the conditions: (i) $\int_{-\pi}^{\pi} \ln f(\lambda; \psi) d\lambda$ is twice differentiable with respect to ψ under the integral sign; (ii) $f(\lambda; \psi)$ and $1/f$ are continuous for all $(\lambda; \psi)$; (iii) the first and second derivatives of $1/f(\lambda; \psi)$ are continuous for all $(\lambda; \psi)$; and (iv) the first and second derivatives of $f(\lambda; \psi)$ are continuous for all $(\lambda; \psi)$.

We now state a limiting result for Whittle estimation under short memory; the corresponding limiting covariance matrix will show up under long memory as well. The following result by Hannan (1973a) assumes that the process under consideration has an MA(∞) representation where the squared coefficients are 1-summable.

Proposition 4.12 *Let $\{x_t\}$ meet Assumption 4.1 with*

$$\sum_{j=0}^{\infty} j c_j^2 < \infty, \quad \text{and } f(0) > 0.$$

---

7 The closure of Ψ is assumed to be compact.

More precisely, let conditions A and B from Hannan (1973a) be met. It then holds for the true parameter vector ψ_0 and a sample of size T that

$$\sqrt{T}(\hat{\psi}_W - \psi_0) \overset{D}{\to} \mathcal{N}_k(0, W^{-1}(\psi_0))$$

with $(i, j \in \{1, \ldots, k\})$

$$w_{ij}(\psi) = \frac{1}{4\pi} \int_{-\pi}^{\pi} \frac{\partial \ln f(\lambda; \psi)}{\partial \psi_i} \frac{\partial \ln f(\lambda; \psi)}{\partial \psi_j} d\lambda \tag{4.33}$$

being the typical element of $W(\psi)$.

Proof: Hannan (1973a, Theorem 2). □

Note that this limiting result does not rely on Gaussianity of the process. Assuming a normal distribution of the data, Whittle (1953a, Eq. (3.7)) had already obtained expression Eq. (4.33) for the covariance matrix. Whittle (1953a) derived $W^{-1}(\psi)$ as Fisher information matrix under Gaussianity, thus showing that $\hat{\psi}_W$ is efficient notwithstanding the fact that it is based only on an approximation of maximum likelihood. In fact, in Section 8.2 we will see that $W(\psi)$ from Eq. (4.33) gives the general form of the limiting covariance matrix under (exact) maximum likelihood estimation.

The 1-summability of c_j^2 excludes long memory in terms of fractional integration as considered in subsequent chapters; similarly, the assumption of a positive spectrum (at the origin) rules out antipersistence in terms of fractional integration. But Proposition 4.12 holds for stationary and invertible ARMA processes. Nevertheless, the asymptotic covariance matrix $W(\psi)$ from Eq. (4.33) will later show up in a long memory environment, too, however sometimes *in disguise*. The following lemma contains two alternative representations that hold without $f(0; \psi) > 0$ and without continuity at the origin.

Lemma 4.3 *Let the spectrum $f(\lambda; \psi) = \sigma^2 T_C(\lambda; \psi)/2\pi$ satisfy (i) and (iii) from Assumption 4.1 and (ii) for $\lambda \neq 0$. It then holds for $w_{ij}(\psi)$ from Eq. (4.33) that*

$$4\pi w_{ij}(\psi) = \int_{-\pi}^{\pi} f^{-2}(\lambda; \psi) \frac{\partial f(\lambda; \psi)}{\partial \psi_i} \frac{\partial f(\lambda; \psi)}{\partial \psi_j} d\lambda \tag{4.34}$$

$$= \int_{-\pi}^{\pi} f(\lambda; \psi) \frac{\partial^2 f^{-1}(\lambda; \psi)}{\partial \psi_i \partial \psi_j} d\lambda, \tag{4.35}$$

where $f^{-1} = 1/f$.

Proof: Follows from Fox and Taqqu (1986), see Appendix to this chapter for details.

The expression of $W(\psi)$ given in Eq. (4.34) was obtained in Whittle (1953b, Theorem 9). In a long memory context, Fox and Taqqu (1986, Theorem 2) and Giraitis and Surgailis (1990, Theorem 4) provided $W(\psi)$ according to Eq. (4.35), while Dahlhaus (1989, Theorem 2.1) gave Eq. (4.33). All three covariance expressions reveal the following nice property: If f and $\tilde{f}$ are the spectra of $\{x_t\}$ and $\{\tilde{x}_t\}$, respectively, and if

$$f_x(\lambda) = T_C(\lambda)\frac{\sigma^2}{2\pi} \quad \text{and} \quad \tilde{f}_x(\lambda) = \frac{1}{T_C(\lambda)}\frac{\sigma^2}{2\pi},$$

then the covariance matrices for $f(\lambda)$ and $\tilde{f}(\lambda)$ are identical. This is clear because

$$\frac{\partial \ln \tilde{f}}{\partial \psi_i} = \frac{\partial \ln T_C^{-1}}{\partial \psi_i} = -\frac{\partial \ln f}{\partial \psi_i}.$$

This is exactly the situation of AR and MA processes with equal lag polynomials. Consider the MA(q) process:

$$x_t = \Theta(L)\varepsilon_t, \quad \text{with} \quad f_x(\lambda) = T_\Theta(\lambda)\frac{\sigma^2}{2\pi},$$

see Proposition 4.7. Assume $\Theta(L)$ to be invertible and consider the AR(q) process $\{\tilde{x}_t\}$,

$$\Theta(L)\tilde{x}_t = \varepsilon_t \quad \text{with} \quad \tilde{f}_x(\lambda) = \frac{1}{T_\Theta(\lambda)}\frac{\sigma^2}{2\pi},$$

see Eq. (4.23).

We close this section by turning to the EXP model under Whittle estimation. According to Eq. (4.26), T_C from Eq. (4.32) then becomes

$$T_C(\lambda_j; \psi) = \exp\left(\sum_{\ell=1}^k \psi_\ell \cos(\ell \lambda_j)\right).$$

Further, the matrix W from Proposition 4.12 is diagonal and independent of the true parameter values. The following corollary makes this more precise.

Corollary 4.2 *Let $\{x_t\}$ follow an EXP model according to Eq. (4.26). It then holds for the matrix W from Eq. (4.33) that*

$$W(\psi) = W = \frac{1}{4} I_k,$$

where I_k stands for the k-dimensional identity matrix.

Proof: See Appendix to this chapter.

We observe that Whittle estimation implies a very simple covariance structure under the exponential (EXP) model, and

$$\frac{\sqrt{T}}{2}(\hat{\psi}_W - \psi_0) \overset{D}{\to} \mathcal{N}_k(0, I_k) \quad \text{or} \quad \sqrt{\frac{M}{2}}(\hat{\psi}_W - \psi_0) \overset{D}{\to} \mathcal{N}_k(0, I_k). \quad (4.36)$$

In particular the variances of the estimators do not depend on the true parameter values; further, they are independent of k, i.e. they do not grow with the number of estimated parameters, which is due to the fact that $\cos(i\lambda)$ and $\cos(j\lambda)$ are orthogonal for $i \neq j$ (cf. the proof). Hence, application of Whittle estimation is particularly convenient for EXP time series models. More generally, the covariance matrix from Eq. (4.33) shows up for exact maximum likelihood estimation and different approximations to maximum likelihood in Chapter 8, such that the EXP model will provide very simple covariance matrices for these estimators.

4.7 Technical Appendix: Proofs

Proof of Proposition 4.1

We first deal with the case of even sample size, $T = 2M + 2 = 2N$, where $N = M + 1$ for simplicity. We define a parameter vector of length $2N$,

$$\beta' = (a_0, a_1, b_1, \dots, a_{N-1} \, b_{N-1}, b_N),$$

and a corresponding vector of regressors,

$$F'_t = (1, \cos(\lambda_1 t), \sin(\lambda_1 t), \dots, \cos(\lambda_{N-1} t), \sin(\lambda_{N-1} t), (-1)^t)$$

to write the regression equation

$$x_t = F'_t \beta + u_t, \quad t = 1, \dots, T,$$

or in matrix notation

$$x = F\beta + u$$

with

$$x' = (x_1, \dots, x_T)$$
$$F' = (F_1, \dots, F_T)$$
$$u' = (u_1, \dots, u_T).$$

The ordinary least squares (OLS) estimator is

$$\hat{\beta} = (F'F)^{-1}F'x$$

where $F'F$ is diagonal by Lemma 4.1

$$F'F = \text{diag}\left(T, \frac{T}{2}, \dots, \frac{T}{2}, T\right).$$

Consequently,

$$
\hat{\beta} = \begin{pmatrix}
\dfrac{1}{T} \sum_{t=1}^{T} x_t \\[2mm]
\dfrac{2}{T} \sum_{t=1}^{T} x_t \, \cos(\lambda_1 t) \\[2mm]
\vdots \\[2mm]
\dfrac{2}{T} \sum_{t=1}^{T} x_t \, \sin(\lambda_{N-1} t) \\[2mm]
\dfrac{1}{T} \sum_{t=1}^{T} x_t (-1)^t
\end{pmatrix}.
$$

Note that $F'F$ has full rank T, such that the square matrix F must have full rank. Hence,

$$
F(F'F)^{-1}F' = FF^{-1}F'^{-1}F' = I_T,
$$

and hence it holds for the residual vector

$$
\hat{u} = x - \hat{x} = x - F\hat{\beta} = 0.
$$

Consequently, $x = \hat{x}$, or

$$
x_t = F_t' \, \hat{\beta}.
$$

This is the required result for $T = 2M + 2 = 2N$. The case $T = 2M + 1$ can be treated along the same lines; see also Hamilton (1994, Proposition 6.2 a)). This completes the proof. □

Proof of Proposition 4.2

With the definition of x, F, and $\hat{\beta}$ from the previous proof, we have the usual variance decomposition,

$$
x'x = \hat{x}'\hat{x} + \hat{u}'\hat{u}
$$

with $\hat{u} = 0$, such that

$$
x'x = \hat{\beta}'F'F\hat{\beta},
$$

where again by Lemma 4.1

$$
F'F = \operatorname{diag}\left(T, \frac{T}{2}, \dots, \frac{T}{2} \right).
$$

Consequently,

$$
x'x = T\overline{x}^2 + \frac{T}{2}\left(\frac{2}{T}\right)^2 \sum_{j=1}^{M}\left[\left(\sum_{t=1}^{T} x_t \cos(\lambda_j t)\right)^2 + \left(\sum_{t=1}^{T} x_t \sin(\lambda_j t)\right)^2\right]
$$

or

$$
\hat{\gamma}_x(0) = \frac{2}{T^2} \sum_{j=1}^{M}\left[\left(\sum_{t=1}^{T} x_t \cos(\lambda_j t)\right)^2 + \left(\sum_{t=1}^{T} x_t \sin(\lambda_j t)\right)^2\right],
$$

which proves the claim. $\qquad\qquad\qquad\qquad\qquad\qquad\qquad\qquad\qquad\square$

Proof of Proposition 4.3

For any stationary process $\{x_t\}$ we have from Brockwell and Davis (1991, Corollary 4.3.1) that

$$
\gamma_x(h) = \int_{-\pi}^{\pi} e^{i\lambda h}\, dF_x(\lambda),
$$

where the so-called spectral distribution function F_x is right continuous, nondecreasing, and bounded with $F_x(-\pi) = 0$. If a so-called spectrum f_x exists, then

$$
F_x(\lambda) = \int_{-\pi}^{\lambda} f_x(\omega)\, d\omega \quad\text{or}\quad dF_x(\lambda) = f_x(\lambda)\, d\lambda.
$$

Further, if the stationary process is from Eq. (3.1) or (4.12), then f_x does exist by Brockwell and Davis (1991, Theorem 5.7.2) with $f_x(\lambda)$ defined in Eq. (4.18). Further, by Eq. (4.11) and symmetry:

$$
\gamma_x(h) = \int_{-\pi}^{\pi} (\cos(\lambda h) + i \sin(\lambda h)) f_x(\lambda)\, d\lambda = \int_{-\pi}^{\pi} \cos(\lambda h) f_x(\lambda)\, d\lambda.
$$

Finally, Eq. (4.19) is from Brockwell and Davis (1991, Theorem 5.8.1), which completes the proof. $\qquad\qquad\qquad\qquad\qquad\qquad\qquad\square$

Proof of Proposition 4.8

We begin with the ARMA case. From Eq. (3.9) we have $x_t = \sum_{j=0}^{\infty} c_j\, \varepsilon_{t-j}$. By Eq. (4.15) one obtains for the spectrum:

$$
f_x(\lambda) = T_C(\lambda)\frac{\sigma^2}{2\pi}
$$

with

$$
T_C(\lambda) = \sum_{j=0}^{\infty} c_j^2 + 2 \sum_{h=1}^{\infty}\sum_{j=0}^{\infty} c_j\, c_{j+h} \cos(\lambda\, h).
$$

By Proposition 3.5 (b) the coefficients $\{c_j\}$ are geometrically bounded; by Lemma 3.4, the sequence $\{c_h h^s\}$ is geometrically bounded, too, for any $s > 0$. Consequently, the spectrum is twice differentiable with

$$T_C'(\lambda) = \frac{\partial T_C(\lambda)}{\partial \lambda} = -2 \sum_{h=1}^{\infty} \sum_{j=0}^{\infty} c_j \, c_{j+h} \, h \, \sin(\lambda \, h),$$

$$T_C''(\lambda) = \frac{\partial T_C'(\lambda)}{\partial \lambda} = -2 \sum_{h=1}^{\infty} \sum_{j=0}^{\infty} c_j \, c_{j+h} \, h^2 \, \cos(\lambda \, h).$$

Both derivatives are clearly bounded, and, in particular, $T_C'(0) = 0$.

Now, we turn to the EXP(k) model from Eq. (4.26) with $f_x(\lambda) = f_x(\lambda; \psi)$. It is obvious that

$$f_x'(\lambda) = -f_x(\lambda) \sum_{\ell=1}^{k} \psi_\ell \ell \, \sin(\lambda \, \ell),$$

such that $f_x'(0) = 0$. The calculation of the second derivative is straightforward,

$$f_x''(\lambda) = f_x(\lambda) \left[\left(\sum_{\ell=1}^{k} \psi_\ell \ell \, \sin(\lambda \, \ell) \right)^2 - \sum_{\ell=1}^{k} \psi_\ell \ell^2 \, \cos(\lambda \, \ell) \right],$$

which is bounded. Hence, the proof is complete. □

Proof of Proposition 4.9

With the definition of $I_\varepsilon(\lambda_j)$ from Eq. (4.3), we have

$$2 \frac{I_\varepsilon(\lambda_j)}{\sigma^2/2\pi} = \left(\frac{\sqrt{2}}{\sigma \sqrt{T}} \sum_{t=1}^{T} \varepsilon_t \cos(\lambda_j t) \right)^2 + \left(\frac{\sqrt{2}}{\sigma \sqrt{T}} \sum_{t=1}^{T} \varepsilon_t \sin(\lambda_j t) \right)^2$$

$$= A_T^2 + B_T^2, \quad \text{say.}$$

Both A_T and B_T are normal by assumption with $E(A_T) = E(B_T) = 0$,

$$\text{Var}(A_T) = \frac{2}{\sigma^2 T} \sum_{t=1}^{T} \sigma^2 (\cos(\lambda_j t))^2 = 1,$$

$$\text{Var}(B_T) = \frac{2}{\sigma^2 T} \sum_{t=1}^{T} \sigma^2 (\sin(\lambda_j t))^2 = 1,$$

because of Lemma 4.1 (a). Hence, the squares follow χ^2 distributions with 1 degree of freedom:

$$A_T^2 \sim \chi^2(1) \quad \text{and} \quad B_T^2 \sim \chi^2(1).$$

Since $\{\varepsilon_t\}$ is free of serial correlation, the covariance of A_T and B_T reduces to

$$E(A_T B_T) = \frac{2}{T} \sum_{t=1}^{T} \cos(\lambda_j t)\, \sin(\lambda_j t),$$

which equals 0 by Lemma 4.1 (a). Due to normality, A_T and B_T are hence independent. Therefore,

$$A_T^2 + B_T^2 \sim \chi^2(2).$$

Now, we apply Corollary 4.1 with

$$X = 2\, \frac{I_\varepsilon(\lambda_j)}{\sigma^2/2\pi} \quad \text{and} \quad c = \frac{\sigma^2}{4\pi}.$$

It follows that $I_\varepsilon(\lambda_j) \sim \mathcal{EXP}(2\pi/\sigma^2)$ as required.

The last statement refers to the periodogram at different frequencies. Let us consider $(j \neq k)$:

$$2\, \frac{I_\varepsilon(\lambda_k)}{\sigma^2/2\pi} = \left(\frac{\sqrt{2}}{\sigma\sqrt{T}} \sum_{t=1}^{T} \varepsilon_t \cos(\lambda_k t) \right)^2 + \left(\frac{\sqrt{2}}{\sigma\sqrt{T}} \sum_{t=1}^{T} \varepsilon_t \sin(\lambda_k t) \right)^2$$

$$= C_T^2 + D_T^2, \quad \text{say.}$$

Obviously, C_T and D_T are standard normal variates, too. Further, as above

$$E(A_T C_T) = \frac{2}{T} \sum_{t=1}^{T} \cos(\lambda_j t)\, \cos(\lambda_k t) = 0, \quad j \neq k,$$

$$E(A_T D_T) = \frac{2}{T} \sum_{t=1}^{T} \cos(\lambda_j t)\, \sin(\lambda_k t) = 0, \quad j \neq k,$$

where Lemma 4.1 (b) was used. Similarly, $E(B_T C_T) = E(B_T D_T) = 0$, such that $A_T, B_T, C_T,$ and D_T are all independent. This completes the proof. □

Proof of Corollary 4.1

This is a simple application of Proposition 4.9. With $c > 0$, the domains of X and Y are

$$D_x = D_y = \{z : z > 0\}.$$

The inverse transformation linking the two variables becomes

$$g^{-1}(y) = \frac{y}{c}, \quad \frac{dg^{-1}(y)}{dy} = \frac{1}{c}.$$

Notice that the density of a $\chi^2(2)$ distribution equals that of an exponential distribution with $\theta = 1/2$. Hence, it holds for the density of Y that

$$\phi_y(y) = \frac{1}{c}\, \frac{1}{2}\, \exp\left\{ -\frac{1}{2}\, \frac{y}{c} \right\},$$

i.e.

$$Y \sim \mathcal{EXP}\left(\frac{1}{2c}\right).$$

This completes the proof. □

Proof of Proposition 4.11

Phillips (2007, Theorem 3.2) proves for the discrete Fourier transforms $w_x(\lambda_j)$ that they form an iid sequence asymptotically. Moreover, they follow a complex normal distribution. For a review of the complex normal distribution, see e.g. Brillinger (1975, Section 4.2): If $Z = A + iB$ follows $\mathcal{N}(0, \sigma^2)$ with $\sigma^2 \in \mathbb{R}$, this means

$$\begin{pmatrix} A \\ B \end{pmatrix} \sim \mathcal{N}_2\left(\begin{pmatrix} 0 \\ 0 \end{pmatrix}, \frac{1}{2}\begin{pmatrix} \sigma^2 & 0 \\ 0 & \sigma^2 \end{pmatrix}\right).$$

This implies that

$$2\frac{Z^2}{\sigma^2} = 2\frac{A^2 + B^2}{\sigma^2}$$

is the sum of two independent, squared real standard normal variates: $2Z^2/\sigma^2 \sim \chi^2(2)$. According to Phillips (2007, p. 118), $w_x(\lambda_j)$ has a limiting complex normal distribution with $\mathcal{N}(0, f_x(0))$ as $T \to \infty$.[8] This completes the proof. □

Proof of Lemma 4.3

The first equality in Lemma 4.3 is obvious from $\partial \ln f = f^{-1}\partial f$. Beginning there, we obtain again straightforwardly (dropping all arguments to save space)

$$\int_{-\pi}^{\pi} f^{-2} \frac{\partial f}{\partial \psi_i} \frac{\partial f}{\partial \psi_j} = \int_{-\pi}^{\pi} f^2 \frac{\partial f^{-1}}{\partial \psi_i} \frac{\partial f^{-1}}{\partial \psi_j} = \int_{-\pi}^{\pi} T_C^2 \frac{\partial T_C^{-1}}{\partial \psi_i} \frac{\partial T_C^{-1}}{\partial \psi_j},$$

where $f = T_C \sigma^2/2\pi$ was used. Now, Fox and Taqqu (1986, Lemma 3) comes in, yielding under the assumptions (i), (ii), and (iii) that

$$\int_{-\pi}^{\pi} T_C^2 \frac{\partial T_C^{-1}}{\partial \psi_i} \frac{\partial T_C^{-1}}{\partial \psi_j} = \int_{-\pi}^{\pi} T_C \frac{\partial^2 T_C^{-1}}{\partial \psi_i \partial \psi_j}.$$

Again by $f = T_C \sigma^2/2\pi$ this establishes the second equality stated in Lemma 4.3, which completes the proof. □

---

8 Note that Phillips (2007, Theorem 3.2) has a typo when giving the normal distribution with $\mathcal{N}(0, 2\pi f_x(0))$; see the proof by Phillips (2007).

Proof of Corollary 4.2

To evaluate the limiting Fisher information behind w_{ij} in (4.33), one requires the partial derivatives of the log-spectrum. Under the EXP model from (4.26) one obtains a diagonal covariance matrix since (for $i \neq j$)[9] :

$$\int_{-\pi}^{\pi} \frac{\partial \ln f(\lambda; \psi)}{\partial \psi_i} \frac{\partial \ln f(\lambda; \psi)}{\partial \psi_j} \, d\lambda = \int_{-\pi}^{\pi} \cos(\lambda \, i) \, \cos(\lambda \, j) \, d\lambda$$

$$= \left[\frac{\sin(i-j)\lambda}{2(i-j)} + \frac{\sin(i+j)\lambda}{2(i+j)} \right]_{-\pi}^{\pi}$$

$$= 0.$$

On the main diagonal one has

$$w_{jj} = \frac{1}{4\pi} \int_{-\pi}^{\pi} \left(\frac{\partial \ln f(\lambda; \psi)}{\partial \psi_j} \right)^2 d\lambda$$

$$= \frac{1}{4\pi} \int_{-\pi}^{\pi} \cos^2(\lambda \, j) \, d\lambda = \frac{1}{2\pi j} \int_{0}^{\pi j} \cos^2(x) \, dx$$

$$= \frac{1}{2\pi j} \left[\frac{x}{2} + \frac{\sin 2x}{4} \right]_{0}^{\pi j} = \frac{1}{4}.$$

This proves the claim. □

---

9 We use the trigonometric identity $2 \cos x \cos y = \cos(x+y) + \cos(x-y)$.

5

Differencing and Integration

We recap the case of integer differencing and integration systematically to pave the way for a generalization to noninteger, or fractional, differencing and integration. Four lemmata on the coefficients of fractional differences or integration are established. At the same time we collect some technical results on the approximation of sequences and functions that will be useful in subsequent chapters.

5.1   Integer Case

The book by Box and Jenkins (1970) popularized the integrated ARMA(p, q) process, in short ARIMA(p, d, q), where typically $d \in \{0, 1, 2\}$. Integrated time series models, $I(d)$, have become a standard tool in econometrics, too, since the seminal papers by Granger (1981) and Engle and Granger (1987).

The random walk $\{y_t^{(1)}\}$ is the most prominent example of an $I(1)$ process. It is nonstationary and has to be differenced once in order to obtain white noise:

$$\Delta y_t^{(1)} = \varepsilon_t, \quad t = 1, \dots, T, \quad y_0^{(1)} = 0.$$

With the starting value assumption that $y_0^{(1)} = 0$, this amounts to

$$y_t^{(1)} = \sum_{j=0}^{t-1} \varepsilon_{t-j} = \varepsilon_t + \varepsilon_{t-1} + \cdots + \varepsilon_1$$
$$\sim I(1).$$

The superscript "(1)" should remind us of the integration order 1 of the random walk. Similarly, one defines the $I(2)$ process $\{y_t^{(2)}\}$ by integrating over a random walk $\{y_t^{(1)}\}$:

$$y_t^{(2)} = \sum_{j=0}^{t-1} y_{t-j}^{(1)}, \quad \text{where } \Delta y_t^{(1)} = \varepsilon_t, \ y_0^{(2)} = y_{-1}^{(2)} = 0.$$

Time Series Analysis with Long Memory in View, First Edition. Uwe Hassler.
© 2019 John Wiley & Sons, Inc. Published 2019 by John Wiley & Sons, Inc.

Consequently, $\Delta y_t^{(2)} = y_t^{(1)} \sim I(1)$, such that differencing once is not enough to obtain stationarity. Differencing twice, however,

$$\Delta^2 y_t^{(2)} = \Delta y_t^{(1)} = \varepsilon_t, \quad t = 1, \dots, T,$$

yields white noise, where we used two starting values equal to 0. Inserting

$$y_k^{(1)} = \sum_{i=0}^{k-1} \varepsilon_{k-i} \sim I(1)$$

for $k = t - j$, the level of the $I(2)$ process is given by a double sum (which justifies to call it integrated of order 2):

$$y_t^{(2)} = \sum_{j=0}^{t-1} \sum_{i=0}^{t-j-1} \varepsilon_{t-j-i}$$
$$= \varepsilon_t + 2\varepsilon_{t-1} + \cdots + (t-1)\varepsilon_2 + t\varepsilon_1$$
$$\sim I(2).$$

Similarly, we may define an $I(-1)$ process as a process that has been differenced once although differencing would not have been required to obtain stationarity:

$$y_t^{(-1)} = \varepsilon_t - \varepsilon_{t-1} \sim I(-1).$$

Finally, the white noise sequence itself is stationary, and it has to be differenced 0 times to obtain stationarity; hence, we call the process integrated of order 0:

$$y_t^{(0)} = \varepsilon_t \sim I(0).$$

At the same time we observe that $\text{Var}(\varepsilon_t) = \sigma^2$ equals the long-run variance from Eq. (2.10): $\omega^2 = \sigma^2 > 0$. Hence, the $I(0)$ process $\{y_t^{(0)}\}$ is not antipersistent in the sense of Definition 2.5.

What do $\{y_t^{(-1)}\}$, $\{y_t^{(0)}\}$, $\{y_t^{(1)}\}$, and $\{y_t^{(2)}\}$ have in common? They all are written as a weighted sum of (past) stationary innovations that are not antipersistent:

$$y_t^{(d)} = \sum_{j=0}^{t-1} \psi_j(d)\varepsilon_{t-j}, \tag{5.1}$$

where

$$\psi_j(d) = \frac{j-1+d}{j}\, \psi_{j-1}(d), \quad j \geq 1, \quad \psi_0(d) = 1.$$

We tabulate the coefficients $\psi_j(d)$ for the integer cases $d = -1, 0, 1, 2$:

	$\psi_0(d)$	$\psi_1(d)$	$\psi_2(d)$	$\psi_3(d)$	$\cdots$	$\psi_{t-1}(d)$
$d = 2$	1	2	3	4	$\cdots$	t
$d = 1$	1	1	1	1	$\cdots$	1
$d = 0$	1	0	0	0	$\cdots$	0
$d = -1$	1	-1	0	0	$\cdots$	0

Similarly, appropriate differencing turns $y_t^{(d)}$ into ε_t for $d = 0, 1, 2$:

$$\Delta^d y_t^{(d)} = \varepsilon_t, \quad t = 1, \dots, T,$$

where

$$\Delta^2 = (1 - L)^2 = 1 - 2L + L^2, \quad \Delta = 1 - L, \quad \Delta^0 = 1.$$

The differences Δ^d, $d \in \{0, 1, 2\}$, are given by

$$\Delta^d = \sum_{j=0}^{d} \pi_j(d) L^j,$$

where

$$\pi_j(d) = \frac{j - 1 - d}{j} \pi_{j-1}(d), \quad j \geq 1, \quad \pi_0(d) = 1. \tag{5.2}$$

In the integer case, $\pi_j(d) = 0$ for $j > d$. In the fractional case we will have more generally

$$\Delta^d = \sum_{j=0}^{\infty} \pi_j(d) L^j. \tag{5.3}$$

How can we justify the coefficients $\pi_j(d)$ showing up in Eq. (5.3) for noninteger (or fractional) values d? Before we answer this question, we make a mathematical digression on the approximation of (certain) sequences and functions that we will encounter repeatedly in what follows.

5.2 Approximating Sequences and Functions

For two real sequences $\{a_n\}$ and $\{b_n\}$, let $a_n \sim b_n$ denote the following, where $b_n \neq 0$:

$$a_n \sim b_n \iff \lim_{n \to \infty} \frac{a_n}{b_n} = 1.$$

Similarly, for real functions $g(x)$ and $h(x)$ with $h(x) \neq 0$

$$g(x) \sim h(x), x \to x_0 \iff \lim_{x \to x_0} \frac{g(x)}{h(x)} = 1.$$

Such approximations are often due to a (first- or second-order) Taylor expansion of $f(x)$ at x_0, which provides a linear or a quadratic approximation. Loosely, we are used to write

$$f(x) \approx f(x_0) + f'(x_0)(x - x_0),$$

or

$$f(x) \approx f(x_0) + f'(x_0)\,(x - x_0) + f''(x_0)\,\frac{(x - x_0)^2}{2}.$$

But what is actually meant by the approximation "$\approx$"? To become precise, we introduce with $o((x - x_0)^n)$ the so-called Landau symbol (also called "little o") signifying that a function is of smaller order than $(x - x_0)^n$ for some $n \geq 0$. More generally, we define for a (deterministic) function $g(x)$ that it is of smaller order than $h(x)$ with $h(x) \neq 0$ in a neighborhood of x_0, if

$$g(x) = o(h(x)) \iff \lim_{x \to x_0} \frac{g(x)}{h(x)} = 0. \tag{5.4}$$

With the identity $h(x) = 1$, we simply express that a function converges to 0:

$$g(x) = o(1) \iff \lim_{x \to x_0} g(x) = 0.$$

Hence, the first-order Taylor expansion actually means that the linear approximation is valid up to an error that is of smaller order than $(x - x_0)$ in a neighborhood of x_0:

$$f(x) = f(x_0) + f'(x_0)\,(x - x_0) + o(x - x_0)$$

or

$$f(x) - f(x_0) \sim f'(x_0)\,(x - x_0)\,, \quad x \to x_0 \quad \text{with } f'(x_0) \neq 0.$$

Similarly, the second-order Taylor expansion provides

$$f(x) = f(x_0) + f'(x_0)\,(x - x_0) + f''(x_0)\,\frac{(x - x_0)^2}{2} + o((x - x_0)^2),$$

or with $f''(x_0) \neq 0$:

$$f(x) - f(x_0) - f'(x_0)\,(x - x_0) \sim f''(x_0)\,\frac{(x - x_0)^2}{2}\,, \quad x \to x_0.$$

Note that a Taylor expansion at $x_0 = 0$ is also often called a Maclaurin expansion. This way we get, for example, a linear approximation of the sine at the origin:

$$\sin(x) - \sin(0) \sim \cos(0)\,(x - 0)\,, \quad x \to 0,$$

i.e.

$$\sin(x) \sim x\,, \quad x \to 0. \tag{5.5}$$

In the same way the cosine may be approximated quadratically at the origin:

$$\cos(x) = 1 - \frac{x^2}{2} + o((x - x_0)^2)\,, \quad x \to 0.$$

or

$$2 - 2\,\cos(x) \sim x^2\,, \quad x \to 0. \tag{5.6}$$

In addition to the *little o* notation, we will also use the *big O* notation to characterize that a sequence or function is at most of a certain order. We define for a (deterministic) function $g(x)$ that it is at most of order $h(x)$ with $h(x) \neq 0$ in a neighborhood of x_0, if there exist positive numbers K and δ such that $\frac{|g(x)|}{|h(x)|} \leq K$ for all x with $|x - x_0| < \delta$:

$$g(x) = O(h(x)) \iff \frac{|g(x)|}{|h(x)|} \leq K \text{ for } |x - x_0| < \delta. \tag{5.7}$$

Hence, $g(x) \sim h(x)$ implies $g(x) = O(h(x))$. Thus, we obtain from Eq. (5.5) that $\sin x = O(x), x \to 0$. Note, however, that the *bound h* does not have to be sharp. From (5.6) we observe

$$2 - 2 \cos(x) = O(x^2), \quad x \to 0,$$

as well as

$$2 - 2 \cos(x) = O(x), \quad x \to 0.$$

Repeatedly, we will use three further trigonometric results. First, there is Euler's celebrated formula:

$$e^{ix} = \cos(x) + i \sin(x), \quad i^2 = -1. \tag{5.8}$$

Second, it is worth remembering the so-called half-angle formula:

$$\sin^2(x) = \frac{1 - \cos(2x)}{2}. \tag{5.9}$$

Third, we recall one of the so-called double-angle formulae:

$$\sin(2x) = 2 \sin(x) \cos(x). \tag{5.10}$$

To use the binomial theorem, we return to real sequences and define as binomial coefficients for real d and $j \in \mathbb{N}_0$:

$$\binom{d}{j} := \frac{d\,(d-1)\,\cdots\,(d-j+1)}{j\,!}, \quad j \geq 1, \quad \binom{d}{0} := 1. \tag{5.11}$$

We then have

$$(1 - L)^d = \sum_{j=0}^{\infty} \binom{d}{j} (-L)^j.$$

In fact, binomial coefficients can be expressed conveniently in terms of the so-called Gamma function. For the rest of this section, we refer to, e.g. Gradshteyn and Ryzhik (2000, Section 8.3). We define

$$\Gamma(x) = \begin{cases} \displaystyle\int_0^\infty t^{x-1}\, e^{-t}\, dt, & x > 0 \\[2mm] \dfrac{\Gamma(x+1)}{x}, & x < 0, \quad x \neq -1, -2, \ldots \end{cases} \tag{5.12}$$

with special values

$$\Gamma(1) = \Gamma(2) = 1 , \quad \Gamma(1/2) = \sqrt{\pi}$$

and minimum at $x \approx 1.46$. The Gamma function often facilitates recursive relations by

$$\Gamma(x + 1) = x \, \Gamma(x) , \tag{5.13}$$

which was used in Eq. (5.12) to extend the definition for $x > 0$ to negative, noninteger arguments. For positive integers, the recursion yields the factorial of n:

$$\Gamma(n + 1) = n!.$$

From Eq. (5.13) we further take

$$\Gamma(x) \sim x^{-1} , \quad x \to 0,$$

which characterizes the simple pole at the origin, sometimes written as $\Gamma(0) = \infty$. In that sense, the recursion Eq. (5.13) may be used to characterize the poles for negative integers: $\Gamma(-1) = -\infty$, $\Gamma(-2) = \infty$, and so on. From now on, we reserve the notion *pole* (at x_p of order n) for singularities of the type

$$f(x) = \frac{g(x)}{(x - x_p)^n} , \quad g(x_p) \neq 0,$$

where $n \in \mathbb{N}$. Further, Eq. (5.13) allows to relate the Gamma function to binomial coefficients for noninteger d:

$$\binom{d}{j} (-1)^j = \frac{\Gamma(j - d)}{\Gamma(j + 1) \, \Gamma(-d)} , \quad d \notin \mathbb{Z}. \tag{5.14}$$

We now turn to approximations of the Gamma function. First, remember Stirling's formula for the factorial of n:

$$n! \sim \sqrt{2\pi n} \, (n/e)^n , \quad n \to \infty. \tag{5.15}$$

It generalizes to (see, e.g. Abramowitz and Stegun (1984, 6.1.39))

$$\Gamma(x) \sim \sqrt{\frac{2\pi}{x}} \left(\frac{x}{e}\right)^x , \quad x \to \infty.$$

With this result one can show[1]

$$\Gamma(n + a) \sim n! \, n^{a-1} , \quad n \to \infty.$$

This implies

$$\frac{\Gamma(n + a)}{\Gamma(n + b)} \sim n^{a-b} , \quad n \to \infty. \tag{5.16}$$

---

1 Use $(1 + a/n)^n \to e^a$.

From Gradshteyn and Ryzhik (2000, 8.334(3)), we learn (*Euler's Reflection*)

$$\Gamma(1 - x)\,\Gamma(x) = \frac{\pi}{\sin(\pi x)}. \tag{5.17}$$

Further, $\ln(\Gamma(x))$ is convex for $x > 0$, implying that $\partial \ln(\Gamma(x))/\partial x$ grows monotonically. In particular, for the so-called psi function,

$$\psi(x) := \frac{\partial \ln(\Gamma(x))}{\partial x} = \frac{\Gamma'(x)}{\Gamma(x)}, \tag{5.18}$$

we have the following recursion (see, e.g. Abramowitz and Stegun (1984, 6.3.5)):

$$\psi(x + 1) = \psi(x) + \frac{1}{x}, \quad x > 0. \tag{5.19}$$

Note in particular that

$$\psi(1) = \Gamma'(1) = -\gamma, \tag{5.20}$$

where γ is again Euler's constant defined in Lemma 3.3.

Finally, the so-called Beta function is related to $\Gamma(\cdot)$, too. Define,

$$B(x, y) = \int_0^1 u^{x-1}(1 - u)^{y-1} du \quad \text{for } x, y > 0.$$

The Beta function is symmetric, $B(x, y) = B(y, x)$. Further, it holds true that

$$B(x, y) = \frac{\Gamma(x)\,\Gamma(y)}{\Gamma(x + y)}. \tag{5.21}$$

This relation is sometimes helpful to simplify ratios given in terms of the Gamma function.

5.3 Fractional Case

We now wish to justify the expansion of the differences $\Delta^d = (1 - L)^d$ given in (5.3). To that end, define

$$D(x) = (1 - x)^d.$$

The derivatives up to order k obviously are

$$D^{(1)}(x) = -d(1 - x)^{d-1},$$
$$D^{(2)}(x) = -d(1 - d)(1 - x)^{d-2},$$
$$D^{(3)}(x) = -d(1 - d)(2 - d)(1 - x)^{d-3},$$
$$\vdots$$
$$D^{(k)}(x) = -d(1 - d) \dots (k - 1 - d)(1 - x)^{d-k}.$$

The Maclaurin expansion, or Taylor expansion about 0, yields for $|x| < 1$ the famous binomial theorem (see, e.g. Knopp (1990, p. 209)):

$$(1 - x)^d = D(0) + D^{(1)}(0)x + \frac{D^{(2)}(0)}{2!}x^2 + \cdots + \frac{D^{(k)}(0)}{k!}x^k + \cdots$$

$$= 1 - dx - \frac{d(1-d)}{2}x^2 - \frac{d(1-d)(2-d)}{3!}x^3 - \cdots$$

$$= \sum_{j=0}^{\infty} \pi_j(d)x^j = \sum_{j=0}^{\infty} \binom{d}{j}(-x)^j.$$

We hence obtain for $\pi_j(d)$ the usual binomial expansion satisfying Eq. (5.2):

$$\pi_j(d) = (-1)^j \binom{d}{j} = \frac{j-1-d}{j}\, \pi_{j-1}(d) = \frac{\Gamma(j-d)}{\Gamma(j+1)\,\Gamma(-d)}. \tag{5.22}$$

It is worthwhile to comment on Eq. (5.22), which is a short expression for the following: The first equality defines the coefficient $\pi_j(d)$ for any d and $j \geq 0$; the recursion (for $j > 0$) behind the second equality follows from the definition of the binomial coefficients (see Eq. (5.11)); the third equality follows from Eq. (5.14), however, only for $d \notin \mathbb{Z}$. Hence, we have in fact two equations behind Eq. (5.22):

$$\pi_j(d) = \frac{j-1-d}{j}\, \pi_{j-1}(d) \quad \text{for all } d\,, \quad j > 0\,, \quad \pi_0(d) = 1,$$

$$\pi_j(d) = \frac{\Gamma(j-d)}{\Gamma(j+1)\,\Gamma(-d)} \quad \text{for } d \notin \mathbb{Z}\,, \quad j \geq 0.$$

The fractional differences $\Delta^d = (1-L)^d$ from Eq. (5.3) are then given by replacing x by L in the expansion of $(1-x)^d$.

We can now prove the following lemma that will turn out to be helpful in the next chapter.

Lemma 5.1 *For $\{\pi_j(d)\}$ from Eq. (5.22) it holds for $d \in \mathbb{R}$ that*

$$\pi_j(d) = \pi_j(d-1) - \pi_{j-1}(d-1) \quad \text{for } j > 0\,, \tag{5.23}$$

$$\sum_{j=0}^{k} \pi_j(d) = \pi_k(d-1) \quad \text{for } k \geq 0\,, \tag{5.24}$$

and

$$\pi_j(d) \sim \frac{j^{-d-1}}{\Gamma(-d)}, \quad j \to \infty\,, \quad d \notin \mathbb{Z}. \tag{5.25}$$

Proof: See Appendix to this chapter.

We learn from Eq. (5.25) that $\pi_j(d)$ converges to 0 with growing j if and only if $d > -1$:

$$\pi_j(d) \xrightarrow{j \to \infty} 0 \Longleftrightarrow d > -1.$$

This motivates the restriction $d > -1$ that we will introduce in the next chapter. Further, conditions for (square) summability may be induced from Eq. (5.25) by means of Lemma 3.1 (d).

Lemma 5.2 *For $\{\pi_j(d)\}$ from Eq. (5.22), it holds that*

$$\sum_{j=0}^{\infty} \pi_j(d) < \infty \quad \textit{if and only if } d \geq 0$$

and

$$\sum_{j=0}^{\infty} \pi_j^2(d) < \infty \quad \textit{if and only if } d > -\frac{1}{2}.$$

Proof: See Appendix to this chapter.

Under summability, the sum over all coefficients is simply 0 as long as d is strictly positive:

$$\sum_{j=0}^{\infty} \pi_j(d) = (1-z)^d \Big|_{z=1} = 0, \quad d > 0.$$

The above results on differencing carry over to the case of integration. Upon inverting, one defines the fractional integration operator,

$$(1-L)^{-d} = \sum_{j=0}^{\infty} \pi_j(-d) L^j.$$

Note that $\pi_j(-d)$ equals what had been called $\psi_j(d)$ before; see Eq. (5.1). Just as in Eq. (5.25), it holds

$$\pi_j(-d) \sim \frac{j^{d-1}}{\Gamma(d)}, \quad j \to \infty, \quad d \notin \mathbb{Z}. \tag{5.26}$$

Replacing d by $-d$, the results from Lemma 5.2 carry over. We hence have the following.

Lemma 5.3 *For $\{\pi_j(-d)\}$ from Eq. (5.22), it holds*

$$\pi_j(-d) \to 0 \quad \textit{if and only if } d < 1, \quad j \to \infty,$$

$$\sum_{j=0}^{\infty} \pi_j(-d) < \infty \quad \textit{if and only if } d \leq 0,$$

and

$$\sum_{j=0}^{\infty} \pi_j^2(-d) < \infty \quad \text{if and only if } d < \frac{1}{2}.$$

Proof: Obvious. □

In particular, for strictly negative d one obtains of course again:

$$\sum_{j=0}^{\infty} \pi_j(-d) = 0 , \quad d < 0.$$

We now turn to the convolution of fractional integration and an absolutely summable filter.

Lemma 5.4 *Let $\{c_j\}_{j=0,1,\dots}$ be absolutely summable, $\sum_{j=0}^{\infty} |c_j| < \infty$, and $\sum_{j=0}^{\infty} c_j \neq 0$. Further, we assume that*

$$\sum_{j=0}^{\infty} j^s |c_j| < \infty \quad \text{for some } s > 1 - d \tag{5.27}$$

and $d < 1$. It then holds for the coefficients given by convolution,

$$b_k(d) = \sum_{j=0}^{k} \pi_{k-j}(-d)c_j, \tag{5.28}$$

that

$$b_k(d) \sim \sum_{j=0}^{\infty} c_j \frac{k^{d-1}}{\Gamma(d)} , \quad d \notin \mathbb{Z}, \tag{5.29}$$

as $k \to \infty$.

Proof: See Appendix to this chapter.

The proposition is an extension of Hassler and Kokoszka (2010) and Hassler (2012) who did not allow for values of d smaller than -1. It guarantees that the harmonic decay of the fractionally integrated filter given in Eq. (5.26) carries over to the convolution with an absolutely summable filter under condition Eq. (5.27) for all noninteger values $d < 1$. Hassler and Kokoszka (2010) and Hassler (2012) give examples that absolute summability of $\{c_j\}$ alone is not enough to ensure Eq. (5.29), which corrects differing claims made in the literature before.

Now we are equipped to define stationary processes in terms of the fractional integration operator $\Delta^{-d} = (1 - L)^{-d}$ and to discuss their stochastic properties, in particular their (long) memory.

5.4 Technical Appendix: Proofs

Proof of Lemma 5.1

We prove the first statement first. Assume the inductive hypothesis $\pi_j(d) = \pi_j(d-1) - \pi_{j-1}(d-1)$, where the base case for $j = 1$ obviously holds true: $\pi_1(d) = -d = -(d-1) - 1$. To complete the inductive step, we proceed as follows:

$$
\pi_{j+1}(d) = \frac{j-d}{j+1}\pi_j(d) = \frac{j-d}{j+1}(\pi_j(d-1) - \pi_{j-1}(d-1))
$$

$$
= \left(\frac{j-d+1}{j+1} - \frac{1}{j+1}\right)\pi_j(d-1) - \frac{j-d}{j+1}\pi_{j-1}(d-1)
$$

$$
= \pi_{j+1}(d-1) - \frac{1}{j+1}\pi_j(d-1) - \frac{j-d}{j+1}\pi_{j-1}(d-1)
$$

$$
= \pi_{j+1}(d-1) - \pi_j(d-1) + R,
$$

where one has for the remainder term R that

$$
R = \frac{j}{j+1}\pi_j(d-1) - \frac{j-d}{j+1}\pi_{j-1}(d-1).
$$

By recursion it follows that $R = 0$, and we have Eq. (5.23).

The second claim in Eq. (5.24) follows elementarily by summing over Eq. (5.23). The third claim simply follows from Eqs. (5.22) by (5.16), where integer values of d are ruled out. This completes the proof. □

Proof of Lemma 5.2

For $d = 0, 1, 2, \ldots$, the summability is obvious since $\pi_j(d) = 0$ for $j > d$. For $d = -1, -2, \ldots$, the sums cannot be finite since $\pi_j(d)$ do not converge to 0 with growing j. For the noninteger case, both claims follow by Eq. (5.25) and Lemma 3.1 (d). □

Proof of Lemma 5.4

(i) We consider the positive case first, $0 < d < 1$. Note that Eq. (5.27) implies that

$$
c_j \, j^{1-d} \to 0, \quad j \to \infty. \tag{5.30}
$$

This condition was shown to be necessary and sufficient for Eq. (5.29) by Hassler and Kokoszka (2010, Proposition 2.1).

(ii) Now, we turn to the negative case. The range $-1 < d < 0$ has been covered by Hassler (2012). We show how his proof extends to smaller values of d. To that end, we will invoke the dominated convergence theorem (DCT). It reads as follows.

Lemma 5.5 *Let the double-indexed sequence* $\{a_{n,j}\}$, $n \in \mathbb{N}$, $j \in \mathbb{N}$, *be absolutely dominated by the summable sequence* $\{d_j\}$:

$$|a_{n,j}| \le d_j \quad \text{for all } j \text{ and } n, \quad \text{where } \sum_{j=1}^{\infty} d_j < \infty.$$

Let further

$$\lim_{n \to \infty} a_{n,j} = \alpha_j \quad \text{for all } j.$$

It then holds that $\{\alpha_j\}$ *is summable, and*

$$\sum_{j=1}^{n} a_{n,j} \to \sum_{j=1}^{\infty} \alpha_j \tag{5.31}$$

as $n \to \infty$.

Proof: Textbooks on measure theory typically contain Lebesgue's DCT for sequences of integrable functions, $f_n(x)$, $n \in \mathbb{N}$. The proof for our case of *sequences of summable sequences* is readily adopted, yielding

$$\lim_{n \to \infty} \sum_{j=1}^{\infty} a_{n,j} = \sum_{j=1}^{\infty} \alpha_j.$$

Since $\sum_{j=n+1}^{\infty} a_{n,j}$ must converge to 0, result Eq. (5.31) follows. $\qquad \square$

We continue with the negative case. To show that condition Eq. (5.27) is sufficient for Eq. (5.29), the coefficients from Eq. (5.28) are split into two sums,

$$b_k(d) = \Sigma_{1,k} + \Sigma_{2,k}$$
$$= \sum_{j=0}^{n_k} \pi_{k-j}(-d)c_j + \sum_{j=n_k+1}^{k} \pi_{k-j}(-d)c_j \, ,$$

where we choose a series of integers $n_k = \left\lfloor k^{\frac{1-d}{s}} \right\rfloor$ with $s > 1 - d$ from Eq. (5.27) such that $n_k \to \infty$ and $n_k = o(k)$. Note that

$$k^{1-d} \pi_{k-j}(-d) \to \frac{1}{\Gamma(d)} \quad \text{for } j \le n_k$$

by Eq. (5.26), such that we can define a constant K with

$$\max_{j=0,\dots,n_k} k^{1-d} |\pi_{k-j}(-d)| \le K \quad \text{for all } k.$$

Now, consider

$$k^{1-d}\Sigma_{1,k} = \sum_{j=0}^{n_k} k^{1-d}\pi_{k-j}(-d)c_j,$$

where the assumptions of the dominated convergence theorem are met,

$$|k^{1-d}\pi_{k-j}(-d)c_j| \le K\,|c_j|,$$

$$k^{1-d}\pi_{k-j}(-d)c_j \to \frac{c_j}{\Gamma(d)} \quad \text{for } k \to \infty,$$

such that by the DCT

$$\lim_{k\to\infty} k^{1-d}\Sigma_{1,k} = \sum_{j=0}^{\infty} \frac{c_j}{\Gamma(d)}.$$

We are left with showing that $\Sigma_{2,k} = o(k^{d-1})$. To do so, consider

$$|\Sigma_{2,k}| \le \sum_{j=n_k+1}^{k} |\pi_{k-j}(-d)||c_j| \le \max_{j=n_k+1,\dots,k} |\pi_{k-j}(-d)| \sum_{j=n_k+1}^{k} |c_j|\,,$$

where $\max_{j=n_k+1,\dots,k}|\pi_{k-j}(-d)| = \pi_0(-d) = 1$. For $j > n_k$ it holds that $j^s > k^{1-d}$. Hence,

$$k^{1-d}|\Sigma_{2,k}| \le \sum_{j=n_k+1}^{k} j^s|c_j|\,,$$

which converges to 0 since $\sum_{j\ge 0} j^s|c_j| < \infty$ by assumption.
   Hence, the proof is complete. $\qquad\qquad\square$

6

Fractionally Integrated Processes

First, we define a stationary fractionally integrated process of order d. In large parts of the book, we will maintain Assumption 6.2 below. Under stationarity ($d < 1/2$), we establish stochastic properties in the frequency and in the time domain. Invertibility in the sense of Definition 3.1 holds as long as $d > -1$. Fractionally integrated processes are strongly persistent and display long memory at the same time whenever $d > 0$, while they are antipersistent and have short memory for $d < 0$. Some examples and special cases of fractional integration are used to illustrate these properties. Then we turn to an alternative definition of fractional integration (often labeled as "type II") and discuss the extension of fractional integration beyond stationarity. The chapter closes with a different set of assumptions that is widely used in the book (Assumption 6.3). It only maintains rather weak conditions for the spectrum in a vicinity of the origin.

6.1 Definition and Properties

The ingredient to define fractionally integrated processes will be the $I(0)$ process $\{x_t\}$, which is not integrated, sometimes also called integrated of order 0. We define it as an MA(∞) process of short memory that is not antipersistent; see Definitions 2.4 and 2.5. This amounts to an absolutely summable process with positive long-run variance.

Assumption 6.1 (*Integration of order 0*) *Let $\{x_t\}$ be an MA(∞) process,*

$$x_t = \sum_{j=0}^{\infty} c_j\, \varepsilon_{t-j}, \quad \varepsilon_t \sim \text{WN}\,(0, \sigma^2), \quad t \in \mathbb{Z},$$

which is absolutely summable, $\sum_{j=0}^{\infty} |c_j| < \infty$, with $\sum_{j=0}^{\infty} c_j \neq 0$.

Time Series Analysis with Long Memory in View, First Edition. Uwe Hassler.
© 2019 John Wiley & Sons, Inc. Published 2019 by John Wiley & Sons, Inc.

Remember the power transfer function defined in terms of impulse responses of $\{x_t\}$ from Eq. (4.13):

$$T_C(\lambda) = \left| \sum_{j=0}^{\infty} c_j e^{-i\lambda j} \right|^2 .$$

It then follows by Propositions 4.3 and 4.4 that $\{x_t\}$ has a continuous spectrum $f_x(\lambda)$ on $[0, \pi]$ given by

$$f_x(\lambda) = \left| \sum_{j=0}^{\infty} c_j e^{-i\lambda j} \right|^2 \frac{\sigma^2}{2\pi} , \tag{6.1}$$

which by Assumption 6.1 is bounded and bounded away from 0 at the origin:

$$0 < f_x(0) = \left(\sum_{j=0}^{\infty} c_j \right)^2 \frac{\sigma^2}{2\pi} = \frac{\omega^2}{2\pi} < \infty ,$$

where ω^2 is the long-run variance of the process.

We now define the fractionally integrated process $\{y_t\}$ as $y_t = (1 - L)^{-d} x_t$.

Definition 6.1 (*Fractionally integrated processes*) *Let $\{y_t\}$ be given by*

$$y_t = (1 - L)^{-d} x_t, \quad t \in \mathbb{Z}, \quad d < \frac{1}{2} , \tag{6.2}$$

where the expansion $(1 - L)^{-d}$ is in terms of $\{\pi_j(-d)\}_{j=0,1,\dots}$ from Eq. (5.22). Further, $\{x_t\}$ satisfies Assumption 6.1 and Eq. (5.27), i.e.

$$\sum_{j=0}^{\infty} j^s |c_j| < \infty \quad \text{for some } s > 1 - d .$$

We call the process $\{y_t\}$ fractionally integrated of order d. In particular, $\{x_t\}$ satisfying Assumption 6.1 is integrated of order 0. If $\{x_t\}$ equals $\{\varepsilon_t\}$ (white noise), then $\{y_t\}$ from Eq. (6.2) is called fractionally integrated noise (FIN).

By definition, the fractionally integrated process possesses an MA representation,

$$y_t = \sum_{k=0}^{\infty} b_k(d) \varepsilon_{t-k} = \sum_{n=0}^{\infty} \pi_n(-d) \left(\sum_{j=0}^{\infty} c_j \varepsilon_{t-n-j} \right) \tag{6.3}$$

with MA or impulse response coefficients given in (5.28):

$$b_k(d) = \sum_{j=0}^{k} \pi_j(-d) c_{k-j} .$$

By Lemma 5.4 it holds that $\{b_k(d)\}$ dies out with rate k^{d-1}:

$$k^{1-d}b_k(d) \quad \rightarrow \quad \frac{\sum_{j=0}^{\infty} c_j}{\Gamma(d)}.$$

Hence, $\{b_k(d)\}$ is square summable if and only if $d < 1/2$, see Lemma 3.1 (d). For this reason, we assumed $d < 1/2$ in Definition 6.1, guaranteeing a stationary process according to Proposition 3.1.

The condition Eq. (5.27) maintained in Definition 6.1 can be weakened to $c_j j^{1-d} \rightarrow 0$ in the case of $d > 0$. Note that $\sum_{j=0}^{\infty} j^s |c_j| < \infty$ for some $s > 1 - d$ implies $c_j j^{1-d} \rightarrow 0$, which was shown to be necessary and sufficient under $d > 0$ for the hyperbolic rate in Eq. (5.26) to carry over to $\{b_k(d)\}$; see Hassler and Kokoszka (2010). The condition $c_j j^{1-d} \rightarrow 0$ is quite intuitive in that it requires the impulse responses of $\{x_t\}$ to die out faster than the hyperbolic rate of the fractional integration filter:

$$\frac{c_j}{\pi_j(-d)} \rightarrow 0, \quad j \rightarrow \infty.$$

Clearly, our Assumption 6.1 and Eq. (5.27) maintained in Definition 6.1 are met by stationary and invertible ARMA processes; see Proposition 3.5. In the early literature on fractionally integrated models by Granger and Joyeux (1980) and Hosking (1981), it was assumed that $\{x_t\}$ is a stationary and invertible ARMA process; see also Brockwell and Davis (1991, Section 13.2). Such processes are called fractionally integrated ARMA processes, or ARFIMA(p, d, q) in short, with p and q denoting the lag order of the AR and MA polynomials, respectively.

Now we have set the stage to derive the first results for fractionally integrated processes.

Proposition 6.1 *The process $\{y_t\}$ defined in Definition 6.1 is stationary with* $E(y_t) = 0$ *for $d < \frac{1}{2}$. Its spectrum is given by*

$$f_y(\lambda) = 4^{-d}\sin^{-2d}\left(\frac{\lambda}{2}\right)f_x(\lambda), \quad \lambda > 0$$

$$\sim \lambda^{-2d}f_x(0), \quad \lambda \rightarrow 0,$$

where $f_x(\lambda)$ is the continuous spectrum of $\{x_t\}$ from Eq. (6.1).

Proof: See Appendix to this chapter.

Next, we further restrict the parameter space to $-1 < d < 1/2$. With this parameter space we follow Palma (2007, Theorem 3.4); see also Palma (2007, Remark 3.1). With the notable exception of Palma (2007), all other authors followed Hosking (1981) and restricted the range of fractional integration to $|d| < 1/2$. Clearly, $d < 1/2$ is required for stationarity since the expansion

$$\Delta^{-d} = \sum_{j=0}^{\infty} \pi_j(-d)L^j$$

is square summable if and only if $d < 1/2$; see Lemma 5.3. Similarly, by Lemma 5.2

$$\Delta^{d} = \sum_{j=0}^{\infty} \pi_j(d)L^j$$

is square summable if and only if $d > -1/2$. Hence, $d > -1/2$ has been maintained to ensure invertibility of $I(d)$ processes. Bondon and Palma (2007), however, proved that this condition is not necessary; see also Bloomfield (1985, p. 231) for the special case of FIN. From Bondon and Palma (2007), we conclude the following invertibility result in the sense of Definition 3.1; see also Bloomfield (1985, Theorem 4) and Palma (2007, Theorem 3.4 c).[1]

Proposition 6.2 *Let $\{x_t\}$ from Assumption 6.1 have a spectrum f_x such that $1/f_x$ is integrable on $[0, \pi]$. It then holds that $\{y_t\}$ from Definition 6.1 is invertible in the sense of Definition 3.1 for $d > -1$.*

Proof: Follows from Bondon and Palma (2007, Theorem 3). □

Note that invertibility does not hold anymore for $d = -1$; see Bondon and Palma (2007, Theorem 4). This is not surprising since $\pi_j(d) \rightarrow 0$ if and only if $d > -1$; see Lemma 5.2. From now on we will assume that the fractionally integrated process is stationary and invertible, i.e. we maintain the assumptions behind Proposition 6.2.

Assumption 6.2 (*Invertibility*) *Let $\{y_t\}$ be a fractionally integrated process of order d with*

$$-1 < d < \frac{1}{2} ,$$

where $\{y_t\}$ is from Definition 6.1. Further, assume for the spectrum f_x of the underlying $I(0)$ process $\{x_t\}$ that $1/f_x$ is integrable on $[0, \pi]$.

It is worthwhile to stress that stationary ARMA processes having all moving average roots outside the unit circle, which is condition Eq. (3.8), have a strictly

---

1 Building on yet another, forecastability-based concept of invertibility by Granger and Andersen (1978), Odaki (1993) established invertibility for $d \in (-1, 1/2)$, too.

positive spectrum (see Eq. (4.23)), such that $1/f_x$ is integrable. Hence, all stationary and invertible ARMA processes meet the assumptions maintained in Assumption 6.2 with respect to $\{x_t\}$.

The case $-1 < d < -1/2$ occurs in practice when differencing a stationary process integrated of order δ, $0 < \delta < 1/2$. Let us call the $I(\delta)$ process $\{z_t\}$ with

$$z_t = \sum_{j=0}^{\infty} \pi_j(-\delta) x_{t-j} .$$

Then define $y_t = \Delta z_t = z_t - z_{t-1}$ resulting in

$$y_t = \sum_{j=0}^{\infty} \pi_j(-\delta) x_{t-j} - \sum_{j=1}^{\infty} \pi_{j-1}(-\delta) x_{t-j}$$

$$= x_t + \sum_{j=1}^{\infty} (\pi_j(-\delta) x_{t-j} - \pi_{j-1}(-\delta)) x_{t-j}$$

$$= \sum_{j=0}^{\infty} \pi_j(-\delta + 1) x_{t-j} ,$$

where Lemma 5.1 was used for the last equality. Hence, $y_t = \Delta z_t$ is integrated of order $d = \delta - 1$ according to Definition 6.1.

For the case $d > 0$, we next translate Proposition 6.1 from the frequency domain into the time domain.

Corollary 6.1 *Under the assumptions of Proposition 6.1, it holds under $0 < d < \frac{1}{2}$ for the autocovariances that*

$$\gamma_y(h) \sim C_{\gamma,d} \, h^{2d-1}, \quad h \to \infty ,$$

where

$$C_{\gamma,d} := 2\pi f_x(0) \frac{\Gamma(1-2d)}{\Gamma(d)\Gamma(1-d)} . \tag{6.4}$$

Proof: See Appendix to this chapter.

Note that Corollary 6.1 contains just the statement by Hassler and Kokoszka (2010, Corollary 2.1). It follows from Proposition 6.1 by Giraitis et al. (2012, Proposition 3.1.1 a) but equally by the following references: Beran (1994, Theorem 2.1), Taniguchi and Kakizawa (2000, Proposition 5.1.2), Palma (2007, Theorem 3.3), and Beran et al. (2013, Theorem 1.3). All these results state that a singularity of the spectrum at the origin of order λ^{-2d} for $d > 0$ translates into a hyperbolic decay h^{2d-1} for the autocovariances at lag h; see Appendix to this chapter for details.

We now wish to carry Corollary 6.1 to the case $d < 0$. Although the statement with respect to the autocovariance behavior will remain the same, the technique of proof is quite different under this new assumption. We first translate the result Eq. (5.23) from Lemma 5.1 to $\{b_k(d)\}$ defined in Eq. (5.28).

Lemma 6.1 *For $\{b_k(d)\}$ from Eq. (5.28) it holds that*

$$b_k(d-1) = b_k(d) - b_{k-1}(d) \quad for\ k > 0 . \tag{6.5}$$

Proof: See Appendix to this chapter.

Now we are able to show the following.

Proposition 6.3 *Under Assumption 6.2 it holds under $-1 < d < 0$ for the autocovariances of the fractionally integrated process that*

$$\gamma_y(h) \sim C_{\gamma,d}\ h^{2d-1}, \quad h \to \infty ,$$

where $C_{\gamma,d}$ is defined in Eq. (6.4).

Proof: See Appendix to this chapter.

In the next section we look at some examples to illustrate the properties established so far.

6.2 Examples and Discussion

Granger (1966) described as "typical spectral shape of an economic variable" that the periodogram estimates tend to grow when approaching the origin: they do not seem to converge to a constant value at $\lambda = 0$. Such a behavior is captured by $f_y(\lambda)$ from Eq. (6.1) for $d > 0$ where the spectrum has a singularity as $\lambda \to 0$; see also the discussion in Granger (1981). For $d < 0$, on the other hand, $f_y(0) = 0$, and the order of integration characterizes the rate with which $f_y(\lambda)$ converges to 0 at the origin. In view of Eq. (4.22) the spectrum contains the same information as the accumulated impulse responses. In terms of Definition 2.5, we hence observe the following: A fractionally integrated process is strongly persistent if and only if $d > 0$ and antipersistent if and only if $d < 0$. If $d = 0$, then $y_t = x_t$ satisfies Assumption 6.1 and is hence moderately persistent with $0 < \omega_x^2 < \infty$.

The persistent behavior for $0 < d < 1/2$ translates into long memory in the sense of Definition 2.4 according to Corollary 6.1. With Lemma 3.3 we observe

$$\sum_{h=1}^{H} \gamma_y(h) \to \infty \quad as\ H \to \infty , \quad d > 0 .$$

With $0 < d < 1/2$, the constant $C_{\gamma,d}$ from Eq. (6.4) is positive. Similarly, under antipersistence, $\gamma_y(h)$ is absolutely summable, and $C_{\gamma,d} < 0$. Hence, $I(d)$ processes have short memory in the sense of Definition 2.4 as long as $d < 0$, while being antipersistent at the same time.

As has been mentioned before, a fully parametric model formulated in the time domain often assumes that $\{x_t\}$ is a stationary and invertible ARMA process of finite orders p and q. Its spectrum at the origin is well known from Eq. (4.24). Consequently, it is straightforward to insert $f_x(0)$ in Propositions 6.1 and 6.3.

As a special case we first consider FIN, also often called fractional noise in short, where $x_t = \varepsilon_t$. A detailed discussion of FIN is contained in Hassler (2016, Chapter 5); see also Hassler and Hosseinkouchack (2014).

Proposition 6.4 *For the FIN(d) process $\{y_t\}$ with $-1 < d < 1/2$ and $x_t = \varepsilon_t$ being white noise with variance σ^2, the autocovariances are*

$$\gamma_y(h) = \sigma^2 \frac{(-1)^h \Gamma(1-2d)}{\Gamma(1-d+h)\,\Gamma(1-d-h)}, \quad h \in \mathbb{N}_0 \tag{6.6}$$
$$= \frac{h-1+d}{h-d}\,\gamma_y(h-1)\,, \quad h > 0\,.$$

Proof: From Hosking (1981, Theorem 1), we have the first equality given in Eq. (6.6). His proof for $|d| < 1/2$ extends to $-1 < d \le -1/2$. With Eq. (5.13) we obtain the recursive relation given by the second equality. Hence, the proof is complete. □

In another special case, we next consider the fractionally integrated MA(1) model, ARFIMA$(0, d, 1)$, or FIMA$(d, 1)$, and adopt the following result from Hosking (1981, Lemma 2).

Proposition 6.5 *For the FIMA(d, 1) process $\{y_t\}$ with $-1 < d < 1/2$ and*

$$x_t = \varepsilon_t + \theta\,\varepsilon_{t-1}, \quad |\theta| < 1,$$

the autocovariances are

$$\gamma_y(h) = c(h) \frac{(1+\theta)^2 h^2 - (1-d)\,((1-d)(1+\theta^2) + 2\theta d)}{h^2 - (1-d)^2}$$

where $\{c(h)\}$ equals the sequence of autocovariances of FIN(d) defined in Eq. (6.6).

Proof: The proof given by Hosking (1981, Lemma 2) for $|d| < 1/2$ extends to $-1 < d \le -1/2$. □

For $\theta = 0$ Proposition 6.5 reproduces the case of FIN of course. The result of Proposition 6.5 has been extended for ARFIMA(p, d, q) processes in terms of the so-called hypergeometric functions by Sowell (1992, Eq. (8)). A computationally simpler expression was obtained by Chung (1994). The explicit autocovariances of ARFIMA(p, d, q) processes are of course relevant when it comes to maximum likelihood estimation in Chapter 8. In particular, for FIMA(d, q) processes, Chung (1994, p. 296) provides formulae that can be verified to be equal to those in Proposition 6.5 for $q = 1$. The result following Eq. (8) in Sowell (1992), however, does not reproduce Proposition 6.5 for $q = 1$. We conclude that the formulae by Sowell (1992) must be used with care.

We complement the above discussion by means of some numerical and graphical examples. We first look at some simulated FIN processes that were generated according to Eq. (5.1). In fact, we simulated samples of size $T + 1000$ and discarded the first 1000 observations from this burn-in period. The effective sample size is $T = 1000$, too. Different approaches to the simulation of fractionally integrated processes are discussed at the end of Section 6.4. In Figure 6.1 we see series that are integrated of integer orders 2 and 1. To make the graphs comparable, we let both time series begin at the origin. Clearly, the $I(2)$ series is much smoother than the $I(1)$ series, displaying a more global trend over time. The $I(1)$ series is a random walk that wanders in the positive

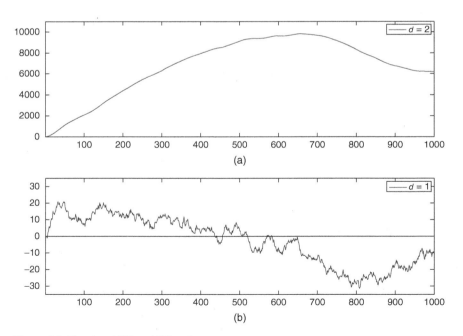

Figure 6.1 Simulated $I(2)$ and $I(1)$ series.

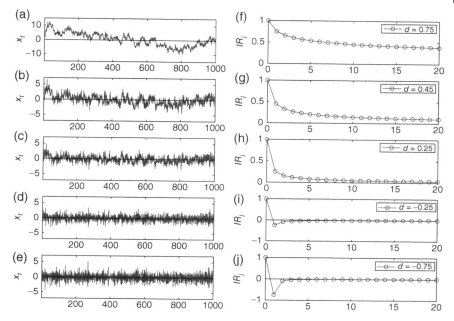

Figure 6.2 Simulated $I(d)$ series and their impulse responses.

range for about half of the time span, only to turn negative then. This series hence has a strong trend component, but not as strong as the $I(2)$ case. In Figure 6.2 we see series of (fractional) integration of smaller order. The $I(0.75)$ series resembles the $I(1)$ series but displays a less trending behavior. Notice the change in scale from the $I(0.75)$ to the $I(0.45)$ and $I(0.25)$ series in Figure 6.2: We thus observe that the drifting behavior is less pronounced the smaller $d > 0$ is. At the same time the impulse responses $\psi_j(d) = \pi_j(-d)$ from Eq. (5.1) converge to 0 but are not summable for $d > 0$. In the antipersistent case ($d < 0$), we find just one dominantly negative impulse response coefficient $\psi_1(d)$, while $\psi_j(d) \approx 0$ for $j > 1$. Consequently, the $I(-0.25)$ and $I(-0.75)$ cases are not very different from the $I(0)$ and $I(-1)$ cases, respectively, as shown in Figure 3.1.

The behavior of the impulse response coefficients is of course reflected by the theoretical autocorrelations plotted in Figure 6.3 for selected values of d in the region of stationarity. The corresponding spectra are hard to plot due to their singularity at the origin for positive d; we hence show $f_y(\lambda)$ only over $(0, \pi/4]$. If $d > 0$, there is a very pronounced peak at frequency 0 highlighting the importance of a trend component for long memory processes under fractional integration; for negative d, the spectrum is quite flat, but $f_y(0) = 0$. When $\ln f_y(\lambda)$ is plotted against $\ln \lambda$, however, we obtain a linear relationship around 0

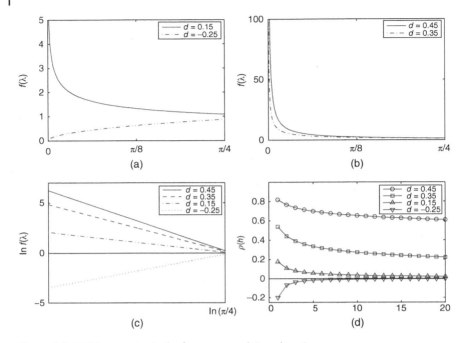

Figure 6.3 FIN(*d*) processes in the frequency and time domain.

according to Proposition 6.1, where the slope is $-2d$; see Figure 6.3c. The fourth panel displays the theoretical autocorrelations. With $d = 0.45$ or $d = 0.35$, the long memory is quite obvious; remember that the autocorrelations do converge to 0 eventually notwithstanding the differing eyeball-impression suggesting at first glance that they converge to a nonzero asymptote.

Next, we illustrate the persistence properties of FIMA(d, 1) models in Figure 6.4. For $\theta = 0$ it simply reproduces the (log) spectrum and autocorrelogram of the FIN process. Positive values of θ have a positive effect on the autocorrelogram, while negative values have a negative influence, and, not surprisingly, the first autocorrelation is particularly strongly affected. Similarly, negative values of θ dampen the spectral singularity at the origin and the other way around for positive θ.

When it comes to estimating d in subsequent chapters, the problem will be to disentangle long memory from a (strong) short memory component. To illustrate this point, we plot spectra from three ARFI(1,d) models over harmonic frequencies $\lambda_j = 2\pi j/1000, j = 1, \ldots, 499$. In particular, Figure 6.5 contains plots for the pure AR(1) process ($d = 0$, $\phi = 0.72$), for the pure FIN model ($d = 0.22$, $\phi = 0$), and for the long memory model with additional short memory ($d = 0.1$, $\phi = 0.5$). The parameter values are picked such that the spectra have (almost) equal values at the first frequency λ_1. Clearly, the pure FIN model

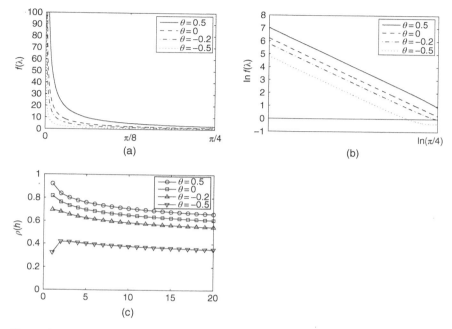

Figure 6.4 FIMA(d, 1) processes with $d = 0.45$ and θ.

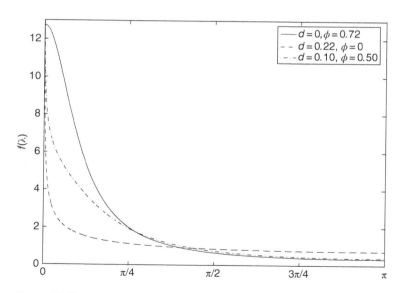

Figure 6.5 Spectra of ARFI(1, d) processes.

has the steepest spectrum at the origin, while only the AR(1) model is bounded (and actually flat) at 0. In practice, where we do not observe spectra but have to estimate them using periodograms computed from finite samples, it will be challenging to discriminate between such models.

Many authors take a hyperbolic decay of $\gamma_y(h)$ as in Corollary 6.1 for $d > 0$ as definition for long Memory; see, e.g. Brockwell and Davis (1991, Section 13.2), Beran (1994, Definition 2.1), or Taniguchi and Kakizawa (2000, Definition 5.1.1). Beran et al. (2013, Definition 1.2) define long-range dependence (corresponding to long memory) in terms of the spectral behavior given in Proposition 6.1 for $d > 0$. Palma (2007, Theorem 3.1) considers the relations between different definitions of long memory in terms of impulse responses, spectra, or autocorrelations. Our definition from Definition 2.4 corresponds most closely to Giraitis et al. (2012, Definition 3.1.2). In the particular framework of fractional integration defined in Eq. (6.2) with Assumption 6.1, we summarize with our definitions of long memory and persistence: A fractionally integrated process has long memory and equivalently strong persistence if and only if $d > 0$, while it has short memory for $d \leq 0$. For $d < 0$ the process is antipersistent, while $d = 0$ ($y_t = x_t$) is the moderately persistent case without antipersistence.

Note, however, that our definition of long memory allows for processes that are not fractionally integrated. Remember Example 3.5 from Section 3.3 with MA(∞) coefficients $c_j = 1/(j+1)$. We saw that the autocovariances $\gamma(h)$ converge to 0 with rate $\ln(h)/h$, which clearly is not hyperbolic, although the sequence is not summable. Further, note that long memory in the sense of Definition 2.4 is not in general equivalent to an unbounded spectrum at the origin, although one may read such statements (see, e.g. Baillie, 1996, p. 10) because many authors implicitly equate long memory with fractional integration. Remember Example 3.4 from Section 3.3 with alternating MA(∞) coefficients $c_j = (-1)^{j+1}/j$; we have seen that the autocovariances are not absolutely summable, while the accumulated impulse responses (and hence the spectrum at frequency 0) are finite.

6.3 Nonstationarity and Type I Versus II

In Section 5.1 we defined the random walk as a leading example of an $I(1)$ process, $\varepsilon_1 + \varepsilon_2 + \cdots + \varepsilon_t$. More generally, for $\{x_j\}$ being $I(0)$ in that this sequence satisfies Assumption 6.1, we define the nonstationary $I(1)$ process as partial sum process

$$S_t = \sum_{j=1}^{t} x_j \quad \sim I(1) .$$

If $\{x_j\}$ is replaced by the stationary $I(d)$ sequence $\{y_j\}$, $|d| < 1/2$, then we define the nonstationary partial sum process $\{S_t\}_{t \in \{1,...,T\}}$ as integrated of order $1 + d$. Alternatively, one may define a nonstationary $I(1 + d)$ process by filtering an $I(0)$ sequence with $\{\pi_j(-1 - d)\}$ from the expansion Δ^{-1-d}; see Lemma 5.3. However, since $\{\pi_j(-1 - d)\}$ is not square summable, the summation has to be truncated in order to get a well-defined process. Such processes are also called fractionally integrated of type II after Marinucci and Robinson (1999), and they are widely used in econometrics:

$$y_t^{II} = \sum_{j=0}^{t-1} \pi_j(-1 - d)x_{t-j} \sim I(1 + d), \quad t = 1, \dots, T.$$

The advantage of type II processes is that the same definition can be applied to higher orders of integration than $1 + d$ but equally to the case d itself with $|d| < 1/2$:

$$y_t^{II} = \sum_{j=0}^{t-1} \pi_j(-d)x_{t-j} \sim I(d), \quad t = 1, \dots, T.$$

Note that the following definition has actually been anticipated in (5.1) when introducing the idea of fractional integration.

Definition 6.2 (Fractional integration of type II) *Let $\{x_t\}$ satisfy Assumption 6.1. With $\{\pi_j(\cdot)\}$ from Eq. (5.22) we define the processes $\{y_t^{II}\}_{t \in \{1,...,T\}}$,*

$$y_t^{II} = \sum_{j=0}^{t-1} \pi_j(-\delta)x_{t-j}, \quad t = 1, \dots, T, \tag{6.7}$$

as type II fractionally integrated of order δ, $y_t^{II} \sim I(\delta)$.

By type I fractional integration for $d < 1/2$, we from now on understand the stationary process considered so far, meeting the restrictions of Assumption 6.2; see Definition 6.1.[2] Nonstationary type I processes are defined by cumulation.[3] For $|d| < 1/2$ we define $\{S_t\}_{t \in \{1,...,T\}}$ with

$$S_t = \sum_{j=1}^{t} y_j \sim I(1 + d), \quad t = 1, \dots, T,$$

---

2 In this and the next chapter, we will distinguish notationally between type I processes $\{y_t\}$ and type II processes $\{y_t^{II}\}$. In later chapters, however, we will use the symbol $\{y_t\}$ for both processes to simplify the notation, and we will add the information whether we mean type I or type II.
3 Repeated cumulation allows for higher orders of integration: The partial sum process of a type I fractional $I(1 + d)$ process, for instance, will be called type I fractionally integrated or order $I(2 + d)$.

as integrated of order $1 + d$ as long as $\{y_j\}$ is $I(d)$. Note that the type I process S_t is difference stationary (by construction) in that the $I(d)$ differences $\Delta S_t = y_t$ are stationary by assumption. This does not hold for the type II process, $y_t^{\text{II}} \sim I(d+1)$, where we observe

$$\Delta y_t^{\text{II}} = \sum_{j=0}^{t-1} \pi_j(-1-d)x_{t-j} - \sum_{j=0}^{t-2} \pi_j(-1-d)x_{t-1-j}$$

$$= x_t + \sum_{j=1}^{t-1} [\pi_j(-d-1) - \pi_{j-1}(-d-1)]x_{t-j}$$

$$= x_t + \sum_{j=1}^{t-1} \pi_j(-d)x_{t-j} \,,$$

where we used Eq. (5.23) from Lemma 5.1 for the last equality. With the stationary $I(d)$ process from Assumption 6.2,

$$\Delta S_t = \sum_{j=0}^{\infty} \pi_j(-d)x_{t-j} \,, \quad \pi_0(-d) = 1, \quad -1 < d < 1/2 \,,$$

it becomes obvious that

$$\Delta y_t^{\text{II}} = \Delta S_t - \sum_{j=t}^{\infty} \pi_j(-d)x_{t-j}$$

is not stationary for finite t. One may call $\{\Delta y_t^{\text{II}}\}$ asymptotically stationary, since the difference between Δy_t^{II} and ΔS_t vanishes with $t \to \infty$ as long as $d < 1/2$. More precisely, we know from Marinucci and Robinson (1999, p. 119) that it holds for two $I(1+d)$ processes that

$$E[(\Delta y_t^{\text{II}} - \Delta S_t)^2] = \text{Var}\left(\sum_{j=t}^{\infty} \pi_j(-d)x_{t-j}\right) = O(t^{2d-1}), \quad |d| < \frac{1}{2} \,;$$

see also Robinson (2005b, Eq. (16)).

There are different ways used in the literature to denote type II integration. They all build on the indicator function $\mathbf{1}_A$ that takes on the value 1 for the argument falling into the set A and is 0 otherwise:

$$\mathbf{1}_A(x) = \begin{cases} 1, & x \in A \\ 0, & x \notin A \end{cases} \,.$$

Hence,

$$\{x_t \mathbf{1}_{\{t>0\}}(t)\}_{t \in \mathbb{Z}} = \{\dots, 0, \dots, 0, x_1, x_2, \dots\} \,,$$

such that a type II process of order δ from Eq. (6.7) becomes

$$y_t^{\text{II}} = \Delta^{-\delta} x_t \mathbf{1}_{\{t>0\}}(t) \,,$$

which is used, e.g. in Robinson (2005b, Eq. (1.5)). Alternatively, we may write, e.g. as in Nielsen (2004a, Eq. (1))

$$\Delta^\delta y_t^{II} = x_t \mathbf{1}_{\{t>0\}}(t) \, .$$

Johansen (2008, Appendix A.4) seems to be the first who formally introduced and discussed the operators $\Delta_+^{-\delta}$ and Δ_+^{δ} defined as

$$\Delta_+^{-\delta} x_t = \Delta^{-\delta} x_t \mathbf{1}_{\{t>0\}}(t) \quad \text{and} \quad \Delta_+^{\delta} y_t = \Delta^\delta y_t \mathbf{1}_{\{t>0\}}(t), \text{ respectively.}$$

We wish to add a further piece of interpretation to type II integration. Let us consider the type I fractional difference equation

$$\Delta^\delta y_t = x_t, \quad t = 1, \dots, T \, .$$

If we now assume that the unobserved past of $\{y_t\}$ is 0,

$$y_0 = y_{-1} = y_{-2} = \cdots = 0,$$

it then holds that

$$\Delta^\delta y_t = \Delta^\delta y_t \mathbf{1}_{\{t>0\}}(t) = \Delta_+^\delta y_t = x_t \, .$$

The last equation can safely be multiplied with the truncated filter $\Delta_+^{-\delta}$ such that

$$y_t = \Delta_+^{-\delta} x_t = y_t^{II}, \quad t = 1, \dots, T \, .$$

Consequently, we find that our type II model Eq. (6.7) arises from the autoregressive representation of type I model under the assumption of zero starting values for the whole past previous to $t = 1$; for a more thorough discussion, see Johansen (2008, Appendix A.5) and the paper by Johansen and Nielsen (2012c).

Let us consider the special case of integer integration. We begin with $I(0)$ processes. Since $\pi_j(0) = 0$ for $j > 0$,

$$y_t^{II} = \sum_{j=0}^{t-1} \pi_j(0) x_{t-j} = x_t = y_t \sim I(0),$$

such that type II and the conventional type I processes coincide for $I(0)$. The same holds true for the $I(1)$ case. Consider

$$y_t^{II} = \sum_{j=0}^{t-1} \pi_j(-1) x_{t-j}$$

with

$$\pi_j(-1) = \frac{j-1+1}{j} \pi_{j-1}(-1) = \cdots = \pi_0(-1) = 1.$$

Consequently,

$$y_t^{II} = \sum_{j=0}^{t-1} \pi_j(-1)x_{t-j} = \sum_{j=0}^{t-1} x_{t-j} = S_t \sim I(1) \ .$$

For noninteger values of d, however, the distinction between type I and type II processes is quite substantial.

6.4 Practical Issues

We now touch upon three practical issues. First, we address how to (fractionally) difference real data. Second, we discuss how to allow for nonzero means. Third, we review competing approaches to simulate fractionally integrated processes that is required for computer experimental exercises.

Let us assume that d is a known or estimated value characteristic of the observed sample $y_1, \ldots, y_T$. If we assume a type II process, $y_t = \Delta_+^{-d} x_t$, this amounts to a past of zero values, $0 = y_0 = y_{-1} = \cdots$. In this case, the appropriate differencing filter is Δ_+^d:

$$\Delta_+^d y_t = \sum_{j=0}^{t-1} \pi_j(d)y_{t-j} \ .$$

The required sequence of coefficients $\{\pi_j(d)\}$ can be computed recursively or by employing a routine for the Gamma function provided by some software; see Eq. (5.22). Alternatively, a very fast way to compute these coefficients is due to Jensen and Nielsen (2014) mentioned at the end of this section.

If we assume a type I process, the infinite difference filter has to be truncated to obtain an approximation for large N:

$$\Delta^d y_t = \sum_{j=0}^{\infty} \pi_j(d)y_{t-j} \approx \sum_{j=0}^{N} \pi_j(d)y_{t-j} \ , \quad t = N+1, \ldots, T \ .$$

In this case, the expansion provides the same precision depending on N at each point in time. Alternatively, one may fix a minimum of N starting values and expand the filter as far as possible:

$$\Delta^d y_t = \sum_{j=0}^{\infty} \pi_j(d)y_{t-j} \approx \Delta_+^d y_t = \sum_{j=0}^{t-1} \pi_j(d)y_{t-j} \ , \quad t = N+1, \ldots, T \ .$$

In the majority of applications, one wants to allow for expected values different from 0. To that end, we read in some papers the model

$$(1-L)^d(y_t - \mu) = x_t \ .$$

This is the standard notation for d being a positive integer. For positive fractional d, this definition is still fine. For negative d, however, we know from Lemma 5.2 that $(1 - L)^d \mu = \mu \sum_{j=0}^{\infty} \pi_j(d)$ is not defined. Therefore, one should allow for nonzero means by generalizing equation (6.2):

$$ y_t = \mu + (1 - L)^{-d} x_t, \quad t \in \mathbb{Z}, \quad d < \frac{1}{2} , \tag{6.8} $$

or

$$ y_t = \mu + \Delta_+^{-d} x_t, \quad t = 1, 2, \dots . \tag{6.9} $$

How should we empirically difference the data if the underlying model is Eq. (6.8) or (6.9)? Under long memory, one may argue that there is no need to estimate μ since it is removed (approximately) by differencing with positive d, where we only consider the type II case for brevity:

$$ \Delta_+^d y_t = \Delta_+^d \mu + x_t \approx x_t . $$

The approximation builds again on

$$ \lim_{t \to \infty} \sum_{j=0}^{t-1} \pi_j(d) = 0 \quad \text{for } d > 0 , $$

and obviously will not work well for small t. Alternatively, one may estimate μ by means of the sample average $\bar{y} = T^{-1} \sum_{t=1}^{T} y_t$, and difference the demeaned series

$$ \Delta_+^d (y_t - \bar{y}) \approx 0 + x_t , $$

where the approximation relies on $\mu - \bar{y} \approx 0$. In the next chapter, however, we learn that the sample average is a poor estimator under long memory; see page 138. A third method to take out the mean of type II process was proposed by Robinson (1994b) (see also page 212 for the case of a mean and a linear time trend): regress the differences $\Delta_+^d y_t$ on the differenced constant intercept defined as $r_t := \sum_{j=0}^{t-1} \pi_j(d)$

$$ \Delta_+^d y_t = \hat{\mu} r_t + \hat{x}_t , $$

and compute the empirical residuals $\hat{x}_t$. At least when assuming a type II model, this last procedure will be most appropriate for removing the mean.

For Monte Carlo experiments, one has to simulate fractionally integrated processes building on so-called pseudorandom numbers. We begin with the stationary type I process. We wish to generate a vector $y' = (y_1, \dots, y_T)$ with autocovariance matrix Σ_y containing all autocovariances,

$$ \Sigma_y = (\gamma_y(|i - j|))_{i,j=1,\dots,T}. $$

One may decompose Σ_y into its Cholesky factors,

$$ \Sigma_y = C \, C'. $$

Given a sample of pseudorandom numbers, $\varepsilon' = (\varepsilon_1, \ldots, \varepsilon_T)$, we obtain

$$y = C\,\varepsilon$$

with the exact required covariance structure. Note that Gaussianity of ε is not assumed for this exercise. Such a procedure requires the autocovariance matrix, which is readily available for FIN; see Proposition 6.4. Still, the $T \times T$ Cholesky decomposition is computationally quite demanding for large T. Alternatively, one may employ the so-called Durbin–Levinson algorithm following Brockwell and Davis (1991, Proposition 5.2.1). In case of FIN, such a procedure can be made faster by employing the known partial autocorrelation structure; see Hosking (1981). Details on the Durbin–Levinson recursion for simulation of FIN are found in Hosking (1984, p. 1900); while Hosking (1984) considered normal FIN, the Gaussianity is not essential to the procedure. The number of operations required to execute the Durbin–Levinson algorithm, however, grows with T^2. A much faster algorithm (of order $T \ln T$) has been proposed by Davies and Harte (1987); however, it is only designed to generate Gaussian data. Moreover, this Davies–Harte algorithm requires a nonnegativity constraint to be met. It was shown to hold for FIN(d) processes with $|d| < 1/2$ by Craigmile (2003, Lemma 2); we omit details.

For a fractionally integrated process of type II, one may explicitly compute the sequence $\{\pi_j(-d)\}$, $j = 0, \ldots, T-1$ and filter some white noise sequence or ARMA sequence $\{x_t\}$ according to Definition 6.2. Jensen and Nielsen (2014) pointed out that such a procedure requires again a number of arithmetic operations of the order T^2. Alternatively, Jensen and Nielsen (2014, Theorem 2) proposed a method of much smaller complexity (namely, again of order $T \ln T$). It has the nice property that it does not rely on Gaussianity. Further, it cannot only be applied to generate a fractionally integrated series, but just as well to compute fractional differences with empirical data by means of $\pi_j(d)$. Technical details and indeed codes in several programming languages are given in Jensen and Nielsen (2014).

6.5 Frequency Domain Assumptions

So far we have worked under Assumption 6.2 or Definition 6.2 in the time domain. In particular for semiparametric modeling, a different set of assumptions is often maintained. It is settled in the frequency domain and semiparametric in that it maintains assumptions on the spectrum only in a vicinity of the origin. Remember the order notation $O(\cdot)$ introduced in Eq. (5.7) in Section 5.2.

Assumption 6.3 (*Integration in the frequency domain*) *Let $\{y_t\}$ be a stationary process with spectrum $f_y(\lambda) = \lambda^{-2d} h(\lambda)$, $d < 1/2$, where h is bounded and* *bounded away from 0 at frequency $\lambda = 0$. It further holds that*

$$h(\lambda) = h(0) + O(\lambda^\beta), \quad \lambda \to 0 \tag{6.10}$$

for some $\beta \in (0, 2]$.

The first assumption that $h(0)$ is bounded and positive is minimal and made in order to identify d. Next, assumption Eq. (6.10) imposes a rate of convergence characterizing the smoothness of the short memory component h around 0.[4] This implies that h is continuous at the origin. The larger β, the smoother is $h(\cdot)$ at 0 in that $h(\lambda) - h(0)$ vanishes faster. It further follows that

$$\frac{h(\lambda) - h(0)}{\lambda} = O(\lambda^{\beta-1}).$$

Hence, h possesses a first derivative at 0 as long as $\beta > 1$ with $h'(0) = 0$. Assumption 6.3 is semiparametric in that only assumptions within an arbitrarily small neighborhood of frequency 0 are made.

Next, we characterize a class of models meeting Assumption 6.3.

Proposition 6.6 *Let $\{y_t\}$ be a stationary process satisfying Assumption 6.2, where we assume that the $I(0)$ component $\{x_t\}$ from $(1 - L)^d y_t = x_t$ possesses a three times continuously differentiable spectrum f_x in a neighborhood of the origin including zero with $0 < f_x(0) < \infty$ and $f_x'(0) = 0$. It then holds that $f_y(\lambda) = \lambda^{-2d} h(\lambda)$, where*

$$h(\lambda) = f_x(0) + \left(f_x(0) \, \frac{d}{6} + f_x''(0) \right) \frac{\lambda^2}{2} + O(\lambda^3)$$

as $\lambda \to 0$.

Proof: See Appendix to this chapter.

Obviously, this function $h(\lambda)$ from Proposition 6.6 with $h(0) = f_x(0)$ meets Assumption 6.3 for $\beta = 2$. Note that $f_x'(0) = 0$ is not an additional assumption. Since f_x is a spectrum, it is an even function; see Proposition 4.4. By symmetry, f_x has a relative extremum at the origin, such that $f_x'(0) = 0$ is an implication, not an assumption.

We also have a fractionally integrated parametric model settled in the frequency domain. If $\{x_t\}$ is an EXP(k) model (see Eq. (4.26)), then we will follow (Beran, 1993) and call $\{y_t\}$ from Eq. (6.2) a FEXP model, or more precisely FEXP(d, k). Another early recommendation of the use of FEXP models was by Robinson (1994b, Eq. (22)). We notice that FEXP models as well as the

---

4 Sometimes, a slightly stronger parametric version is maintained:

$$h(\lambda) = b_0 + b_1 \lambda^\beta + o(\lambda^\beta).$$

more familiar ARFIMA models satisfy Assumption 6.3. For FEXP or ARFIMA models, we have of course $f_y(\lambda) = \left(4\sin^2\lambda/2\right)^{-d} f_x(\lambda)$; see Proposition 6.1. If $\{x_t\}$ is an ARMA or EXP process, one can derive that the third derivative f_x''' exists continuously, following the arguments proving Proposition 4.8. Hence, the assumptions of Proposition 6.6 on f_x are met as long as the ARMA process is invertible, which implies Assumption 6.3 with $\beta = 2$. Consequently, the reader may think of the (invertible) ARFIMA or FEXP models as simple parametric fractionally integrated examples meeting the more abstract frequency domain Assumption 6.3.

Some papers maintain semiparametric models that are more general than what we have considered so far. Remember that the model of fractional integration from Eq. (6.2) with Assumption 6.1 amounts to (Proposition 6.1)[5]

$$f_y(\lambda) \sim \lambda^{-2d} f_x(0),$$

where

$$\lim_{\lambda \to 0} f_x(\lambda) = f_x(0) =: \ell > 0 .$$

This means that $\tilde{f}_x(\omega) := f_x(\omega^{-1})$ has an asymptote as $\omega \to \infty$:

$$\lim_{\omega \to \infty} \tilde{f}_x(\omega) = f_x(0) = \ell .$$

This kind of behavior can be generalized by allowing for slowly varying functions replacing the number ℓ. Here, we only consider the (natural) logarithm as an example. For a general treatment of slowly varying functions in the context of long memory; see Palma (2007, Section 3.1), Giraitis et al. (2012, Section 2.3) and in particular the detailed exposition in Beran et al. (2013, Section 1.3.1). So, our leading example for a slowly varying function is

$$\tilde{L}(x) = \ln(x) ,$$

such that $\tilde{L}$ is positive for large enough arguments,

$$\tilde{L}(x) > 0 \quad \text{for} \quad x > 1 .$$

Further, for any $u > 0$ it holds

$$\frac{\tilde{L}(ux)}{\tilde{L}(x)} = 1 + \frac{\ln(u)}{\ln(x)} \to 1, \quad x \to \infty.$$

A function $\tilde{L}$ with such a behavior for large x is called slowly varying at infinity (in Karamata's sense). Consequently,

$$L(\lambda) := \tilde{L}\left(\frac{1}{\lambda}\right) = \ln\left(\frac{1}{\lambda}\right), \quad \lambda > 0 ,$$

---

5 Similarly, we have by Assumption 6.3 that $f_y(\lambda) \sim \lambda^{-2d} h(0)$.

is slowly varying at the origin in that for any $u > 0$

$$\frac{L(u\lambda)}{L(\lambda)} = 1 + \frac{\ln(u)}{\ln(\lambda)} \to 1, \quad \lambda \to 0 .$$

Hence, although $L(\lambda)$ does not converge to a constant value as $\lambda \to 0$, it *varies only very little*. Some papers replace our $f_x(\lambda)$ (or $h(\lambda)$ from Assumption 6.3) by a function $L(\cdot)$ that is slowly varying at the origin,

$$f_y(\lambda) = \lambda^{-2d} L(\lambda) .$$

Note that few papers define slowly varying in Zygmund's sense, which is a special case of Karamata's sense defined above; see Beran et al. (2013, p. 22–24). For the rest of the book, however, we will not allow for slowly varying functions.

6.6 Technical Appendix: Proofs

Proof of Proposition 6.1

Under Definition 6.1, we know from Lemma 5.4 that $\{y_t\}$ has a square summable MA(∞) representation as long as $d < \frac{1}{2}$ (see Theorem 3.1),

$$\sum_{k=0}^{\infty} b_k^2(d) < \infty$$

and $\{y_t\}$ is stationary for $d < \frac{1}{2}$; see also Proposition 3.1. By Proposition 4.3 it holds

$$f_y(\lambda) = T_B(\lambda) \frac{\sigma^2}{2\pi}$$

with the power transfer function T_B of

$$
\begin{aligned}
B_d(\lambda) &= \sum_{k=0}^{\infty} b_k(d) \, e^{-i\lambda k} \\
&= (1 - e^{-i\lambda})^{-d} \, C(e^{-i\lambda}) .
\end{aligned}
$$

Hence,

$$T_B(\lambda) = |1 - e^{-i\lambda}|^{-2d} |C(e^{-i\lambda})|^2 ,$$

and

$$f_y(\lambda) = |1 - e^{-i\lambda}|^{-2d} f_x(\lambda) .$$

Note that

$$
\begin{aligned}
|1 - e^{-i\lambda}|^{-2d} &= [(1 - e^{-i\lambda})(1 - e^{i\lambda})]^{-d} \\
&= [2 - (e^{-i\lambda} + e^{i\lambda})]^{-d}
\end{aligned}
$$

$$= [2 - 2\cos(\lambda)]^{-d}$$

$$= \left[4\sin^2\left(\frac{\lambda}{2}\right)\right]^{-d},$$

where the trigonometric half-angle formula was used,

$$2\sin^2(x) = 1 - \cos(2x);$$

see (5.9). Remember the first-order Taylor expansion of the sine function from (5.5): $\sin\lambda \sim \lambda$. It then holds that

$$\left[4\sin^2\left(\frac{\lambda}{2}\right)\right]^{-d} \sim \left[4\left(\frac{\lambda}{2}\right)^2\right]^{-d} = \lambda^{-2d}, \quad \lambda \to 0,$$

which completes the proof. $\qquad\qquad\qquad\qquad\qquad\qquad\qquad\qquad$ □

Proof of Corollary 6.1

The corollary follows from Giraitis et al. (2012, Proposition 3.1.1 a). In their notation it holds

$$C_{\gamma,d} = f_x(0)\, 2\, \Gamma(1 - 2d)\sin(\pi d).$$

Remember the so-called reflection formula Eq. (5.17):

$$\Gamma(1 - x)\Gamma(x) = \frac{\pi}{\sin(\pi x)}.$$

Inserting $\sin(\pi d)$ in $C_{\gamma,d}$ yields the expression given in Eq. (6.4), which completes the proof. $\qquad\qquad\qquad\qquad\qquad\qquad\qquad$ □

Proof of Lemma 6.1

The proof follows essentially by definition. Consider

$$b_k(d) - b_{k-1}(d) = c_k + \sum_{j=1}^{k} c_{k-j}(\pi_j(-d) - \pi_{j-1}(-d))$$

$$= c_k + \sum_{j=1}^{k} c_{k-j}\pi_j(1 - d),$$

where Lemma 5.1 was used. Consequently,

$$b_k(d) - b_{k-1}(d) = \sum_{j=0}^{k} c_{k-j}\pi_j(1 - d) = b_k(d - 1),$$

which completes the proof. $\qquad\qquad\qquad\qquad\qquad\qquad\qquad\qquad\qquad$ □

Proof of Proposition 6.3

The result uses Bondon and Palma (2007, Theorem 1). To that end, we have to check two conditions for the coefficients $\{b_k(d)\}$ from Eq. (5.28):

(1) $\sum_{k=0}^{\infty} b_k(d) = 0$,

(2) $b_k(d) - b_{k-1}(d) \sim \sum_{j=0}^{\infty} c_j \frac{k^{d-2}}{\Gamma(d-1)}$, $\quad k \to \infty$.

The first condition clearly holds under $d < 0$:

$$\sum_{k=0}^{\infty} b_k(d) = (1-z)^{-d}|_{z=1} C(1) = 0 \ .$$

For the second condition, we use Lemma 6.1 with Eq. (5.28):

$$b_k(d) - b_{k-1}(d) = b_k(d-1) \sim \sum_{j=0}^{\infty} c_j \frac{k^{d-2}}{\Gamma(d-1)} \ ,$$

such that Condition (2) is met. By Bondon and Palma (2007, Theorem 1), we have

$$\gamma_y(h) \sim h^{2d-1} \frac{\left(\sum_{j=0}^{\infty} c_j\right)^2 \sigma^2 \Gamma(1-2d) \sin(\pi d)}{\pi} \ .$$

Using Eq. (5.17) again, we complete the proof. □

Proof of Proposition 6.6

By assumption, we have from Proposition 6.1 that $f_y(\lambda) = \left(4 \sin^2 \frac{\lambda}{2}\right)^{-d} f_x(\lambda)$. Hence, h from $f_y(\lambda) = \lambda^{-2d} h(\lambda)$ is defined by

$$h(\lambda) = g(\lambda) f_x(\lambda) \quad \text{with } g(\lambda) = \left(\frac{\sin \frac{\lambda}{2}}{\frac{\lambda}{2}}\right)^{-2d} \ .$$

Taylor expansions yield

$$g(\lambda) = 1 + d \, \frac{\lambda^2}{12} + O(\lambda^4) \ ,$$
$$f_x(\lambda) = f_x(0) + f'_x(0)\lambda + f''_x(0)\frac{\lambda^2}{2} + R(\lambda) \ ,$$

where R is Lagrange's remainder:

$$R(\lambda) = \int_0^{\lambda} f'''_x(u) \frac{(\lambda - u)^2}{2} du = \frac{f'''_x(\lambda^*)}{3!} \lambda^3 \ .$$

where the last equality holds for some λ^* from the mean value theorem. Hence,

$$f_x(\lambda) = f_x(0) + f_x''(0)\frac{\lambda^2}{2} + O(\lambda^3) \ .$$

This completes the proof. □

7

Sample Mean

Under long memory the sample mean does not converge at the standard rate $\sqrt{T}$. This chapter addresses related issues. A general central limit theorem (CLT) ensuring limiting normality is given in Theorem 7.1. In order to apply it to fractionally integrated processes of type I, one has to determine the variance of the arithmetic mean. It turns out that the mean converges all the more slowly, the longer the memory is. After an extension to a *functional* CLT (for type I and type II processes), we are able to compare the limiting variance of the sample mean under the assumption of type I and type II, respectively. After discussing some further aspects of inference about the mean, we turn to sample autocorrelations. They strongly underestimate the true theoretical memory, which is a motivation to rather estimate long memory in the frequency domain.

7.1   Central Limit Theorem for $I(0)$ Processes

CLTs are the reason behind limiting normality of sample means. Classical CLTs, however, have to be adjusted to meet the special needs of time series analysis where serial correlation is rather the rule than the exception. We now assume that $\{x_t\}$ is a moving average process of infinite order (see (3.1)):

$$x_t = \mu + \sum_{j=0}^{\infty} c_j \, \varepsilon_{t-j} \, , \quad \varepsilon_t \sim \mathrm{WN}(0, \sigma^2) \, . \tag{7.1}$$

The arithmetic mean from a sample $x_1, \dots, x_T$ of size T is denoted as

$$\bar{x} = \frac{1}{T} \sum_{t=1}^{T} x_t \, .$$

Properties of $\bar{x}$ are typically inferred from properties of the innovation sequence with mean $\bar{\varepsilon} = \frac{1}{T} \sum_{t=1}^{T} \varepsilon_t$. Let "$\xrightarrow{D}$" stand for convergence in distribution. Then

Time Series Analysis with Long Memory in View, First Edition. Uwe Hassler.
© 2019 John Wiley & Sons, Inc. Published 2019 by John Wiley & Sons, Inc.

we assume the following:

$$\sqrt{T}\,\bar{\varepsilon} \;\overset{D}{\to}\; \mathcal{N}(0,\ \sigma^2) \quad \text{as } T \to \infty . \tag{7.2}$$

The next proposition contains a result for absolutely summable moving average processes.

Proposition 7.1 *Let* $\{x_t - \mu\}$ *be an* $I(0)$ *process from Assumption 6.1,*

$$x_t = \mu + \sum_{j=0}^{\infty} c_j\, \varepsilon_{t-j}\,, \quad \sum_{j=0}^{\infty} |c_j| < \infty\,,$$

with $\sum_{j=0}^{\infty} c_j \neq 0$, *such that the long-run variance* ω_x^2 *from Eq. (2.10) or (3.5) is positive. Further, assume limiting normality of the innovation mean, Eq. (7.2). It then holds that*

$$\sqrt{T}\,(\bar{x} - \mu) \;\overset{D}{\to}\; \mathcal{N}(0,\ \omega_x^2) \tag{7.3}$$

as $T \to \infty$.

Proof: Fuller (1996, Theorem 6.3.3). □

We want to briefly discuss two situations where the assumption Eq. (7.2) behind Proposition 7.1 holds. First, for iid innovations, Eq. (7.2) is just the classical CLT due to Lindeberg–Lévy; see, e.g. Davidson (1994, Proposition 23.3) or White (2001, Proposition 5.2). Consequently, Proposition 7.1 is a classical time series result if the innovations are iid; see, e.g. Anderson (1971, Theorem 8.4.1) or Brockwell and Davis (1991, Theorem 7.1.2). Second, we relax the iid assumption and maintain a strictly stationary, ergodic martingale difference sequence (MDS)(with finite variance σ^2). The following lemma ensures a CLT for MDS.

Lemma 7.1 *Let* $\{\varepsilon_t\}$ *be a strictly stationary, ergodic martingale difference sequence with finite variance* σ^2. *Then Eq. (7.2) holds as* $T \to \infty$.

Proof: Davidson (1994, p. 385) argued that a stationary ergodic MDS with finite variance meets the assumptions of Davidson (1994, Theorem 24.3) (see also White, 2001, Theorem 5.24). Then the result follows. □

The innovation variance can be estimated consistently

$$\frac{1}{T}\sum_{t=1}^{T} \varepsilon_t^2 \;\overset{a.s.}{\to}\; \sigma^2\,, \tag{7.4}$$

which follows from the assumption of ergodicity; see Theorem 2.2. The consistent estimation of the long-run variance ω_x^2 showing up in Eq. (7.3) is a more delicate issue. Hamilton (1994, Section 10.5) contains a recommendable overview; see also Section 7.4.

Proposition 7.1 shows that the *standard* textbook case of a limiting normal distribution at rate $\sqrt{T}$ for stationary time series is actually rather special: It only holds for $0 < \omega_x^2 < \infty$. First, we have to rule out long memory in that ω_x^2 has to be finite. Second, we may not allow for *negative short memory* or *antipersistence* where $\omega_x^2 = 0$. In what follows we will see how Eq. (7.3) has to be generalized to account for violations of $0 < \omega_x^2 < \infty$.

7.2 Central Limit Theorem for *I(d)* Processes

To motivate a generalization of Proposition 7.1, we consider a reformulation using Proposition 2.3, where we observed under short memory that

$$\mathrm{Var}(\bar{x}) \sim \frac{\omega_x^2}{T} \; .$$

Hence, we have

$$\frac{\bar{x} - \mu}{\sqrt{\mathrm{Var}(\bar{x})}} \overset{D}{\to} \mathcal{N}(0, \, 1) \tag{7.5}$$

as $T \to \infty$. This statement is known to hold true for linear processes under the very weak assumption that the variance $\mathrm{Var}\left(\sum_{t=1}^{T} x_t\right) = T^2 \, \mathrm{Var}(\bar{x})$ grows with the sample size; see Ibragimov and Linnik (1971, Theorem 18.6.5). The following theorem provides a generalization by Peligrad and Utev (2006) and Abadir et al. (2014). We formulate it in terms of $\{y_t\}$, since it will also cover $I(d)$ processes as in (6.2).

Theorem 7.1 (CLT) *Let $\{y_t\}$ be a stationary MA(∞) process,*

$$y_t = \mu + \sum_{j=0}^{\infty} \gamma_j \varepsilon_{t-j} \, , \quad \sum_{j=0}^{\infty} \gamma_j^2 < \infty \, ,$$

with

$$\mathrm{Var}\left(\sum_{t=1}^{T} y_t\right) = T^2 \, \mathrm{Var}(\bar{y}) \; \to \; \infty \, .$$

Let $\{\varepsilon_t\}$ be a stationary, ergodic martingale difference sequence or satisfy Abadir et al. (2014, Assumption 2.1). It then holds that

$$\frac{\bar{y} - \mu}{\sqrt{\mathrm{Var}(\bar{y})}} \overset{D}{\to} \mathcal{N}(0, 1)$$

as $T \to \infty$.

Proof: Peligrad and Utev (2006, Proposition 4) and Abadir et al. (2014, Theorem 2.1). □

Note that this theorem holds without assumptions with respect to higher moments of $\{\varepsilon_t\}$, only requiring a finite variance. The original result by Ibragimov and Linnik (1971, Theorem 18.6.5) was established for linear processes, i.e. $\varepsilon_t \sim \mathrm{iid}(0, \sigma^2)$. Peligrad and Utev (2006, Proposition 4) relaxed this assumption allowing for stationary and ergodic MDS; see also the proof in Abadir et al. (2014, Theorem 2.1). Alternatively, Abadir et al. (2014) allowed for conditionally heteroskedastic MDS meeting certain requirements with respect to conditional moments; see the discussion in Abadir et al. (2014, Assumption 2.1, Section 4.1).

To establish limiting normality by means of Theorem 7.1, one essentially has to show that the variance of the sample sum diverges. This is where the following proposition comes in. We only treat type I processes:

$$y_t = (1 - L)^{-d} x_t .$$

The case of type II integration will be dealt with in Section 7.3.

Proposition 7.2 *Let $\{y_t\}$ satisfy Assumption 6.2, and define*

$$S_T = \sum_{t=1}^{T} y_t \quad \text{with } \sigma_T^2 := \mathrm{Var}(S_T) .$$

It then holds

(a) if $|d| < 1/2$ that

$$\sigma_T^2 \sim C_{S,d}^2 T^{1+2d},$$

where

$$C_{S,d}^2 := \frac{\omega_x^2 \Gamma(1 - 2d)}{\Gamma(d + 1)\Gamma(1 - d)(1 + 2d)} ; \qquad (7.6)$$

(b) if $d = -1/2$ that

$$\sigma_T^2 \sim \frac{2\omega_x^2}{\pi} \ln T ;$$

(c) if $-1 < d < -1/2$ that

$$\sigma_T^2 < \infty ,$$

as $T \to \infty$, where $\omega_x^2 = 2\pi f_x(0)$ is the long-run variance of $\{x_t\}$ from Assumption 6.1.

Proof: See Appendix to this chapter.

Note that $C_{S,d}^2$ is linked to $C_{\gamma,d}$ from Eq. (6.4) for $d \neq 0$:

$$C_{S,d}^2 = \frac{C_{\gamma,d}}{d(1 + 2d)}, \quad |d| < 1/2.$$

By Theorem 7.1 and Proposition 7.2, we have the following result for type I processes; again, the case of type II integration is postponed to Section 7.3.

Corollary 7.1 *Let $\{y_t - \mu\}$ satisfy Assumption 6.2, where $\{\varepsilon_t\}$ meets the restrictions of Theorem 7.1. It then holds*

(a) for $|d| < 1/2$ that

$$T^{1/2-d}(\bar{y} - \mu) \overset{D}{\to} \mathcal{N}(0, C_{S,d}^2)$$

with $C_{S,d}^2$ from Eq. (7.6),
(b) and for $d = -1/2$ that

$$\frac{T}{\sqrt{\ln T}}(\bar{y} - \mu) \overset{D}{\to} \mathcal{N}\left(0, \frac{2\omega_x^2}{\pi}\right),$$

where ω_x^2 is again the long-run variance of $\{x_t\}$

as $T \to \infty$.

Proof: Theorem 7.1 and Proposition 7.2. □

We have now restricted for the first time the parameter range, which covered $(-1, 1/2)$ before. The range $|d| < 1/2$ is standard in the long memory literature; in Corollary 7.1, we still are able to cover $d = -1/2$, too. The case $-1 < d < -1/2$ does not result in limiting normality. This fact is easy to see. Assume $\{z_t\}$ to be $I(d)$, $0 < d < 1/2$, with differences $y_t = \mu + \Delta z_t$ as in the proof of Proposition 7.2 (c). It is clear that $\bar{y} = (\mu T + z_T - z_0)/T$ does not converge when normalized with the square root of the variance, since $\sigma_T^2 = \text{Var}(\sum_t y_t) < \infty$:

$$\frac{\bar{y} - \mu}{\sqrt{\text{Var}(\bar{y})}} = \frac{z_T - z_0}{\sigma_T}.$$

Inferential procedures relying on Corollary 7.1 will be discussed in Section 7.4.

7.3   Functional Central Limit Theory

In this section we will provide so-called functional CLTs for type I and type II processes. As a by-product, we will obtain a CLT for type II processes, paralleling Corollary 7.1 (a). This is of applied interest if one assumes a deterministic component behind a type II process,

$$y_t^{\text{II}} = \mu + \sum_{j=0}^{t-1} \pi_j(-\delta)x_{t-j}, \quad t = 1, \dots, T,$$

and wishes to make inference about the expected value μ. In view of Eq. (6.3) it is not surprising that the sample mean $\bar{y}^{\text{II}}$ converges at the same rate as $\bar{y}$. As in Proposition 7.2 (a), it holds

$$\text{Var}(\bar{y}^{\text{II}}) = \text{E}\left[\left(\sum_{j=0}^{t-1} \pi_j(-d)x_{t-j}\right)^2\right] = O(T^{2d-1}), \quad |d| < 1/2 ;$$

see also Shimotsu (2010, Eq. (6)). The limiting variance of $T^{1/2-d}\bar{y}^{\text{II}}$, however, differs from that of $T^{1/2-d}\bar{y}$, as we will see in Proposition 7.3. To that end, we will consider

$$T\bar{y} = S_T = \sum_{t=1}^{T} y_t, \quad y_t \sim I(d),$$

$$T\bar{y}^{\text{II}} = S_T^{\text{II}} = \sum_{t=1}^{T} y_t^{\text{II}}, \quad y_t^{\text{II}} \sim I(d).$$

In fact, we want more than just a CLT. This time we are interested in asymptotic distributions of suitably scaled versions of

$$S_{\lfloor rT \rfloor} = \sum_{t=1}^{\lfloor rT \rfloor} y_t \quad \text{and} \quad S_{\lfloor rT \rfloor}^{\text{II}} = \sum_{t=1}^{\lfloor rT \rfloor} y_t^{\text{II}}, \quad y_t^{\text{II}}, y_t \sim I(d), \quad r \in [0, 1] ,$$

with $|d| < 1/2$, where $\lfloor x \rfloor$ truncates real positive numbers x such that $\lfloor x \rfloor$ is the largest integer with $\lfloor x \rfloor \le x$. Note that $S_{\lfloor rT \rfloor}$ and $S_{\lfloor rT \rfloor}^{\text{II}}$ are stochastic step functions constant on the T intervals

$$r \in \left[\frac{j-1}{T}, \frac{j}{T}\right), \quad j = 1, \dots, T,$$

which are open on the right. They map the domain $[0, 1]$ into a range of step values, where the length of the steps, $j/T - (j-1)/T = 1/T$, decreases with the sample size T. In the limit one obtains a stochastic process with continuous paths depending on $r \in [0, 1]$. For a well-defined limit to be achieved, $S_{\lfloor rT \rfloor}$ and $S_{\lfloor rT \rfloor}^{\text{II}}$ have to be normalized appropriately with the sample size. The required rate is obvious from Corollary 7.1: $T^{-1/2-d}$. Note, however, that $T^{-1/2-d}S_{\lfloor rT \rfloor}$ and $T^{-1/2-d}S_{\lfloor rT \rfloor}^{\text{II}}$, $r \in [0, 1]$, are not random variables, but random functions.

Hence, we do not have convergence in distribution, "$\xrightarrow{D}$," as in Corollary 7.1, but rather so to speak infinitely many such CLTs, one for each $r \in [0, 1]$. They are *summarized* in a functional central limit theorem (FCLT), where the corresponding weak convergence is denoted by "$\Rightarrow$." The following result for $T^{-1/2-d}S^{II}_{\lfloor rT \rfloor}$ is taken from Marinucci and Robinson (2000, Theorem 1); for $T^{-1/2-d}S_{\lfloor rT \rfloor}$ the most recent result is by Abadir et al. (2014, Corollary 4.1). Earlier results under different assumptions with respect to $\{\varepsilon_t\}$ are discussed at the end of this section.

Proposition 7.3 *Let*

$$y_t = \sum_{j=0}^{\infty} \pi_j(-d)x_{t-j} \quad and \quad y_t^{II} = \sum_{j=0}^{t-1} \pi_j(-d)x_{t-j}$$

be I(d) processes of type I and II, respectively, where $\{x_t\}$ with $x_t = \sum_{j=0}^{\infty} c_j \varepsilon_{t-j}$, $t \in \mathbb{Z}$, is I(0) satisfying Assumption 6.1 with long-run variance ω_x^2 from Eq. (3.5). Further, assume

$$E(|\varepsilon_t|^q) < \infty \quad for\ some \quad q > \max\left(2, \frac{2}{2d+1}\right). \tag{7.7}$$

It then holds for $r \in [0, 1]$ as $T \to \infty$

(a) that

$$T^{-1/2-d}S_{\lfloor rT \rfloor} \Rightarrow C_{S,d}B_d(r), \quad C_{S,d} = \left(\frac{\omega_x^2 \Gamma(1-2d)}{\Gamma(d+1)\Gamma(1-d)(2d+1)}\right)^{1/2},$$

as long as $|d| < 1/2$ and $\{\varepsilon_t\}$ is a strictly stationary and ergodic MDS, where $C_{S,d}$ is from Eq. (7.6),
(b) and further that

$$T^{-1/2-d}S^{II}_{\lfloor rT \rfloor} \Rightarrow C^{II}_{S,d}B^{II}_d(r), \quad C^{II}_{S,d} := \left(\frac{\omega_x^2}{\Gamma^2(d+1)(2d+1)}\right)^{1/2},$$

as long as $d > -1/2$ and $\{\varepsilon_t\}$ is iid.

Here $B_d(\cdot)$ and $B^{II}_d(\cdot)$ are so-called standard fractional Brownian motions (fBm) of type I and type II, respectively, which are defined below.

Proof: Abadir et al. (2014, Corollary 4.1) for (a) (type I) and Marinucci and Robinson (2000, Theorem 1) for (b) (type II). □

Note that $\{S_t^{II}\}$ is a type II process integrated of order $d + 1$ in the sense of Eq. (6.7):

$$
\begin{aligned}
S_t^{II} &= x_1 + [x_2 + \pi_1(-d)x_1] + \cdots \\
&\quad + [x_t + \pi_1(-d)x_{t-1} + \cdots + \pi_{t-1}(-d)x_1] \\
&= x_1 \sum_{k=0}^{t-1} \pi_k(-d) + x_2 \sum_{k=0}^{t-2} \pi_k(-d) + \cdots + \pi_0(-d)x_t \\
&= \sum_{j=0}^{t-1} \pi_j(-d-1)x_{t-j} \sim I(d+1),
\end{aligned}
$$

where Eq. (5.24) from Lemma 5.1 was used in the last step.

To understand the limiting processes in Proposition 7.3, we first consider the case of $d = 0$, where $y_t = y_t^{II} = x_t$, $S_t = S_t^{II}$, and $C_{S,d} = C_{S,d}^{II} = \omega_x$. Then we know from Davidson (1994, Chapter 29) or White (2001, Section 7.3) under several sets of assumptions that[1]

$$
T^{-1/2} S_{\lfloor rT \rfloor} \Rightarrow \omega_x B_0(r), \tag{7.8}
$$

where B_0 is a standard Brownian motion or Wiener process. The standard Brownian motion is characterized by three properties, which we review rather informally here: (i) At time $t = 0$ it holds that $B_0(0) = 0$ with probability 1; (ii) nonoverlapping increments, $B_0(t_2) - B_0(t_1)$ and $B_0(t_4) - B_0(t_3)$ for $t_1 \leq t_2 \leq t_3 \leq t_4$, are independent; and (iii) the increments are stationary and follow a normal distribution, $B_0(t) - B_0(s) \sim \mathcal{N}(0, t - s)$ for $s < t$. Consequently, $B_0(r) \sim \mathcal{N}(0, r)$ such that Eq. (7.8) provides for $r = 1$ that

$$
T^{-1/2} S_T = T^{1/2} \bar{x} \Rightarrow \omega_x B_0(1) \sim \mathcal{N}(0, \omega_x^2),
$$

which reproduces Proposition 7.1 of course or Corollary 7.1 for $d = 0$.

Now, we turn to a type I fBm. Let $B_d(t)$ be a Gaussian process, $t \in \mathbb{R}$, $|d| < 1/2$, with $E(B_d(t)) = 0$ and $B_d(0) = 0$ with probability 1, and let

$$
E(B_d(s)B_d(t)) = \frac{1}{2}(|s|^{2d+1} + |t|^{2d+1} - |s-t|^{2d+1}), \quad s, t \in \mathbb{R}. \tag{7.9}
$$

Then $B_d(\cdot)$ is called a fBm; see, e.g. Giraitis et al. (2012, Definition 3.4.4). Further, we call $B_d(\cdot)$ a standard fBm, since $\mathrm{Var}(B_d(1)) = 1$. In much of the probability literature, the fBm is typically denoted as $B_H(\cdot)$, where H with $0 < H < 1$ is often called the Hurst coefficient. We follow Pipiras and Taqqu (2003, p. 176) and write fBm as $B_d(\cdot)$ with $d = H - 1/2 \in (-1/2, 1/2)$. The fBm can be represented as integral in terms of the standard Brownian motion B_0 (see, e.g. Mandelbrot and Van Ness (1968, Eq. (2.1)), Taqqu (2003, p. 29), or Pipiras and

---

1 An intuitive introduction to FCLTs is provided by Hassler (2016, Chapter 14).

Taqqu (2003, Eq. (3.1))):

$$B_d(t) = \frac{1}{C(d)} \left[\int_{-\infty}^0 ((t-s)^d - (-s)^d) dB_0(s) + \int_0^t (t-s)^d dB_0(s) \right] \qquad (7.10)$$

for $0 \le t$ and $|d| < 1/2$, where

$$C^2(d) := \left[\frac{1}{2d+1} + \int_0^\infty ((1+s)^d - s^d)^2 ds \right] .$$

A convenient closed-from expression for this latter integral is available from Taqqu (2003, Eq. (9.3)):

$$C^2(d) = \frac{\Gamma^2(d+1)}{\Gamma(2d+2)\sin((d+1/2)\pi)} . \qquad (7.11)$$

Marinucci and Robinson (1999) called B_d a type I fBm in order to distinguish it from the type II process to be introduced below. Note that (7.10) yields $B_d(0) = 0$ at the origin and

$$B_0(t) = \int_0^t dB_0(t) \quad \text{with} \quad C(0) = 1, \quad d = 0 .$$

Generally, $B_d(t)$ is a process with stationary Gaussian increments and autocovariance function Eq. (7.9). From Eq. (7.9) it follows that

$$\text{Var}(B_d(t)) = t^{2d+1}, \quad \text{Var}(B_d(t) - B_d(s)) = (t-s)^{2d+1}, \quad 0 \le s \le t.$$

A different expression for $C^2(d)$ is widely used in econometrics. It was given by Davidson and Hashimzade (2008, Lemma 5.1) and is reproduced in Eq. (7.12). Since the equality of Eqs. (7.11) and (7.12) is not obvious, we establish it in the following lemma.

Lemma 7.2 *For $C^2(d)$ from Eq. (7.11) it holds that*

$$C^2(d) = \frac{1}{2d+1} \frac{\Gamma(1-2d)\Gamma(d+1)}{\Gamma(1-d)} . \qquad (7.12)$$

Proof: See Appendix to the chapter.

Mandelbrot and Van Ness (1968, p. 424) briefly mention an integral originally discussed by Lévy (1953). Marinucci and Robinson (1999) called it type II fBm:

$$B_d^{II}(t) = \sqrt{2d+1} \int_0^t (t-s)^d dB_0(s), \quad t \ge 0, \ d > -1/2. \qquad (7.13)$$

While B_d and B_d^{II} share the same variance (and mean equal to 0),

$$\text{Var}(B_d(t)) = \text{Var}(B_d^{II}(t)) = t^{2d+1},$$

the increments of B_d^{II} are not stationary; for details, see Marinucci and Robinson (1999). This mirrors of course that the differences $\{\Delta y_t^{II}\}$ from Eq. (6.7) are not stationary. Further, it is noteworthy that B_d^{II} is well defined for any value $d > -1/2$, which parallels Definition 6.2.

We now discuss conditions under which results like in Proposition 7.3 were established in the past. Let us begin with the more classical type I result. Davydov (1970) proved the FCLT for $S_{\lfloor rT \rfloor}$ for $\{x_t\} = \{\varepsilon_t\} \sim \text{iid}(0, \sigma^2)$. This was an extension of Donsker's theorem; Donsker (1951) dealt with the case $d = 0$ under the moment condition

$$E(|\varepsilon_t|^q) < \infty \quad \text{for some } q \geq 2 .$$

For $d \in [0, 1/2)$, Taqqu (1975) maintained this restriction, too, while employing the stricter assumption Eq. (7.7) in case of antipersistence ($-1/2 < d < 0$):

$$q > \frac{2}{2d + 1} .$$

This moment condition relaxed even stricter sufficient conditions employed by Davydov (1970). Generally, it is not possible[2] to further relax $q > 2/(2d + 1)$, which was shown to be necessary by Johansen and Nielsen (2012a, Theorem 2). This accomplished work by Davidson and de Jong (2000, Theorem 3.1), who had missed $q > 2/(2d + 1)$. For values of d close to $-1/2$, this moment condition becomes increasingly restrictive, requiring more control over the tails of the distribution with d getting small. Most recently, Abadir et al. (2014) proved a FCLT under type I fractional integration. They allowed for considerable dependence in

$$x_t = \sum_{j=0}^{\infty} c_j \varepsilon_{t-j} ,$$

only requiring $\{\varepsilon_t\}$ to form a stationary and ergodic MDS. It is very instructive to follow from Abadir et al. (2014, Theorem 2.1) to Abadir et al. (2014, Theorem 3.1): One thus learns that the moment condition Eq. (7.7) is not required for a CLT (see Corollary 7.1), and it is not needed for a finite-dimensional convergence result; see Abadir et al. (2014, Proposition 3.1). Condition Eq. (7.7) only comes in to establish (tightness in order to prove) the FCLT; see also Abadir et al. (2014, Corollary 4.1).

The FCLT for type II processes has attracted less attention. Our result from Proposition 7.3 is adopted from Marinucci and Robinson (2000, Theorem 1). The authors assume a linear process $\{x_t\}$. Again, the now familiar moment condition kicks in when $d \in (-1/2, 0)$.

---

2 A mild relaxation allowing for equality, $q \geq 2/(2d + 1)$, is possible under the particular model of fractional integration that we maintain in Eq. (6.2); see Johansen and Nielsen (2012a, Theorem 1).

For $r = 1$, Proposition 7.3 provides the limiting distribution of the arithmetic mean of type II processes. Under Eq. (6.9) it holds that

$$T^{1/2-d}(\bar{y}^{II} - \mu) \xrightarrow{D} C_{S,d}^{II} B_d^{II}(1) \sim \mathcal{N}(0, (C_{S,d}^{II})^2), \quad d > -\frac{1}{2}. \tag{7.14}$$

The same limiting result has been given by Tanaka (1999, Corollary 2.3). It further follows from Tanaka (1999, Corollary 2.2) for $d = -1/2$ that

$$\frac{T}{\sqrt{\ln T}}(\bar{y}^{II} - \mu) \xrightarrow{D} \mathcal{N}\left(0, \frac{\omega_x^2}{\pi}\right), \quad d = -\frac{1}{2}. \tag{7.15}$$

This parallels the type I result from Corollary 7.1 (b). Note, however, that the limiting variance is twice as large for a type I process than for $\bar{y}^{II}$.

We next evaluate the relative efficiency for the whole range of d:

$$\text{eff}(d) := \lim_{T\to\infty} \frac{\text{Var}(\bar{y})}{\text{Var}(\bar{y}^{II})} = \begin{cases} C_{S,d}^2/(C_{S,d}^{II})^2, & |d| < 1/2 \\ 2, & d = -1/2 \end{cases}.$$

The variance terms $C_{S,d}^2$ and $(C_{S,d}^{II})^2$ are both defined in Proposition 7.3. For $d \to -1/2$ they diverge; however, their singularities cancel, such that $\text{eff}(d)$ is continuous, leaving 2 in the limit. Further, it is straightforward to verify that

$$\text{eff}(d) = \frac{\Gamma(1-2d)\Gamma(d+1)}{\Gamma(1-d)}, \quad |d| < 1/2.$$

This expression can be shown to be bounded from below by 1.

Proposition 7.4 *For eff(d) = $\Gamma(1-2d)\Gamma(d+1)/\Gamma(1-d)$, it holds with $|d| < 1/2$ that eff(d) $\geq$ 1.*

Proof: See Appendix to the chapter.

For a more detailed comparison, we plot the ratio of the type I variance relative to that of type II. From Figure 7.1 we learn that this ratio is smaller than 2 for $d \in (-0.5, 0.35]$. For $d > 0.45$ the value of eff(d) is larger than 5, showing that the variance under type I fractional integration is at least five times as large as under type II. For $d \to 1/2$, this variance ratio diverges, which is obvious from the definition.

There is an important issue associated with the estimation of expected values under long memory, namely, the rate of convergence. From Corollary 7.1 (a) or Eq. (7.14), we learn that for $d = -1/2 + \epsilon$, $\epsilon > 0$, the rate of convergence of the arithmetic mean is $T^{1-\epsilon}$, in that the bias $\bar{y} - \mu$ vanishes so fast that a multiplication with $T^{1-\epsilon}$ still renders a well-defined limiting distribution. For

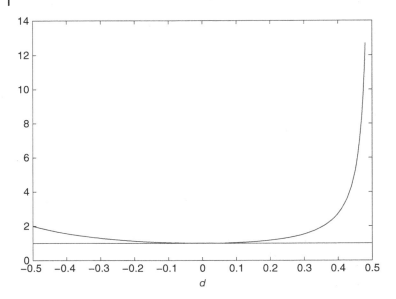

Figure 7.1 $\mathrm{eff}(d) = C_{S,d}^2/(C_{S,d}^{\parallel})^2$.

$d = -1/2$ the rate of convergence is even faster (see Corollary 7.1 (b) and Eq. (7.15)), since[3]

$$\frac{T/\sqrt{\ln T}}{T^{1-\epsilon}} \to \infty \, , \quad \epsilon > 0 \, .$$

For $d = 0$ the standard rate of convergence, $\sqrt{T}$, is of course reproduced. For positive d the rate is all the slower, the larger d is, and $T^{1/2-d}$ is very slow for $d \to 1/2$. This is intuitively clear, since we have seen that the long memory case ($d > 0$) means *reversing trends* (see, e.g. Figure 6.2), which inevitably makes the estimation of the theoretical mean difficult.

In fact, the slow rate of convergence under fractional integration with long memory is not a property of the arithmetic mean exclusively. This point was made more rigorously by Samarov and Taqqu (1988) under the assumption that a process is fractionally integrated noise (of type I). They compared the relative efficiency of the best linear unbiased estimator (BLUE) of the expected value μ relative to the arithmetic mean. Building on previous work by Adenstedt (1974),

---

3 For $d < -1/2$ the arithmetic mean converges even mildly faster with rate T as we know from Proposition 7.2 for type I processes:

$$\mathrm{Var}(\bar{y}) = T^{-2}\sigma_T^2 = O(T^{-2}) \, , -1 < d < -1/2 \, .$$

However, limiting normality no longer holds; see the discussion following Corollary 7.1.

Samarov and Taqqu (1988, Eq. (3.3)) proved under fractional integration for FIN(d) with $0 \neq |d| < 1/2$ that

$$\lim_{T\to\infty} \frac{\text{Var}(\widehat{\mu}_{\text{BLU}})}{\text{Var}(\overline{y})} = \frac{(1+2d)\Gamma(2-2d)\Gamma(1+d)}{\Gamma(1-d)} \tag{7.16}$$

where $\widehat{\mu}_{\text{BLU}}$ is the BLUE of $E(y_t) = \mu$. Hence, the slow rate of convergence of $\overline{y}$ under long memory cannot be overcome by the BLUE. More interesting than the formula itself is the numerical evaluation performed by Samarov and Taqqu (1988, Figure 1), from which we learn that

$$0.98 < \lim_{T\to\infty} \frac{\text{Var}(\widehat{\mu}_{\text{BLU}})}{\text{Var}(\overline{y})} \leq 1, \quad 0 < d < \frac{1}{2}. \tag{7.17}$$

Consequently, only very marginal efficiency gains are available when replacing $\overline{y}$ by $\widehat{\mu}_{\text{BLU}}$. All in all, mean estimation is a notoriously difficult problem under long memory. Further aspects of inference about the mean are discussed in the following section.

7.4 Inference About the Mean

We now turn to the issue of computing confidence intervals or performing tests about μ according to Corollary 7.1 under $|d| < 1/2$ or according to Eq. (7.14). This topic was pioneered by Beran (1989). If $C_{S,d}$ or $C_{S,d}^{\text{II}}$ were known, the approximate $1 - \alpha$ confidence interval for the unknown μ would become

$$\left[\overline{y} \pm z_{1-\alpha/2}\frac{C_{S,d}}{T^{1/2-d}}\right] \quad \text{or} \quad \left[\overline{y}^{\text{II}} \pm z_{1-\alpha/2}\frac{C_{S,d}^{\text{II}}}{T^{1/2-d}}\right], \tag{7.18}$$

with $z_{1-\alpha/2}$ denoting the $(1 - \alpha/2)$ quantile of the standard normal distribution. Similarly, one could perform a significance test on $H_0 : \mu = \mu_0$ by t-type statistics

$$t = \sqrt{T^{1-2d}}\frac{\overline{y} - \mu_0}{C_{S,d}} \sim \mathcal{N}(0,1), \quad T \to \infty, \tag{7.19}$$

if $C_{S,d}$ was known; just as well one might wish to standardize with $C_{S,d}^{\text{II}}$ in the case of a type II process. See also Beran (1989, Eq. (2.1)). For convenience, we ignore the type II case for a moment. The constant $C_{S,d}^2$ is made up by two components: First, by a function of d, say,

$$G(d) = \frac{\Gamma(1-2d)}{\Gamma(d+1)\Gamma(1-d)(1+2d)},$$

and second by the long-run variance ω_x^2 of $x_t = \Delta^d y_t$, where d enters through Δ^d, too. The (consistent) estimation of d will be treated in the next two chapters.

For now, we assume the existence of an estimator $\widehat{d}$ converging to the true value of d faster than with logarithmic rate, i.e.

$$\ln T(\widehat{d} - d) \xrightarrow{p} 0.$$

This provides a consistent plug-in estimator for $G(d)$:

$$G(\widehat{d}) = \frac{\Gamma(1 - 2\widehat{d})}{\Gamma(\widehat{d} + 1)\Gamma(1 - \widehat{d})(1 + 2\widehat{d})} .$$

In principle, one could next compute

$$\widehat{x}_t := \Delta^{\widehat{d}} y_t$$

by fractionally differencing $\{y_t\}$ with the estimated memory parameter, and then use $\{\widehat{x}_t\}$ to estimate ω_x^2. This, however, required the arbitrary truncation of the infinite filter $\Delta^{\widehat{d}}$. Alternatively, we follow a proposal made by Robinson (2005a) and elaborated on by Abadir et al. (2009).[4] We note from Proposition 6.1 that

$$\lambda^{2d} f_y(\lambda) \sim \frac{\omega_x^2}{2\pi} , \quad \lambda \to 0 .$$

This suggests to estimate ω_x^2 from the periodogram $I_y(\cdot)$ of $\{y_t\}$ (see (4.3) or equivalent definitions), which is evaluated at the first harmonic frequencies,

$$\lambda_j = \frac{2\pi j}{T} , \quad j = 1, \ldots, m ,$$

such that

$$\widehat{\omega}_{x,m}^2 := \frac{2\pi}{m} \sum_{j=1}^{m} \lambda_j^{2\widehat{d}} I_y(\lambda_j) . \tag{7.20}$$

Again, this requires a consistent estimator $\widehat{d}$ of d. We now define

$$\widehat{C}_{S,\widehat{d}}^2 := G(\widehat{d})\, \widehat{\omega}_{x,m}^2 , \tag{7.21}$$

which is a special case of the more general proposal by Robinson (2005a). This estimator has been called MAC (memory and autocorrelation consistent), because $\widehat{C}_{S,\widehat{d}}^2$ is consistent for $C_{S,d}^2$ in the presence of memory in the form of fractional integration and in the presence of additional short memory autocorrelation behind $\{x_t\}$. In view of Eq. (7.18) or (7.19), $\widehat{C}_{S,\widehat{d}}$ is also called MAC standard error. It not only requires $(\ln T)$-consistent estimation of d in

---

4 A different proposal has been made by McElroy and Politis (2012) who follow the so-called fixed-b approach.

a first step but also an adequate choice of m in Eq. (7.20) growing with the sample size, but more slowly than T. From Abadir et al. (2009) we adopt

$$\frac{1}{m} + \frac{\ln^2 T}{T} m \;\to\; 0 \text{ as } T \to \infty .$$

Under some additional assumptions, Abadir et al. (2009, Theorem 3.1) established that the MAC estimation from Eq. (7.21) is consistent for $C_{S,d}^2$, which allows for asymptotic standard normal inference about the unknown mean μ. To make MAC work, one further has to replace $T^{1/2-d}$ by $T^{1/2-\hat{d}}$ in Eq. (7.18) or (7.19).

Note that the same way one may construct MAC standard errors for the type II case by replacing $G(d)$ by

$$G^{\mathrm{II}}(d) = \frac{1}{\Gamma^2(d+1)\,(2d+1)} ,$$

in order to obtain

$$(\widehat{C}_{S,\hat{d}}^{\mathrm{II}})^2 = G^{\mathrm{II}}(\hat{d})\,\hat{\omega}_{x,m}^2 .$$

This leaves open the problem of how to discriminate between type I and type II processes. In practice, one typically does not try to do so. It depends on prior beliefs, on the assumption of the modeler, whether he or she should standardize the arithmetic mean with estimates of $C_{S,d}$ or $C_{S,d}^{\mathrm{II}}$. In view of Proposition 7.4, one may conclude that the confidence intervals will be larger when using $C_{S,d}$ in comparison with the usage of $C_{S,d}^{\mathrm{II}}$. Or in other words, the use of $C_{S,d}$ will result in a conservative test if the true process is of type II, while $C_{S,d}^{\mathrm{II}}$ will cause size distortions in the sense of overrejection if the true process is of type I. Hence, one may recommend a standardization with $C_{S,d}$ for practical purposes, as long as overrejections are to be avoided.

7.5 Sample Autocorrelation

Sample autocorrelations are averages, too, and hence their treatment comes in naturally in this chapter. A general condition for consistent autocovariance estimation has been given previously in Proposition 3.3. It relies on iid innovations entering the MA(∞) representation of the stationary process. Here, we strive for more, namely, rates of convergence for the standard estimator,

$$\hat{\rho}(h) = \frac{\sum_{t=1}^{T-h}(y_t - \bar{y})(y_{t+h} - \bar{y})}{\sum_{t=1}^{T}(y_t - \bar{y})^2} , \quad h = 0, 1, \dots ,$$

where the lag h is a fixed number and $\rho(h) = \gamma(h)/\gamma(0)$. First, we focus on $\mathrm{E}(\hat{\rho}(h))$ for $0 \le d < 1/2$.

Proposition 7.5 *Let $\{y_t\}$ be from Definition 6.1: $y_t = (1 - L)^{-d} x_t$ and $x_t = \sum_{j=0}^{\infty} c_j \, \varepsilon_{t-j}$. Further, assume that $\{\varepsilon_t\}$ is iid with constant finite fourth moments. For $0 \le d < 1/2$ it holds that*

$$\mathrm{E}(\hat{\rho}(h)) = \rho(h) + O(T^{2d-1})$$

as $T \to \infty$.

Proof: Fuller (1996, Theorem 6.2.3) for $d = 0$ and Hosking (1996, Theorem 6) for $0 < d < 1/2$. □

Obviously, the larger the d, the more slowly vanishes the bias, and for d close to $1/2$, the convergence is very slow. More precisely, we have from Hosking (1996, Eq. (18)) that

$$\mathrm{E}(\hat{\rho}(h)) = \rho(h) - \frac{1 - \rho(h)}{\gamma(0)} C_{S,d}^2 T^{2d-1} + o(T^{2d-1}),$$

with $C_{S,d}^2$ from Eq. (7.6) in Proposition 7.2. For the special FIN case, where $x_t = \varepsilon_t$, this yields

$$\mathrm{E}(\hat{\rho}(h)) \approx \rho(h) - \frac{1 - \rho(h)}{d\,(1 + 2d)} \frac{\Gamma(1 - d)}{\Gamma(d)} T^{2d-1}. \tag{7.22}$$

We used the right-hand side of Eq. (7.22) to produce graphs illustrating how long memory affects the (sample) autocorrelograms.[5] Figure 7.2 displays $\rho(h)$ and $\mathrm{E}(\hat{\rho}(h))$ for FIN of order $d = 0.2$, while Figure 7.3 plots the corresponding figures for $d = 0.45$. In the latter case the negative bias is still very large even for $T = 1000$. This illustrates that the sample autocorrelation is not a very reliable estimator of the theoretical autocorrelation under long memory, and this is all the more true the larger d is. It hence has been recommended to identify potential long memory by frequency domain methods instead of in the time domain.

Limiting distributions are available, too. Assuming a Gaussian sequence $\{\varepsilon_t\}$ behind the process in Proposition 7.5, Hosking (1996, Theorem 7) established the standard result for $0 < d < 1/4$; *standard* means that $\sqrt{T}(\hat{\rho}(h) - \rho(h))$ converges to a normal distribution with mean 0 and the variance given by Bartlett's formula; see Brockwell and Davis (1991, Theorem 7.2.2) or Fuller (1996, Corollary 6.3.6.1) for $d = 0$. For $d = 1/4$, Hosking (1996, Theorem 7) established limiting normality, however at the unconventional rate of $\sqrt{T / \ln T}$, while limiting normality fails altogether for $1/4 < d < 1/2$. Further asymptotic findings without the assumption of Gaussianity and also without finite fourth moments were provided by Horváth and Kokoszka (2008).

Now, we turn to the behavior of the sample autocorrelations under nonstationarity, where $\hat{\rho}(h)$ converges to one. The case $d = 1$ has been pioneered by

---

5 Early numerical evidence on this issue had been provided by Newbold and Agiakloglou (1993).

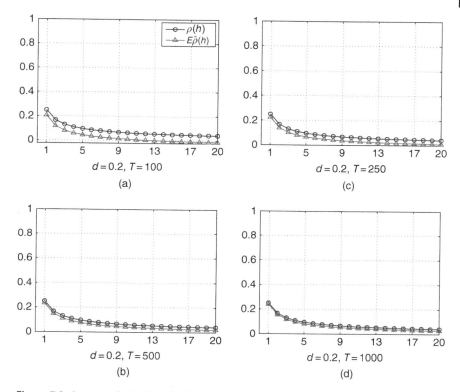

Figure 7.2 Autocorrelation bias for $d = 0.2$.

Box and Jenkins (1970, p. 200) in the framework of an ARIMA$(0, 1, 1)$ model; see also Wichern (1973). Hasza (1980) obtained for ARIMA$(p, 1, q)$ models the limiting distribution,

$$T\,(\widehat{\rho}(h) - 1) \xrightarrow{D} \mathcal{L}(h)\,,$$

where the limit $\mathcal{L}(h)$ depends on the ARMA(p, q) parameters. For ARIMA$(0, 1, 1)$ this specializes to

$$T\,(\widehat{\rho}(h) - 1) \xrightarrow{D} -h\,\mathcal{L}_1 + \mathcal{L}_2\,,$$

with $\mathcal{L}_1$ and $\mathcal{L}_2$ defined appropriately, and $\mathcal{L}_1 \geq 0$. Hence,

$$\widehat{\rho}(h) \approx 1 - h\,\frac{\mathcal{L}_1}{T} + \frac{\mathcal{L}_2}{T}\,. \tag{7.23}$$

Consequently, the sample autocorrelations tend to decline linearly in h. Hasza (1980) argued that this continues to hold for ARIMA$(0, 1, q)$ models as long as

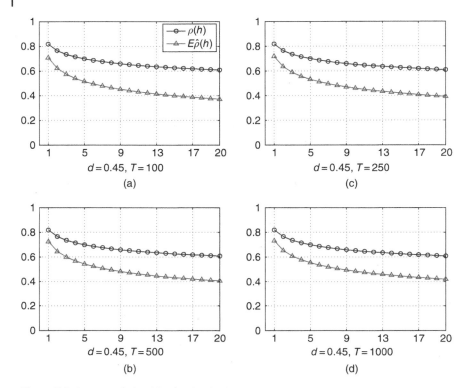

Figure 7.3 Autocorrelation bias for $d = 0.45$.

$h \geq q$. Further, in the ARIMA$(0, 1, 0)$ case of a random walk, Eq. (7.23) becomes

$$\widehat{\rho}(h) \approx 1 - h\, \frac{\mathcal{L}_1}{T} \tag{7.24}$$

since $\mathcal{L}_2$ is forced to 0 for $q = 0$. An analogous result has been established by Hassler (1997, Corollary 3) for $I(d)$ processes with $1 < d < 3/2$ under conditions maintained in Proposition 7.3:

$$T\,(\widehat{\rho}(h) - 1) \overset{D}{\to} -h\, \mathcal{L}_d\,, \quad 1 < d < 3/2\,,$$

where $\mathcal{L}_d$ depends on d and is nonnegative. Hence,

$$\widehat{\rho}(h) \approx 1 - h\, \frac{\mathcal{L}_d}{T}\,, \quad 1 < d < 3/2\,. \tag{7.25}$$

The linear decay of $\widehat{\rho}(h)$ is well documented in Figure 7.4 for $d = 1.25$ and also for the random walk case $d = 1$; Figure 7.4 displays the sample median and sample average plus the 5% and 95% quantiles from 10 000 Monte Carlo

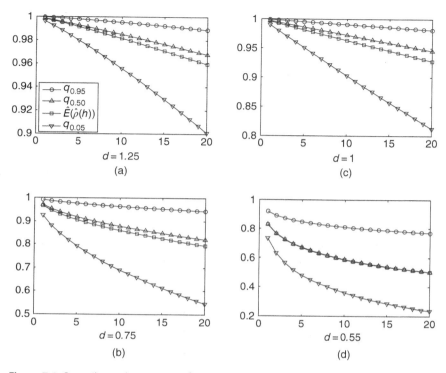

Figure 7.4 Quantiles and averages of $\hat{\rho}(h)$ for FIN(d), $T = 1000$.

replications[6] of fractionally integrated noise series with $T = 1000$ observations. We also generated nonstationary FIN sequences integrated of order 0.75 and 0.55, where the decline of $\hat{\rho}(h)$ is no longer linear in h.

7.6 Technical Appendix: Proofs

Proof of Proposition 7.2

In order to prove (a), we may rely on Giraitis et al. (2012, Proposition 3.3.1). For $d = 0$, they give the special case of Proposition 7.2 with

$$C_{S,0}^2 = \omega_x^2.$$

For $|d| < 1/2$, we note that our assumptions guarantee Proposition 6.1, such that Giraitis et al. (2012, Eq. (3.1.8)) applies. We obtain from Giraitis et al. (2012,

---

6 We simulated type II processes following Jensen and Nielsen (2014).

Proposition 3.3.1) that

$$C_{S,d}^2 = \frac{2\Gamma(1-2d)\sin(\pi d)}{d(1+2d)} f_x(0)$$

$$= \frac{\Gamma(1-2d)2\pi f_x(0)}{d(1+2d)\Gamma(1-d)\Gamma(d)} \, ,$$

where (5.17) was used once more for the second equality, such that by (5.13)

$$C_{S,d}^2 = \frac{\Gamma(1-2d)\omega_x^2}{\Gamma(1+d)\Gamma(1-d)(1+2d)}$$

as required.

To prove (b) we define as partial sum of the autocovariances

$$G_j := \sum_{k=-j}^{j} \gamma_y(k)$$

such that

$$\text{Var}(S_T) = T\gamma_y(0) + 2\sum_{j=1}^{T-1}(T-j)\gamma_y(j)$$

$$= \gamma_y(0) + \sum_{j=1}^{T-1} G_j.$$

Since $d < 0$ (with $f_y(0) = 0$ or $\sum_{k=-\infty}^{\infty} \gamma_y(k) = 0$), we rewrite

$$G_j = G_j - \sum_{k=-\infty}^{\infty} \gamma_y(k) = -2\sum_{k=j+1}^{\infty} \gamma_y(k)$$

$$\sim -2\sum_{k=j+1}^{\infty} C_{\gamma,-1/2} k^{-2}, \quad j \to \infty,$$

where Proposition 6.3 was used for $j \to \infty$. Notice the following:[7]

$$\sum_{k=j}^{\infty} k^{-p} \sim \frac{j^{1-p}}{p-1}, \quad p > 1. \tag{7.26}$$

With $p = 2$ we thus obtain

$$G_j \sim -2C_{\gamma,-1/2}\, j^{-1}.$$

Hence, σ_T^2 grows at logarithmic rate by Lemma 3.3:

$$\text{Var}(S_T) \sim -2C_{\gamma,-1/2} \ln T \, .$$

---

7 This result is clear by

$$\int_j^{\infty} x^{-p}\, dx = \frac{j^{1-p}}{p-1}, \quad p > 1.$$

Further, by definition of $C_{\gamma,d}$ from Eq. (6.4),

$$C_{\gamma,-1/2} = \frac{\omega_x^2 \Gamma(2)}{\Gamma(-1/2)\Gamma(3/2)} = \frac{\omega_x^2}{-\Gamma(1/2)\Gamma(1/2)}$$

where $\Gamma(2) = 1$ and $x\Gamma(x) = \Gamma(1 + x)$ were used again. Since $\Gamma(1/2) = \sqrt{\pi}$, we find that

$$\text{Var}(S_T) \sim \frac{-2\omega_x^2}{-\pi} \ln T$$

as required to show (b).

To prove (c), define the type I process $\{z_t\}$ as integrated of order δ, $0 < \delta < 1/2$, such that the differences $y_t = \mu + \Delta z_t$ are $I(d)$ with $-1 < d = \delta - 1 < -1/2$; see page 107. By construction, $S_T = \sum_{t=1}^{T} y_t = \mu T + z_T - z_0$ and

$$\sigma_T^2 = \text{Var}(S_T) = 2\,\gamma_z(0) - 2\,\gamma_z(T)\,,$$

where $\gamma_z(\cdot)$ is the autocovariance sequence of $\{z_t\}$. Therefore, $\sigma_T^2 < \infty$ for all T as required.

Hence, the proof is complete. □

Proof of Lemma 7.2

Let us call the right-hand side of Eq. (7.11) A, while the right-hand side of Eq. (7.12) is abbreviated as B. Consider

$$\frac{A}{B} = \frac{\Gamma(d+1)\Gamma(1-d)(2d+1)}{\Gamma(2d+2)\sin((d+1/2)\pi)\Gamma(1-2d)}$$

$$= \frac{\Gamma(d)\Gamma(1-d)}{2\Gamma(2d)\sin((d+1/2)\pi)\Gamma(1-2d)}$$

$$= \frac{\sin(2\pi d)}{2\sin(\pi d)\sin((d+1/2)\pi)}$$

$$= \frac{\sin(2\pi d)}{2\sin(\pi d)\cos(\pi d)}\,,$$

where the usual recursion Eq. (5.13) was used twice for the second equality and the reflection formula Eq. (5.17) was used twice for the third equality, while the fourth one is obvious. Because of the trigonometric double-angle formula Eq. (5.10), this ratio equals 1, and the proof is complete. □

Proof of Proposition 7.4

Obviously, eff(0) = 1. For positive d we get

$$\frac{\Gamma(1-2d)\Gamma(d+1)}{\Gamma(1-d)} = d\,\frac{\Gamma(1-2d)\Gamma(d)}{\Gamma(1-d)}$$

$$= d\,B(1-2d,d) = d\int_0^1 u^{-2d}(1-u)^{d-1}\mathrm{d}u\,,$$

where Eq. (5.21) was used for the beta function. Since $u^{-2d} \geq 1$ on $(0, 1)$ we have that

$$\text{eff}(d) \geq d \int_0^1 (1 - u)^{d-1} du = 1 .$$

For negative d we argue similarly:

$$\frac{\Gamma(1 - 2d)\Gamma(d + 1)}{\Gamma(1 - d)} = -2d \, \frac{\Gamma(-2d)\Gamma(1 + d)}{\Gamma(1 - d)}$$

$$= -2d \, B(-2d, 1 + d) = -2d \int_0^1 u^{-2d-1}(1 - u)^d du$$

$$\geq -2d \int_0^1 u^{-2d-1} du = 1 .$$

Hence, the proof is complete. □

8

Parametric Estimators

After a discussion of parametric assumptions, we turn to exact maximum likelihood (ML) estimation. Theorem 8.1 characterizes the limiting normal distribution. The same asymptotic theory will continue to hold under approximations to exact ML. The approximations are in the time domain (conditional sum of squares (CSS)) or in the frequency domain (Whittle). In the end, we briefly consider the special parametric case of a fractional EXP model and discuss the log-periodogram regression that is not rooted in the ML principle, but bridges the gap to Chapter 9.

8.1 Parametric Assumptions

In the last two chapters, we distinguished notationally between type I processes $\{y_t\}$ and type II processes $\{y_t^{II}\}$, because for both types of fractional integration, different (functional) CLTs emerged and have been discussed. In this and subsequent chapters, we will present techniques of inference that have been designed for type I or for type II. We will then tell the reader whether we mean type I or type II, but for brevity the same symbol $\{y_t\}$ will be used for both processes.

We maintain that $\{y_t\}$ is given by fractional differencing in terms of $\{x_t\}$ satisfying Assumption 6.1:

$$\Delta^d y_t = x_t, \quad t = 1, \dots, T.$$

Depending on whether $y_0 = y_{-1} = y_{-2} = \dots = 0$ or not, we account for type I as well as type II fractional integration without distinguishing between the two cases notationally. In this chapter we add the assumption that $\{x_t\}$ is generated from a finite-dimensional parametric model.

Assumption 8.1 (*Parametric* $I(0)$)   *Let* $\{x_t\}$ *from Assumption 6.1 meet the Assumption 4.1 for a parametric model with a k-dimensional parameter vector* ψ, *and spectrum*

$$f_x(\lambda) = f_x(\lambda; \psi) = \frac{\sigma^2}{2\pi} T_C(\lambda; \psi),$$

Time Series Analysis with Long Memory in View, First Edition. Uwe Hassler.
© 2019 John Wiley & Sons, Inc. Published 2019 by John Wiley & Sons, Inc.

where the power transfer function T_C according to Eq. (4.13) depends on $\{c_j\} = \{c_{j,\psi}\}$ from $x_t = \sum_{j=0}^{\infty} c_{j,\psi} \varepsilon_{t-j}$.

The most prominent parametric model for $I(0)$ processes is of course the stationary and invertible ARMA(p, q) model from Eq. (3.6), where $k = p + q$ with

$$\psi' = (\phi_1, \ldots, \phi_p, \theta_1, \ldots, \theta_q),$$

and $\{c_{j,\psi}\}$ are given by the expansion Eq. (3.9):

$$x_t = \frac{\Theta(L)}{\Phi(L)} \varepsilon_t = \sum_{j=0}^{\infty} c_{j,\psi} \varepsilon_{t-j}.$$

Alternatively, one may start with a spectral assumption following the Bloomfield model EXP(k) from Eq. (4.26),

$$f_x(\lambda; \psi) = \frac{\sigma^2}{2\pi} \exp\left(\sum_{\ell=1}^{k} \psi_\ell \cos(\ell \lambda) \right),$$

such that $\{c_{j,\psi}\}$ are given in Eq. (4.27). Assuming a type I process, the spectrum of $\{y_t\}$ becomes (see Proposition 6.1),

$$f_y(\lambda; \varphi) = \left[4 \sin^2\left(\frac{\lambda}{2}\right) \right]^{-d} f_x(\lambda; \psi), \tag{8.1}$$

where we define the $(k + 1)$-dimensional parameter vector

$$\varphi' = (d, \psi'). \tag{8.2}$$

The true order of fractional integration, d_0, is contained in $\varphi'_0 = (d_0, \psi'_0)$.

8.2 Exact Maximum Likelihood Estimation

In this section, we assume that $\{y_t\}$ is a stationary, fractionally integrated process of type I following a normal distribution. It then holds for the vector $y' = (y_1, \ldots, y_T)$ that

$$y \sim \mathcal{N}_T(0, \Sigma_y),$$

where Σ_y contains all autocovariances,

$$\Sigma_y = (\gamma_y(|i - j|))_{i,j=1,\ldots,T},$$

depending on the parameter vector φ. The joint normal density is

$$(2\pi)^{-T/2} |\Sigma_y|^{-1/2} \exp\left\{ -\frac{1}{2} y' \Sigma_y^{-1} y \right\}.$$

ML estimation amounts to minimizing the negative logarithm thereof, which becomes upon neglecting additive constants:

$$\frac{1}{2}\ln|\Sigma_y| + \frac{1}{2}y'\Sigma_y^{-1}y.$$

In practice, a nonzero mean may have been removed from the level of the data, where we choose the arithmetic mean, $\hat{\mu} = \bar{y}$. Let us define the demeaned vector

$$\underline{y} = (y_1 - \bar{y}, \dots, y_T - \bar{y})',$$

and plug it into the negative log-likelihood defined now as

$$\ell(\underline{y}; \varphi, \sigma^2) := \frac{1}{2}\ln|\Sigma_y| + \frac{1}{2}\underline{y}'\Sigma_y^{-1}\underline{y}.$$

Note that Σ_y depends on d, ψ, and σ^2, where ψ and σ^2 enter through the spectrum f_x. Hence, $\ell(\cdot; \varphi, \sigma^2)$ depends on both φ and σ^2. The ML estimator is given by

$$(\hat{\varphi}'_{ML}, \hat{\sigma}^2_{ML})' = \arg\ \min\ \ell(\underline{y}; \varphi, \sigma^2).$$

In most practical situations one is primarily interested in $\varphi' = (d, \psi')$. Hence, the following theorem does not cover $\hat{\sigma}^2_{ML}$, although this could be incorporated; see Dahlhaus (1989). Note that the limiting covariance matrix is defined as in Eq. (4.33) but now in terms of $f_y(\lambda; \varphi)$.

Theorem 8.1 (ML)   *Let $\{y_t - \mu\}$ be a stationary, Gaussian $I(d_0)$ process of type I with $|d_0| < 1/2$, where $\{x_t\}$ meets Assumption 8.1. Under some further conditions given in Dahlhaus (1989, 2006) and Lieberman et al. (2012), it holds that $(T \to \infty)$*

$$\sqrt{T}(\hat{\varphi}_{ML} - \varphi_0) \xrightarrow{D} \mathcal{N}_{k+1}(0, W^{-1}(\varphi_0)),$$

where

$$W(\varphi) = (w_{ij}(\varphi))_{i,j=1,\dots,k+1}$$

with $(i, j \in \{1, \dots, k+1\})$

$$w_{ij}(\varphi) = \frac{1}{4\pi}\int_{-\pi}^{\pi} \frac{\partial \ln f_y(\lambda; \varphi)}{\partial \varphi_i} \frac{\partial \ln f_y(\lambda; \varphi)}{\partial \varphi_j} d\lambda \tag{8.3}$$

in terms of $f_y(\lambda; \varphi)$ given in Eq. (8.1).

Proof: Dahlhaus (1989, Theorem 3.2) for $0 < d_0 < 1/2$ and Lieberman et al. (2012, Theorem 1) for $d_0 \leq 0$. □

For $d_0 = 0$, this result was first established by Hannan (1973a); see also Proposition 4.12. A result corresponding to Dahlhaus (1989) under long memory was given earlier by Yajima (1985, Theorem 4.1) under the more restrictive assumption that $\{x_t\} = \{\varepsilon_t\} \sim \text{iid}(0, \sigma^2)$. Only recently, Lieberman et al. (2012) extended Theorem 8.1 to allow for antipersistence ($d < 0$), and even for the case $d \leq -1/2$, which is often called *noninvertible*, notwithstanding Proposition 6.2. To that end, Lieberman et al. (2012, Assumption 5) assumed to demean the data with $\hat{\mu}$ satisfying

$$T^{1/2-d_0-\epsilon}(\hat{\mu} - \mu_0) \xrightarrow{p} 0 \quad \text{for} \quad \epsilon > 0, \quad \mu_0 = \text{E}(y_t).$$

By Corollary 7.1, we know that $\bar{y}$ meets this condition as long as $|d_0| < 1/2$, which is why we maintain $d_0 > -1/2$ in Theorem 8.1. An extension to $d_0 \leq -1/2$ for a generalized least squares estimator of μ is discussed in Lieberman et al. (2012).

Note that $\hat{\varphi}_{\text{ML}}$ converges with the usual standard rate $\sqrt{T}$ toward the true parameter vector irrespective of the true order of integration d_0. The rate of $\bar{y}$ is different as long as $d_0 \neq 0$, which is why $\hat{\varphi}_{\text{ML}}$ is independent of $\bar{y}$ asymptotically. In other words, minimizing $\ell(y; \varphi, \sigma^2)$ for known mean $\mu = 0$ instead of $\ell(\underset{\sim}{y}; \varphi, \sigma^2)$ results in the same estimator with the same properties as the sample size diverges.

From the covariance matrix W^{-1} from Theorem 8.1, we can derive the variance of $\hat{d}$, which is the first component on the diagonal. It turns out to be particularly simple for the exponential model defined in Eq. (4.26).

Corollary 8.1 *Partition the matrix $W(\varphi)$ according to $\varphi' = (d, \psi')$,*

$$W(\varphi) = W = \begin{pmatrix} w_{11} & w'_{12} \\ w_{12} & W_{22} \end{pmatrix},$$

and maintain the assumptions of Theorem 8.1.

(a) It then holds that

$$T \, \text{Var}(\hat{d}_{\text{ML}}) \to \left(\frac{\pi^2}{6} - w'_{12} W_{22}^{-1} w_{12} \right)^{-1}$$

as $T \to \infty$.

(b) Let $\{x_t\}$ follow an EXP(k) model according to Eq. (4.26). It then holds that

$$w'_{12} W_{22}^{-1} w_{12} = \sum_{\ell=1}^{k} \frac{1}{\ell^2}.$$

Proof: See Appendix of this chapter.

Note the particular case of fractionally integrated noise, where $\{x_t\} = \{\varepsilon_t\} \sim$ WN. In this case $\varphi = d$ and $w_{11} = \pi^2/6$, such that

$$\sqrt{T}(\hat{d}_{\text{ML}} - d_0) \xrightarrow{D} \mathcal{N}\left(0, \frac{6}{\pi^2}\right).$$

It has been demonstrated in Eq. (4.36) that the limiting covariance matrix according to Eq. (4.33) becomes particularly simple for EXP models. This holds true for FEXP models, too. We evaluate the limiting variance for a FEXP(d,k) model with $k = 1, 2, 3$:

k	1	2	3
$\left(\frac{\pi^2}{6} - \sum_{\ell=1}^{k} \frac{1}{\ell^2}\right)^{-1}$	1.5505	2.5321	3.5233

This shows how $\lim_{T\to\infty} T \operatorname{Var}(\widehat{d}_{\mathrm{ML}})$ grows with k for FEXP models. For $k \to \infty$, $\sum_{\ell=1}^{k} \frac{1}{\ell^2}$ converges to $\pi^2/6$ (see Lemma 3.1) such that the variance explodes for $k \to \infty$.

Under Gaussianity the ML estimation becomes efficient in that the limiting covariance matrix is the inverse of the Fisher information; see Dahlhaus (1989).

Proposition 8.1 *Let $y \sim \mathcal{N}_T(0, \Sigma_y)$ satisfy the assumptions behind Theorem 8.1. Then $W(\varphi_0)$ is the asymptotic Fisher information matrix.*

Proof: Dahlhaus (1989, Theorem 4.1). $\qquad\qquad\qquad\qquad\qquad\qquad\qquad$ □

All the parametric estimators discussed in this chapter share the feature that the true model is assumed to be known. The difficult problem of choosing an appropriate parametric model for the short memory component $\{x_t\}$ in a long memory environment will be addressed in Section 11.1. On top of this problem, exact ML has four drawbacks:

(1) First, the determination of Σ_y entering $\ell(y; \varphi, \sigma^2)$ may be computationally demanding. If $\{x_t\}$ is assumed to be an ARMA(p,q) process – and this assumption is maintained in the majority of applications of ML – then Sowell (1992) provided a general closed-form expression for $\gamma_y(\cdot)$ in terms of the so-called hypergeometric function; notice, however, our warning following Proposition 6.5. The computational burden was reduced by the simplifications due to Chung (1994). A widely used ML approximation uses partial autocorrelations of ARFIMA$(0, d, 0)$ models instead of ARFIMA(p, d, q); see Haslett and Raftery (1989, p. 12). Still, for inferential purposes, the consistent estimation of $W(\varphi_0)$ by $W(\widehat{\varphi}_{ML})$ inevitably involves numerical methods. While $W(\widehat{\varphi}_{ML})$ is easily constructed for EXP models, the computation of Σ_y becomes more challenging, building possibly on numerical integration according to Eq. (4.17).

(2) A second serious limitation of exact ML in practical situations is the restriction to stationarity, $d < 1/2$. Clearly, many economic and financial time series display persistence just around the border $d = 1/2$, and nonstationarity cannot be ruled out realistically.

(3) Third, we have learned in the previous chapter that the arithmetic mean converges to the true mean μ_0 at a very slow rate under long memory, all the more slowly, the larger d is. This problem cannot be overcome by more efficient estimators; see Eq. (7.17). Hence, a strong bias in $\hat{\varphi}_{ML}$ has been reported experimentally when μ_0 is not known a priori but has to be estimated under long memory; see, e.g. Cheung and Diebold (1994) and Hauser (1999).

(4) Finally, the assumption that the data follows a normal distribution, which was maintained in Theorem 8.1, is often hard to justify in practice.

In the next two sections, we present two approximations to exact ML that are designed to overcome the drawbacks of exact ML. The first approximation is settled in the time domain, often called method of CSS. The second approximation is settled in the frequency domain, and it is simply the Whittle estimation introduced previously in Section 4.6.

8.3 Conditional Sum of Squares

The assumption of Gaussianity behind $\{y_t\}$ amounts to a normal distribution of the innovations:

$$\varepsilon = (\varepsilon_1, \ldots, \varepsilon_T)' \sim \mathcal{N}_T(0, \sigma^2 I_T).$$

Inverting the MA(∞) process, $x_t = C(L)\varepsilon_t$, we have with $\{a_j\}$ from

$$\varepsilon_t = \frac{x_t}{C(L)} = \sum_{j=0}^{\infty} a_j x_{t-j}$$

the AR(∞) representation for $\{y_t\}$ from Eq. (6.2),

$$\varepsilon_t = \frac{\Delta^d}{C(L)} y_t = \sum_{k=0}^{\infty} \alpha_k(d) y_{t-k},$$

where as in Eq. (5.28) it holds that

$$\alpha_k(d) = \sum_{j=0}^{k} \pi_j(d) a_{k-j} \tag{8.4}$$

$$\sim \frac{1}{\sum_{j=0}^{\infty} c_j} \frac{k^{-d-1}}{\Gamma(-d)}, \quad k \to \infty.$$

Assuming a type II process, i.e. $y_0 = y_{-1} = \cdots = 0$, we truncate the infinite AR expansion and define

$$\varepsilon_t(\varphi) := \sum_{k=0}^{t-1} \alpha_k(d) y_{t-k},$$

noting that $\{\alpha_k(d)\}$ depends on $\varphi' = (d, \psi')$ with the k-dimensional vector ψ behind $C(L) = C(L; \psi)$. The conditional likelihood in terms of $\{\varepsilon_t\}$ becomes

$$(2\pi)^{-T/2}(\sigma^2)^{-T/2} \exp\left(-\frac{1}{2\sigma^2} \sum_{t=1}^{T} \varepsilon_t^2(\varphi) \right)$$

or in logarithmic terms upon dropping additive constants:

$$-\frac{T}{2} \ln \sigma^2 - \frac{1}{2\sigma^2} \sum_{t=1}^{T} \varepsilon_t^2(\varphi).$$

Once we concentrate out σ^2, a maximization amounts to minimizing

$$\mathrm{CSS}(\varphi) := \sum_{t=1}^{T} \varepsilon_t^2(\varphi),$$

which has been called the CSS. We define the estimator as

$$\hat{\varphi}_{\mathrm{CSS}} = (\hat{d}_{\mathrm{CSS}}, \hat{\psi}'_{\mathrm{CSS}})' = \arg\ \min \mathrm{CSS}(\varphi).$$

An estimator of the innovation variance hence is

$$\hat{\sigma}^2_{\mathrm{CSS}} = \frac{1}{T} \mathrm{CSS}(\hat{\varphi}_{\mathrm{CSS}}).$$

In a long memory framework, CSS estimation has been pioneered by Li and McLeod (1986) and Chung and Baillie (1993). In particular, Beran (1995) stressed that CSS is applicable to any type II process where the order of fractional integration is $d > -\frac{1}{2}$. The advantage over exact ML hence is that d may exceed the nonstationarity bound of $d = 1/2$. Moreover, we have the following result without assuming Gaussianity of the process; Gaussianity was only the motivation to set up the objective function $\mathrm{CSS}(\varphi)$, but it is not required for the limiting normality of $\hat{\varphi}_{\mathrm{CSS}}$.

Proposition 8.2 *Let $\{y_t\}$ be type II fractionally integrated with $\mathrm{E}(y_t) = 0$. Let $x_t = \sum_{j=0}^{\infty} c_{j,\psi} \varepsilon_{t-j}$ be from Assumption 8.1, where $\{\varepsilon_t\}$ is stationary and ergodic with finite fourth moments, and $C(z; \psi) = \sum_{j=0}^{\infty} c_{j,\psi} z^j$ meets certain regularity conditions; see Hualde and Robinson (2011) and Nielsen (2015). Further, the true parameter d_0 is contained in $D = [D_1, D_2]$ with*

$$-\infty < D_1 < D_2 < \infty.$$

Then $\hat{\varphi}_{\mathrm{CSS}}$ follows the limiting distribution given in Theorem 8.1.

Proof: Hualde and Robinson (2011, Theorem 2.2) or Nielsen (2015, Theorem 2).   □

The limiting covariance matrix $W^{-1}(\varphi_0)$ was already given in Beran (1995, Eq.(18)). Velasco and Robinson (2000), however, observed that the argument establishing limiting normality in Beran (1995, p. 670) was circular. Robinson (2006) fixed the proof under the restrictive assumption of stationary long memory, $0 < d < 1/2$. The general result in Proposition 8.2 was established by Hualde and Robinson (2011) and independently with quite a different proof by Nielsen (2015); see also Nielsen (2011). Note that all assumptions behind Proposition 8.2 are exclusively formulated in the time domain, in terms of $\{\varepsilon_t\}$ and $\{c_j\}$ behind $\{x_t\}$. Consequently, Hualde and Robinson (2011) and Nielsen (2015) give the limiting covariance matrix not in terms of $\ln f_y(\lambda; \varphi)$, but in terms of $\{c_k\}$ and $\partial a_k / \partial \varphi_j$. However, since the covariance matrix is obtained as the limit of the inverse of the Fisher information matrix (under Gaussianity), it must equal $W^{-1}(\varphi_0)$ from Theorem 8.1; see Proposition 8.1. Consequently, under Gaussianity of the data, CSS is efficient.

An obvious drawback of Proposition 8.2 is the assumption that $\{y_t\}$ has a mean of 0; note that $\varepsilon_t(\varphi)$ is computed without demeaning the data. To overcome this problem, Beran (1995) suggested to compute integer differences Δ^D if $d = D + \delta$, $|\delta| < 1/2$, where $D \in \mathbb{N}_0$ is $D = 0$ or $D = 1$ in most applications. Then,

$$z_t := \Delta^D y_t,$$

such that $z_t \sim I(\delta)$ for $y_t \sim I(d)$. Now, $\{z_t\}$ can be demeaned with $\bar{z}$. With

$$\alpha_k^*(d - D) = \alpha_k^*(\delta), \quad \varphi^* = (d - D, \psi')',$$

Beran (1995, Eq. (12)) suggested to build the CSS estimator on

$$\varepsilon_t(\varphi^*) := \sum_{k=0}^{t-1} \alpha_k^*(d - D)(z_{t-k} - \bar{z}).$$

While this is a simple proposal, it requires a priori knowledge of D. Further, according to Hualde and Robinson (2011, p. 3178), the asymptotic justification of CSS accounting for deterministic regressors such as polynomials of time, or simply a constant, still has to be completed.

8.4 Parametric Whittle Estimation

A natural way to circumvent the difficulties encountered with a potential mean different from 0, which is hard to estimate under long memory, is to settle the estimation of the other parameters φ in the frequency domain. The reason is given in Eq. (4.6): The periodogram as sample counterpart to the spectrum is invariant with respect to an additive constant like the mean as long as $I(\cdot)$ is evaluated at harmonic frequencies. In other words, a periodogram analysis does not require demeaning. This is where the Whittle estimator as frequency domain ML estimator comes in. With ψ behind $x_t = \sum_{j=0}^{\infty} c_{j,\psi} \varepsilon_{t-j}$ and

$\varphi = (d, \psi')'$ behind $f_y(\lambda; \varphi)$ from Eq. (8.1), the Whittle estimator is defined as in Eq. (4.32) as

$$\hat{\varphi}_W = \arg \min Q(\varphi),$$

where $Q(\varphi)$ is given in terms of the periodogram I_y of $\{y_t\}$ and of $f_y(\lambda; \varphi)$ from (8.1). More precisely, $T_C(\lambda_j; \psi)$ in Eq. (4.32) has to be replaced by $2\pi f_y(\lambda_j; \varphi)/\sigma^2$. One notable contribution by Velasco and Robinson (2000) consists in allowing for nonstationarity, $d \geq 1/2$. In this case, no spectrum exists, and one has to work with the *pseudo-spectrum*. The latter we define as generalization of (4.25) as

$$f_y(\lambda) = \frac{1}{2 - 2\cos(\lambda)} f_{\Delta y}(\lambda) = \left(2\sin\frac{\lambda}{2}\right)^{-2} f_{\Delta y}(\lambda), \quad d \in \left[\frac{1}{2}, \frac{3}{2}\right),$$

where $f_{\Delta y}$ is the spectrum given in Proposition 6.1 for the stationary process $\{\Delta y_t\}$ that is $I(d - 1)$. Consequently, the pseudo-spectrum f_y is of order λ^{-2d} at the origin; see Eq. (5.6).

Whittle estimation under long memory has been pioneered by Fox and Taqqu (1986). Limiting normality of $\hat{\varphi}_W$ has been established under stationary long memory and Gaussianity by Fox and Taqqu (1986, Theorem 2); see also Dahlhaus (1989, Theorem 2.1). Giraitis and Surgailis (1990, Theorem 4) replaced the Gaussianity assumption by assuming a stationary linear process driven by iid innovations under long memory. Velasco and Robinson (2000) have crucially extended the range of the memory parameter. They allow for antipersistence as well as nonstationarity up to $d < 3/4$. Further, they got rid of the iid assumption with respect to the innovations, and they do not require Gaussianity of the process. In that sense their assumptions parallel a lot those by Hannan (1973a), see also Proposition 4.12, however, with the crucial step of allowing for fractional integration.

Proposition 8.3   *Let $\{y_t - \mu\}$ or $\{\Delta y_t\}$ be a stationary type I process satisfying*

$$y_t = \mu + \Delta^{-d_0} x_t, \quad |d_0| < 1/2,$$

or

$$\Delta y_t = \Delta^{-(d_0-1)} x_t, \quad 1/2 \leq d_0 < 3/4,$$

where $x_t = \sum_{j=0}^{\infty} c_{j,\psi} \varepsilon_{t-j}$ meets Assumption 8.1 and some further restrictions by Velasco and Robinson (2000); in particular, $\{\varepsilon_t\}$ has to be conditionally homoskedastic. Then $\hat{\varphi}_W$ follows the limiting distribution given in Theorem 8.1, where $f_y(\lambda; \varphi)$ is defined as pseudo-spectrum for $d \geq 1/2$. Further, the estimation is consistent on the interval $3/4 \leq d_0 < 1$.

Proof: Velasco and Robinson (2000, Theorem 1, Theorem 2).   □

Under nonstationarity, Velasco and Robinson (2000) rule out $E(\Delta y_t) \neq 0$, which still allows for $E(y_t) \neq 0$. While Velasco and Robinson (2000, Theorem 1) establish consistency of $\hat{\varphi}_W$ for $3/4 \leq d \leq 1$, limiting normality fails to hold for $d \geq 3/4$. Moreover, Velasco and Robinson (2000) note that up to $d < 3/4$, $W(\hat{\varphi}_W)$ building on $\partial \ln f_y(\lambda; \hat{\varphi}_W)/\partial \varphi_j$ is consistent for $W(\varphi_0)$, such that approximate confidence intervals can be constructed the usual way, and asymptotic significance tests are valid. Finally, we wish to remind the reader of the different expressions of $W(\varphi_0)$ encountered in the literature that are linked by Lemma 4.3.

The range of allowed values of d is still too limited in Proposition 8.3 to meet all practical needs. We would like to have consistency and limiting normality beyond $I(1)$ processes. To extend the validity of the Whittle estimation beyond $d < 1$, Velasco and Robinson (2000) proposed the technique of tapering. A taper $\{h_t\}$ is a data window down-weighting observations at the beginning and end of the sample. Tapering has been suggested to reduce the periodogram bias, in particular at peaks and troughs; for a general introduction see Hannan (1970, pp. 265), Priestley (1981, Section 7.4.1), or Bloomfield (2000, Section 6.2). Velasco and Robinson (2000) discussed the properties of several data windows in the context of Whittle estimation, e.g. the rather simple so-called (full) cosine bell taper

$$h_t = \frac{1}{2}\left(1 - \cos\left(\frac{2\pi t}{T}\right)\right), \quad t = 1, \ldots, T. \tag{8.5}$$

The tapered periodogram is defined as

$$I_y^h(\lambda_j) := \frac{1}{2\pi \sum_{t=1}^T h_t^2} \left|\sum_{t=1}^T h_t\, y_t\, e^{i\lambda_j t}\right|^2, \tag{8.6}$$

with, e.g. $\sum_{t=1}^T h_t^2 = 3\,T/8$ for the cosine bell taper. Replacing the periodogram by I_y^h, Velasco and Robinson (2000) show that the tapered Whittle estimator remains consistent and limiting normal for any degree of nonstationarity as long as an adequate taper is applied.[1] This nice property, however, comes at the price of an inflated limiting variance.

The usage of tapering under nonstationary fractional integration had been pioneered earlier than Velasco and Robinson (2000) in the context of semiparametric estimators discussed in the next chapter. In particular, the cosine bell taper had been applied and discussed (with the local log-periodogram regression) by Hurvich and Ray (1995) and Velasco (1999a). With the local Whittle estimator from Section 9.2, Velasco (1999b) considered more general data windows as in Velasco and Robinson (2000). A new class of tapers resulting in a

---

1 Note that the summation in the Whittle criterion function Q from Eq. (4.32) is no longer over all harmonic frequencies, but rather only for $j \in \{p, 2p, 3p, \ldots\}$, where p is the so-called order of the data taper; see Velasco and Robinson (2000, Eq. (5)).

smaller limiting variance inflation had been suggested and discussed by Hurvich and Chen (2000); still, tapering will inflate the variance of estimators in general.

There are at least two possible extensions of Whittle estimation beyond Velasco and Robinson (2000). First, it is desirable to allow for conditional heteroskedasticity in the martingale difference sequence (MDS) $\{\varepsilon_t\}$; second and maybe more relevant, one may wish to extend the results to a wider range of nonstationarity without having to taper the data. Both aims have been achieved by Shao (2010). Under conditional heteroskedasticity he proved that the covariance matrix of the limiting normal distribution is affected, while the rate of convergence from Proposition 8.3 remains unaffected, implying that $\hat{\varphi}_W$ is still consistent. To achieve consistency and limiting normality beyond $d = 3/4$, Shao (2010) suggested to extend the periodogram. Such an extension had been discussed earlier by Phillips (1999) and had been applied with the so-called local Whittle estimator by Abadir et al. (2007); see our exposition in Chapter 9. To cover persistence around $d = 1$, which is of particular interest in the econometric unit root literature, we assume again

$$d = D + \delta, \quad D \in \{0, 1\}, \quad |\delta| < 1/2.$$

Note that Shao (2010) allowed more generally for $D \in \mathbb{N}_0$, while Abadir et al. (2007) even included $D = -1$; we restrict the exposition to $D \in \{0, 1\}$ for simplicity. The extended periodogram is defined as

$$I_y(\lambda_j; d) := |w_y(\lambda_j; d)|^2, \quad w_y(\lambda_j; d) = w_y(\lambda_j) + k(\lambda_j; d). \tag{8.7}$$

Here, $w_y(\lambda_j)$ is the usual discrete Fourier transform defined in Eq. (4.7),

$$w_y(\lambda_j) = \frac{1}{\sqrt{2\pi T}} \sum_{t=1}^{T} y_t e^{it\lambda_j}.$$

The extension $w_y(\lambda_j; d)$ depends on the nonzero correction term $k(\lambda_j; d)$ for $d \in [1/2, 3/2]$:

$$k(\lambda_j; d) = \begin{cases} 0, & d \in \left[-\dfrac{1}{2}, \dfrac{1}{2}\right) \\ \dfrac{e^{i\lambda_j}}{\sqrt{2\pi T}}(1 - e^{i\lambda_j})^{-1}(y_T - y_0), & d \in \left[\dfrac{1}{2}, \dfrac{3}{2}\right) \end{cases}. \tag{8.8}$$

Note that we now assume observed data from y_0 through y_T, where y_0 enters only through $k(\cdot; \cdot)$, while the standard DFT $w_y(\cdot)$ builds on $y_1, \dots, y_T$ only. The extended periodogram has an alternative, more intuitive expression in terms of the standard periodogram of the differenced data $I_{\Delta y}$ for $d \geq 1/2$. To become precise, it follows from Corbae et al. (2002, Eq. (51)) that

$I_y(\lambda_j; d) = |1 - e^{i\lambda_j}|^{-2} I_{\Delta y}(\lambda_j)$ with our standard notation

$$I_{\Delta y}(\lambda_j) = \left| \frac{1}{\sqrt{2\pi T}} \sum_{t=1}^{T} \Delta y_t e^{it\lambda_j} \right|^2 ;$$

see also Abadir et al. (2011, p. 190). Hence,

$$I_y(\lambda_j; d) = \begin{cases} I_y(\lambda_j), & d \in \left[-\frac{1}{2}, \frac{1}{2}\right) \\ \dfrac{1}{2 - 2\cos(\lambda_j)} I_{\Delta y}(\lambda_j), & d \in \left[\frac{1}{2}, \frac{3}{2}\right) \end{cases},$$

where the branch under nonstationarity parallels the definition of the pseudo-spectrum above. Now, the Whittle objective function is defined in terms of the extended periodogram divided by the (pseudo) spectrum

$$Q_{EW}(\varphi) := \frac{2\pi}{M} \sum_{j=1}^{M} \frac{I_y(\lambda_j; d)}{\left[4 \sin^2 \left(\frac{\lambda_j}{2}\right)\right]^{-d} T_C(\lambda_j; \psi)},$$

where $T_C(\lambda_j; \psi)$ is the power transfer function of $x_t = \sum_{j=0}^{\infty} c_{j,\psi} \varepsilon_{t-j}$. The extended Whittle estimator is obtained by minimization:

$$\widehat{\varphi}_{EW} = \arg \min Q_{EW}(\varphi).$$

From Shao (2010) we take the remarkable result that the limiting theory from Proposition 8.3 carries over to the nonstationarity-extended Whittle estimator, where the value $d = \frac{1}{2}$ is excluded:

$$d \in \left(-\frac{1}{2}, \frac{1}{2}\right) \cup \left(\frac{1}{2}, \frac{3}{2}\right).$$

Proposition 8.4 *Let $\{y_t\}$, $t = 0, 1, \ldots, T$, be generated from*

$$\Delta^D y_t = \mu(1 - D) + \Delta^{-\delta_0} x_t, \quad D \in \{0, 1\}, \quad |\delta_0| < \frac{1}{2},$$

where $x_t = \sum_{j=0}^{\infty} c_j \varepsilon_{t-j}$ meets Assumption 8.1 and some further restrictions by Shao (2010); in particular, we assume $\{\varepsilon_t\}$ to be conditionally homoskedastic. Then $\widehat{\varphi}_{EW}$ follows the limiting distribution given in Theorem 8.1 for $d_0 = D + \delta_0$.

Proof: Shao (2010, Theorem 3.2). □

To sum up, we find that the (extended) Whittle estimation overcomes all pitfalls of exact ML listed at the end of Section 8.2. First, $\widehat{\varphi}_W$ or $\widehat{\varphi}_{EW}$ are very simple to compute, only requiring the (extended) periodogram and the (pseudo) spectrum for a given model. Second, $\widehat{\varphi}_W$ remains valid up to a certain degree of nonstationarity: consistent up to $d < 1$ and limiting normal without any changes in the covariance formula up to $d < 3/4$. And $\widehat{\varphi}_{EW}$ has these nice properties even for $3/4 \leq d < 3/2$. Third, a mean different from 0 is automatically accounted for and does not have to be estimated, since the procedure is settled in the frequency domain. Fourth, the limiting distribution is obtained without assuming normality of the data. Still, also the Whittle estimation as an approximation to exact ML assumes the true parametric model to be known.

8.5   Log-periodogram Regression of FEXP Processes

We now consider specifically a fractionally integrated EXP(k) model, FEXP(d, k), where f_x is the spectrum of the short memory EXP process:

$$y_t = \Delta^{-d} x_t, \quad f_x(\lambda) = \exp\left(\sum_{\ell=0}^{k} \psi_\ell \cos(\ell\lambda)\right), \quad |d| < \frac{1}{2}. \tag{8.9}$$

The process $\{y_t\}$ is stationary, i.e. a type I process ($t \in \mathbb{Z}$). Given this model we obtain for the spectrum f_y at the harmonic frequencies $\lambda_j = 2\pi j/T$ upon taking logarithm:

$$\ln f_y(\lambda_j) = -2dX_j + \sum_{\ell=0}^{k} \psi_\ell \cos(\ell\lambda),$$

where according to Proposition 6.1

$$X_j = \ln\left(2\sin\frac{\lambda_j}{2}\right). \tag{8.10}$$

Let us define

$$I_j = I_y(\lambda_j), \quad f_j = f_y(\lambda_j),$$

where $I_y(\cdot)$ is the usual periodogram computed from $y_1, \ldots, y_T$. It then holds under the FEXP model that

$$\ln I_j = \psi_0 - 2dX_j + \sum_{\ell=1}^{k} \psi_\ell \cos(\ell\lambda_j) + \ln\frac{I_j}{f_j},$$

where $j = 1, \ldots, M = \lfloor\frac{T-1}{2}\rfloor$. This suggests to estimate d by an ordinary least squares (OLS) regression of the log-periodogram, $\ln I_j$, on a constant, the trigonometric regressors $\cos(\ell\lambda_j)$ and on $-2X_j$, where M is the number of harmonic frequencies involved. We call the resulting estimator, $\widehat{d}_{Exp}^{(k)}$, the log-periodogram regression estimator (assuming a FEXP(d, k) model). This estimator was originally proposed by Janacek (1982), however, without establishing asymptotic results rigorously; see also Robinson (1994a, Section 3.3). Further, Kashyap and Eom (1988) assumed $k = 0$, where FEXP($d, 0$)

turns into fractionally integrated noise, FIN(d). Then the estimation amounts to a regression on a constant and on $-2X_j$, which is the case treated in the next chapter.[2] A generalization of $\widehat{d}^{(k)}_{\text{Exp}}$ has been proposed by Beran (1993), allowing for more general functions $h_\ell(\lambda)$ replacing $\cos(\ell\,\lambda)$; see Beran (1993, Definition 1).

To motivate a log-periodogram regression, Kashyap and Eom (1988), Geweke and Porter-Hudak (1983), and Beran (1993) argued that the errors $\ln I_j - \ln f_j$ turn into an iid sequence under appropriate assumptions. In fact, if $d = 0$ and $\{y_t\}$ is Gaussian white noise, then we know from Proposition 4.9 that $\{I_j\}, j = 1, \dots, M$, forms an iid sequence with

$$2\,\frac{I_j}{f_j} \sim \chi^2(2).$$

Consequently, one can show with Lemma 4.2 (see also the proof of Lemma 9.2 below) under this restrictive assumption that

$$\mathrm{E}\left(\ln\left(\frac{I_j}{f_j}\right)\right) = -\gamma,$$

where γ is again Euler's constant. This suggests to define as regression error U_j with mean 0,

$$U_j = \ln\left(\frac{I_j}{f_j}\right) + \gamma, \tag{8.11}$$

such that (with $c = \psi_0 - \gamma$)

$$\ln I_j = c - 2dX_j + \sum_{\ell=1}^{k} \psi_\ell \cos(\ell\,\lambda_j) + U_j,$$

resulting in an OLS regression,

$$\ln I_j = \widehat{c} - \widehat{d}^{(k)}_{\text{Exp}}\, 2\,X_j + \sum_{\ell=1}^{k} \widehat{\psi}_\ell \cos(\ell\,\lambda_j) + \widehat{U}_j, \tag{8.12}$$

where $j = 1, \dots, M = \left\lfloor\frac{T-1}{2}\right\rfloor$. Next, we step beyond the case of Gaussian white noise, i.e. $y_t = \varepsilon_t \sim \mathcal{N}(0, \sigma^2)$. If $d = 0$ and $\{y_t\}$ is a linear process, then Proposition 4.10 justifies $2\,I_j/f_j \sim$ iid with a $\chi^2(2)$ distribution for a collection of p errors at least asymptotically, where p might even grow with the sample size

---

2 Geweke and Porter-Hudak (1983) considered $k = 0$, too, however, regressing only in a neighborhood of the origin, that is $j = 1, \dots, m$ with $m/T \to 0$, which turns the estimator into a semiparametric one; see Section 9.1. Hurvich and Brodsky (2001) considered $\widehat{d}^{(k)}_{\text{Exp}}$ for k growing (more slowly than T) to infinity. They did not assume a true parametric EXP model for $\{x_t\}$, but more generally that the spectrum of this short memory process has a convergent infinite Fourier expansion; this case falls into the category of semiparametric estimators dealt with in Section 9.6.

according to Proposition 4.11. This, however, does not carry over to the case $d \neq 0$. First, $\{U_j\}$ is not an independent sequence for $d \neq 0$, and, second, it does not follow an identical distribution, in particular, $E(U_j)$ is not 0 but depends on j in general. This is true under much more general assumptions than maintained in Eq. (8.9). We report the relevant results from the literature next.

Lemma 8.1 *Let $\{y_t\}$ with $y_t = (1 - L)^{-d} x_t$, $|d| < 1/2$, be a type I process where the spectrum of $\{x_t\}$ is positive, continuous, bounded and bounded away from 0. It then holds for the regression errors from Eq. (8.11) for fixed j and k that (as $T \to \infty$)*

(a) U_j and U_k are not independent if $d \neq 0$, $j \neq k$,

(b) $E(U_j) = \begin{cases} 0, & \text{for } d = 0 \\ \mu_j \neq 0, & \text{for } d \neq 0 \end{cases}$,

where μ_j is characterized in Hurvich and Beltrao (1993, Theorem 7).

Proof: The first result was given independently by Hurvich and Beltrao (1993, Theorem 5) and Robinson (1995a, Theorem 1). The second result is taken from Hurvich and Beltrao (1993, Theorem 7); see also Künsch (1986). □

At first glance, Lemma 8.1 (a) seems to contradict Yajima (1989) who proved for $0 < d < 1/2$ that

$$\frac{I_y(\omega_j)}{f_y(\omega_j)} \quad \text{and} \quad \frac{I_y(\omega_k)}{f_y(\omega_k)}, \quad \text{where } \omega_j, \omega_k \text{ are fixed},$$

are asymptotically independent. Note, however, that ω_j and ω_k are assumed to be fixed by Yajima (1989), which rules out harmonic frequencies $2\pi j/T$ and $2\pi k/T$ varying with T, where only j and k are fixed. It is not straightforward to extend Proposition 4.10 to the reign of fractional integration. Giraitis et al. (2012, Theorem 5.3.1), however, proved under fractional integration for harmonic frequencies, λ_j and λ_k, $j \neq k$, with $j, k \to \infty$ as $T \to \infty$, that U_j and U_k are asymptotically independent; see also Lemma 9.2 below; again, this does not contradict Lemma 8.1 (a) where we assumed finite j and k. Similarly, Hurvich et al. (1998, Lemma 8) established that $E(U_j)$ converges to 0 with growing j, more precisely:

$$E(U_j) = O\left(\frac{\ln j}{j}\right), \quad \ln^2 m \leq j \leq m,$$

where m passes to infinity more slowly than T; further details are given in Lemma 9.1 below. Hence, there is hope to get a consistent estimator from an OLS regression of Eq. (8.12).[3] In fact, we have the following result.

---

3 The theoretical analysis is further complicated by the fact that the regressors from (8.10) are unbounded for growing T, e.g. $X_1 = \ln\left(2 \sin \frac{\pi}{T}\right) \sim \ln\left(\frac{2\pi}{T}\right)$.

Proposition 8.5 *Let $\{y_t\}$ be a FEXP process from Eq. (8.9) with finite order k; further assume a normal distribution. With $\widehat{d}_{\text{Exp}}^{(k)}$ from Eq. (8.12) and $M = \lfloor (T-1)/2 \rfloor$ it holds for the true parameter d_0 with $|d_0| < 1/2$ that*

$$E(\widehat{d}_{\text{Exp}}^{(k)} - d_0) \to 0 \quad and \quad T \, \text{Var}(\widehat{d}_{\text{Exp}}^{(k)}) \to V_{\text{Exp}}^{(k)} = \frac{1}{1 - \frac{6}{\pi^2} \sum_{\ell=1}^{k} \frac{1}{\ell^2}},$$

as $T \to \infty$.

Proof: Follows from Hurvich and Brodsky (2001); for details, see Appendix to this chapter.

A numerical evaluation shows that the variance $V_{\text{Exp}}^{(k)}$ increases fast with k:

$$V_{\text{Exp}}^{(1)} = 2.551, \quad V_{\text{Exp}}^{(2)} = 4.165, \quad V_{\text{Exp}}^{(3)} = 5.796.$$

Not surprisingly, the log-periodogram regression produces an estimator that is asymptotically less efficient than the one from the ML principle. Remember the limiting variance obtained for the ML estimator $\widehat{d}_{\text{ML}}$ in Corollary 8.1 for FEXP models. It just turns out to be a multiple of $V_{\text{Exp}}^{(k)}$:

$$\lim_{T \to \infty} T \, \text{Var}(\widehat{d}_{\text{ML}}) = \frac{1}{\frac{\pi^2}{6} - \sum_{\ell=1}^{k} \frac{1}{\ell^2}} = \frac{6}{\pi^2} V_{\text{Exp}}^{(k)}.$$

From that we observe that the relative efficiency of ML over the log-periodogram regression equals $6/\pi^2 = 60.79\%$, irrespective of k.

Hurvich and Brodsky (2001, Remark 2), allowing for $k \to \infty$, were not able to establish asymptotic normality. In the light of Proposition 8.5, it is not worth striving for limiting normality here, since the parametric log-periodogram regression is inferior to ML estimators in terms of variance. Hence, if one is willing to assume the FEXP model Eq. (8.9), then ML, and in particular the Whittle approximation to ML, is the superior choice.

8.6 Fractionally Integrated Noise

We now consider a very special parametric model, namely, the fractionally integrated noise of order d, FIN(d). On top we maintain the assumption of Gaussianity, i.e.

$$y_t = (1 - L)^{-d} \varepsilon_t, \quad \varepsilon_t \sim \mathcal{N}(0, \sigma^2), \quad |d| < \frac{1}{2}, \tag{8.13}$$

where $\{\varepsilon_t\}$ denotes an iid sequence. Note that FIN is a special case of FEXP with $k = 0$. Hence, the results from the previous section carry over. As a special case of the log-periodogram regression Eq. (8.12), we obtain

$$\ln I_j = c - 2dX_j + U_j, \quad j = 1, \dots, M = \left\lfloor \frac{T-1}{2} \right\rfloor,$$

with the resulting OLS estimator $\widehat{d}^{(0)}_{\text{Exp}}$. With the usual convention that a summation from 1 up to 0 equals 0, Proposition 8.5 provides under Eq. (8.13) that for a true parameter value d_0 it holds asymptotically

$$E(\widehat{d}^{(0)}_{\text{Exp}} - d_0) \to 0 \quad \text{and} \quad T \text{ Var}(\widehat{d}^{(0)}_{\text{Exp}}) \to 1.$$

Note that Kashyap and Eom (1988) originally also stated limiting normality of the log-periodogram regression,

$$\sqrt{T}(\widehat{d}^{(0)}_{\text{Exp}} - d_0) \xrightarrow{D} \mathcal{N}(0, 1) \quad \text{or} \quad \sqrt{M}(\widehat{d}^{(0)}_{\text{Exp}} - d_0) \xrightarrow{D} \mathcal{N}\left(0, \frac{1}{2}\right).$$

Since their arguments rely on asymptotic independence of the regression errors $\{U_j\}$, they do not seem to amount to a rigorous proof. Here, we do not strive for establishing limiting normality, since the parametric log-periodogram regression is inferior to ML estimators in terms of variance, as we observed in the previous section.

Before we close this section, we mention yet another very simple parametric estimator for the model Eq. (8.13). If the true process is FIN(d), we obtain from Proposition 6.4 for the autocorrelation at lag one

$$\rho(1) = \frac{d}{1 - d} \quad \text{or} \quad d = \frac{\rho(1)}{1 + \rho(1)}.$$

This renders itself to a method of moment estimator; see also Anděl (1986, p. 109),

$$\widehat{d}_{\text{MM}} = \frac{\widehat{\rho}(1)}{1 + \widehat{\rho}(1)},$$

which is consistent for d as long as the sample autocorrelation is chosen as consistent, e.g.

$$\widehat{\rho}(1) = \frac{\sum_{t=1}^{T-1}(y_t - \bar{y})(y_{t+1} - \bar{y})}{\sum_{t=1}^{T}(y_t - \bar{y})^2}.$$

Conditions for consistent autocorrelation estimation have been discussed in Section 7.5. Since in particular for $d > 1/4$, limiting normality fails (see Hosking (1996, Theorem 7)), a unified distributional framework for $\widehat{d}_{\text{MM}}$ over the whole parameter range $0 < d < 1/2$ is out of reach.

8.7 Technical Appendix: Proofs

Proof of Corollary 8.1

(a) According to Eq. (8.1) the log-spectrum is

$$\ln f_y(\lambda; \varphi) = -d \ln \left(4 \sin^2\left(\frac{\lambda}{2}\right)\right) + \ln f_x(\lambda; \psi)$$

such that

$$
4\pi w_{11} = \int_{-\pi}^{\pi} \left[\ln\left(4\sin^2\left(\frac{\lambda}{2}\right)\right)\right]^2 d\lambda
$$

$$
= 2 \int_{0}^{\pi} \left[\ln\left(4\sin^2\left(\frac{\lambda}{2}\right)\right)\right]^2 d\lambda
$$

$$
= 4 \int_{0}^{\frac{\pi}{2}} [\ln(4\sin^2(\omega))]^2 d\omega
$$

$$
= 16 \int_{0}^{\frac{\pi}{2}} [\ln 2 + \ln \sin \omega]^2 d\omega.
$$

Using the binomial formula, this integral falls into three terms:

$$
4\pi w_{11} = I_1 + I_2 + I_3,
$$

where

$$
I_1 = 16 \frac{\pi}{2} \ln^2 2,
$$

$$
I_2 = 32 \ln 2 \int_{0}^{\frac{\pi}{2}} \ln \sin \omega \, d\omega,
$$

$$
I_3 = 16 \int_{0}^{\frac{\pi}{2}} \ln^2(\sin \omega) \, d\omega.
$$

For the last two integrals, we have from Gradshteyn and Ryzhik (2000) Eqs. (4.224(3)) and (4.224(7)), respectively, that

$$
\int_{0}^{\frac{\pi}{2}} \ln \sin \omega \, d\omega = -\frac{\pi}{2} \ln 2,
$$

$$
\int_{0}^{\frac{\pi}{2}} \ln^2(\sin \omega) \, d\omega = \frac{\pi}{2} \left(\ln^2 2 + \frac{\pi^2}{12} \right).
$$

It hence follows that

$$
4\pi w_{11} = \frac{2}{3} \pi^3 \quad \text{or} \quad w_{11} = \frac{\pi^2}{6}.
$$

Now, it is well known how to compute the inverse of a partitioned matrix like W; see, e.g. Dhrymes (1978, Proposition 32):

$$
\begin{pmatrix} w_{11} & w_{12}' \\ w_{12} & W_{22} \end{pmatrix}^{-1} = \begin{pmatrix} a & b' \\ b & C \end{pmatrix},
$$

where in particular

$$
a = (w_{11} - w_{12}' W_{22}^{-1} w_{12})^{-1},
$$

which establishes (a).

(b) For a FEXP(d, k) model, we have with $\varphi' = (d, \psi_1, \ldots, \psi_k)$ that

$$\frac{\partial \ln f_y(\lambda; \varphi)}{\partial \varphi_j} = \cos(j-1)\lambda, \quad j = 2, \ldots, k+1,$$

such that W_{22} is diagonal with $1/4$ on the main diagonal by Corollary 4.2. Further, by Eq. (8.1)

$$\frac{\partial \ln f_y(\lambda; \varphi)}{\partial d} = -\ln\left[4 \sin^2\left(\frac{\lambda}{2}\right)\right].$$

Hence, w_{12} consists of k elements $w_{12,j}$ of the following form ($j = 1, \ldots, k$):

$$
\begin{aligned}
w_{12,j} &= \frac{1}{4\pi} \int_{-\pi}^{\pi} \frac{\partial \ln f_y(\lambda; \varphi)}{\partial d} \frac{\partial \ln f_y(\lambda; \varphi)}{\partial \psi_j} \, d\lambda \\
&= \frac{1}{2\pi} \int_0^\pi -\ln\left[4 \sin^2\left(\frac{\lambda}{2}\right)\right] \cos(\lambda j) \, d\lambda \\
&= -\frac{1}{\pi} \left[\int_0^\pi \ln(2) \cos(\lambda j) \, d\lambda + \int_0^\pi \ln\left[\sin\left(\frac{\lambda}{2}\right)\right] \cos(\lambda j) \, d\lambda\right] \\
&= -\frac{1}{\pi} \left[2 \int_0^{\pi/2} \ln[\sin(x)] \cos(2 j x) \, dx\right] \\
&= -\frac{2}{\pi} \left[-\frac{\pi}{4 j}\right],
\end{aligned}
$$

where the latter equality follows from Gradshteyn and Ryzhik (2000, Eq. (4.384.7)). Consequently,

$$w'_{12} = \left(\frac{1}{2}, \ldots, \frac{1}{2k}\right),$$

which establishes (b). Hence, the proof is complete. □

Proof of Proposition 8.5

The proof follows and modifies steps given in Hurvich and Brodsky (2001). They rewrite $\widehat{d}_{\text{Exp}}^{(k)}$ as

$$\widehat{d}_{\text{Exp}}^{(k)} = -\frac{1}{2} \frac{\sum_{j=1}^M r_j \ln I_j}{\sum_{j=1}^M r_j^2},$$

where r_j are the least squares residuals from a regression of X_j on $1, \cos(\lambda_j), \cos(2\lambda_j), \ldots, \cos(k\lambda_j)$:

$$X_j = \widehat{b}_0 + \sum_{\ell=1}^k \widehat{b}_\ell \cos(\ell \lambda_j) + r_j, \quad j = 1, \ldots, M.$$

For $k = 0$, we stick to the convention that $\sum_{\ell=1}^{0} \hat{b}_\ell \cos(\ell\,\lambda_j) = 0$; then $\hat{b}_0 = \overline{X} = \frac{1}{M} \sum_{j=1}^{M} X_j$, and $r_j = X_j - \overline{X}$. Further, by Hurvich and Brodsky (2001, Eq. (9)),

$$\hat{d}_{\text{Exp}}^{(k)} = d - \frac{\sum_{j=1}^{M} r_j \sum_{\ell=1}^{k} \psi_\ell \cos(\ell\,\lambda_j)}{2 \sum_{j=1}^{M} r_j^2} - \frac{\sum_{j=1}^{M} r_j U_j}{2 \sum_{j=1}^{M} r_j^2}, \tag{8.14}$$

where we learn from the proof of Hurvich and Brodsky (2001, Lemma 5) for finite k that

$$\sum_{j=1}^{M} r_j^2 = \frac{T}{4} \left[\frac{\pi^2}{6} - \sum_{\ell=1}^{k} \frac{1}{\ell^2} \right] + O(\ln^2 T). \tag{8.15}$$

Now, we turn to the asymptotic bias. From the proofs of Hurvich and Brodsky (2001, Lemma 6) and Hurvich and Brodsky (2001, Lemma 9), we draw, respectively,

$$\sum_{j=1}^{M} r_j \sum_{\ell=1}^{k} \psi_\ell \cos(\ell\,\lambda_j) = O(\ln T), \tag{8.16}$$

$$E\left(\sum_{j=1}^{M} r_j U_j \right) = O(\ln^3 T). \tag{8.17}$$

From Eqs. (8.14) to (8.17) it follows that

$$E\left(\hat{d}_{\text{Exp}}^{(k)} - d \right) = O\left(\frac{\ln^3 T}{T} \right).$$

Next, we turn to the variance. From the proof of Hurvich and Brodsky (2001, Theorem 2), we learn for

$$\text{Var}\left(\hat{d}_{\text{Exp}}^{(k)} \right) = \frac{\text{Var}\left(\sum_{j=1}^{M} r_j U_j \right)}{4 \left(\sum_{j=1}^{M} r_j^2 \right)^2}$$

that

$$\text{Var}\left(\sum_{j=1}^{M} r_j U_j \right) = \sum_{j=1}^{M} r_j^2 \text{Var}(U_j) + o(T),$$

$$\sum_{j=1}^{M} r_j^2 \text{Var}(U_j) = \frac{\pi^2}{6} \sum_{j=1}^{M} r_j^2 + O(\ln^4 T).$$

Hence, with Eq. (8.15),

$$\text{Var}\left(\hat{d}_{\text{Exp}}^{(k)} \right) = \frac{\frac{\pi^2}{6} + O\left(\frac{\ln^4 T}{T} \right) + o(1)}{T \left[\frac{\pi^2}{6} - \sum_{\ell=1}^{k} \frac{1}{\ell^2} + O\left(\frac{\ln^2 T}{T} \right) \right]},$$

from which the limiting variance follows. This completes the proof. $\qquad\square$

9

Semiparametric Estimators

The methods treated in this chapter do not assume a parametric model for the short memory component $\{x_t\}$. The only parameter that is estimated is d from Assumption 6.2 or 6.3. All estimators are settled in the frequency domain and rely on the periodogram. Two estimators, the log-periodogram regression (PR) and the local Whittle (LW) estimation, become semiparametric in that the periodogram is evaluated only locally, in a vicinity of the origin, where the spectral behavior is characterized by d alone. The crucial tuning parameter in such a setting is the so-called bandwidth m, determining how small or how large the *vicinity* to the origin is, relative to the sample size. There also exist *global* estimators evaluating the periodogram over the whole frequency range. They are semiparametric in that they do not require a parametric specification for $\{x_t\}$; instead they assume that the short memory component is well captured by an AR(k) model with growing k. The critical choice is then of course how to pick k in practice.

9.1 Local Log-periodogram Regression

To begin with, we stick for the moment to Assumption 6.2 as underlying model. From Eq. (6.3) we have for the spectrum

$$f_y(\lambda) = \left(2 \sin \frac{\lambda}{2}\right)^{-2d} f_x(\lambda), \quad \lambda > 0,$$

such that

$$\ln I_j = c_j - 2dX_j + U_j, \quad j = 1, \ldots, m, \tag{9.1}$$

where I_j, X_j, and U_j are defined as in Eq. (8.12),

$$I_j = I_y(\lambda_j), \quad X_j = \ln\left(2 \sin \frac{\lambda_j}{2}\right), \quad U_j = \ln\left(\frac{I_y(\lambda_j)}{f_y(\lambda_j)}\right) + \gamma,$$

Time Series Analysis with Long Memory in View, First Edition. Uwe Hassler.
© 2019 John Wiley & Sons, Inc. Published 2019 by John Wiley & Sons, Inc.

and

$$c_j = \ln f_x(\lambda_j) - \gamma \to c = \ln f_x(0) - \gamma.$$

Note that c_j turns into a constant intercept asymptotically as long as

$$\frac{m}{T} \to 0 \quad \text{as} \quad T \to \infty.$$

Under such a condition on the so-called bandwidth m, Geweke and Porter-Hudak (1983) suggested to treat c_j as constant and to estimate d from an ordinary least squares (OLS) estimation of a simple log-periodogram regression (PR):

$$\widehat{d}_{\text{PR}}^{(m)} = -\frac{1}{2} \frac{\sum_{j=1}^{m} (X_j - \overline{X}_m) \ln I_j}{\sum_{j=1}^{m} (X_j - \overline{X}_m)^2}, \quad \overline{X}_m = \frac{1}{m} \sum_{j=1}^{m} X_j. \tag{9.2}$$

If m was equal to M, then $\widehat{d}_{\text{PR}}^{(m)}$ would turn into the parametric estimator $\widehat{d}_{\text{Exp}}^{(0)}$ from Section 8.6. Assuming that the regression errors U_j form an iid sequence, Geweke and Porter-Hudak (1983, Theorem 2) claimed for a true parameter value $d_0 < 0$ that $\sqrt{m}(\widehat{d}_{\text{PR}}^{(m)} - d_0)$ converges to a normal distribution as long as m diverges fast enough but more slowly than T. In Lemma 8.1, however, we have seen that $\{U_j\}$ does not become iid asymptotically, such that the proof by Geweke and Porter-Hudak (1983, Theorem 2) was incomplete.

Hurvich et al. (1998) added the assumption of normality of the data and extended the range of memory at the same time, $|d| < 1/2$. They established the following result.[1]

Proposition 9.1 *Let $\{y_t - \mu\}$ from Assumption 6.2 be fractionally integrated of type I,*

$$y_t - \mu = \Delta^{-d_0} x_t, \quad |d_0| < 1/2,$$

where $\{x_t\}$ is a Gaussian process with spectrum f_x. Assume that f_x is three times continuously differentiable at zero and in a neighborhood of 0, and $f_x'(0) = 0$.[2] If

$$\frac{1}{m} + \frac{m \ln m}{T} \to 0,$$

then

$$\text{E} \left(\widehat{d}_{\text{PR}}^{(m)} - d_0 \right) \to 0 \quad \text{and} \quad m \, \text{Var} \left(\widehat{d}_{\text{PR}}^{(m)} \right) \to \frac{\pi^2}{24},$$

---

1 We extend Eq. (6.2) from Definition 6.1 to the case of a fractionally integrated process with a nonzero mean as in Eq. (6.8), $y_t = \mu + \Delta^{-d_0} x_t$, since the periodogram is not affected by this; see Eq. (4.6).

2 See also Andrews and Guggenberger (2003, Footnote 2) for a comment on these assumptions. In particular, $f_x'(0) = 0$ necessarily holds since the spectrum f_x is even; see the discussion following Proposition 6.6.

as $T \to \infty$. And if

$$\frac{m}{T^{4/5}} + \frac{\ln^2 T}{m} \to 0 \tag{9.3}$$

then (as $T \to \infty$)

$$\sqrt{m}\left(\widehat{d}_{\mathrm{PR}}^{(m)} - d_0\right) \xrightarrow{D} \mathcal{N}\left(0, \frac{\pi^2}{24}\right).$$

Proof: Hurvich et al. (1998, Theorems 1 and 2). □

The reason behind this result is given in the following lemma on the error terms from Eq. (9.1). Under the additional assumptions of Proposition 9.1, it adds to Lemma 8.1 by characterizing the behavior of U_j and U_k for j and k getting large.

Lemma 9.1 *Let $\{y_t\}$ satisfy the assumptions from Proposition 9.1. It then holds for the regression errors from Eq. (9.1) under*

$$\ln^2 m \le k < j \le m$$

that

(a) $\mathrm{Cov}(U_j, U_k) = O\left(\frac{\ln^2 j}{k^2}\right)$,

(b) $\mathrm{E}(U_j) = O\left(\frac{\ln j}{j}\right)$,

(c) $\mathrm{Var}(U_j) = \frac{\pi^2}{6} + O\left(\frac{\ln j}{j}\right)$,

as $m \to \infty$ and $(m \ln m)/T \to 0$.

Proof: The first result follows from Hurvich et al. (1998, Lemmas 2 and 3); the second and the third ones are given in Hurvich et al. (1998, Lemma 6) and Hurvich et al. (1998, Lemma 7), respectively. □

We have one further lemma building on Giraitis et al. (2012). It is stronger than Lemma 9.1 in that it provides the limiting distribution and asymptotic independence of the regression errors at harmonic frequencies λ_{j_i} where the indices $j_i = j_i(T)$ may vary with T; it is more restrictive at the same time in that it is only given for a finite collection of p harmonic frequencies. It thus carries the earlier Proposition 4.10 to the fractionally integrated case.

Lemma 9.2 *Let $\{y_t\}$ satisfy Assumption 6.2 where $\{x_t\}$ from Assumption 6.1 is a linear process. Consider harmonic frequencies λ_{j_i}, $i = 1, \dots, p$, with*

$$L_T \le j_1 < \cdots < j_p < \frac{T}{2}.$$

Further, let L_T diverge with $T \to \infty$.

(a) The ratios $I_y(\lambda_{j_i})/f_y(\lambda_{j_i})$ are asymptotically independent, $i = 1, \ldots, p$, and

$$2\,\frac{I_y(\lambda_{j_i})}{f_y(\lambda_{j_i})} \xrightarrow{D} \chi^2(2).$$

(b) For $G_{j_i} := -U_{j_i} + \gamma = \ln f_y(\lambda_{j_i}) - \ln I_y(\lambda_{j_i})$ with regression errors U_{j_i} from Eq. (9.1), it holds under $|d| < 1/2$ that G_{j_1} through G_{j_p} are asymptotically independent and that G_{j_i} converges to a Gumbel distribution with mean γ, variance $\pi^2/6$, and density $\phi_g(g) = e^{-g}\exp(-e^{-g})$, $g \in \mathbb{R}$.

Proof: Giraitis et al. (2012, Theorem 5.3.1) show (a), and (b) is an immediate consequence; see Appendix to this chapter for details.

Let us return to Proposition 9.1. For consistency of the log-periodogram regression estimator, $\widehat{d}_{PR}^{(m)}$, we only require that $m \to \infty$ at a sufficiently slower rate than $T \to \infty$. For limiting normality, the bandwidth m has to grow sufficiently fast, faster than $\ln^2 T$, but more slowly than $T^{4/5}$. Note that under $m/T \to 0$, it holds for the regressor that

$$X_j = \ln\left(2\sin\frac{\lambda_j}{2}\right) \sim \ln\lambda_j, \quad j = 1, \ldots, m,$$

because (see Eq. (5.5))

$$2\sin\frac{\lambda}{2} = \lambda(1 + o(1)), \quad \lambda \to 0,$$

such that

$$\ln\left(2\sin\frac{\lambda}{2}\right) = \ln\lambda + \ln(1 + o(1)),$$

i.e.

$$\frac{\ln\left(2\sin\frac{\lambda}{2}\right)}{\ln\lambda} \to 1, \quad \lambda \to 0.$$

Consequently, $-2X_j$ may be replaced by $-2\ln\lambda_j$ in the log-periodogram regression (now with index "LP")

$$\ln I_j = \widehat{c} + \widehat{d}_{LP} Z_j + \widehat{U}_j, \quad j = 1, \ldots, m \tag{9.4}$$

where

$$Z_j = -2\ln\lambda_j.$$

Proposition 9.1 continues to hold for $\widehat{d}_{LP}$ instead of $\widehat{d}_{PR}^{(m)}$; see also Andrews and Guggenberger (2003, p. 680). From now on, we will mostly use $\widehat{d}_{LP}$ from the regression equation (Eq. (9.4)) and suppress the dependence on the bandwidth m for notational convenience. In the present semiparametric context, the symbol $\hat{d}_{\mathrm{LP}}$ may remind us of both *local periodogram* regression and *log-periodogram* regression.

From the above discussion, it is not surprising but noteworthy that Proposition 9.1 continues to hold when replacing Assumption 6.2 by a corresponding frequency domain assumption. If we maintain Assumption 6.3 where $f_y(\lambda) = \lambda^{-2d}h(\lambda)$ and restrict h to be three times continuously differentiable close to 0, then the limiting normality of Proposition 9.1 follows from Andrews and Guggenberger (2003, Theorem 2); see also their bias approximation given in Proposition 9.8.

A slightly different route to the log-periodogram regression has been taken by Robinson (1995a). In addition to the bandwidth parameter m, he introduced the so-called trimming parameter ℓ and defined

$$\overline{d}_{\mathrm{LP}}^{(\ell)} = \frac{\sum_{j=\ell+1}^{m}\left(Z_j - \overline{Z}^{(\ell)}\right)\ln I_j}{\sum_{j=\ell+1}^{m}\left(Z_j - \overline{Z}^{(\ell)}\right)^2}, \quad \overline{Z}^{(\ell)} = \frac{1}{m-\ell}\sum_{j=\ell+1}^{m}Z_j. \tag{9.5}$$

Without trimming, $\ell = 0$, we have

$$\overline{d}_{\mathrm{LP}}^{(0)} = \hat{d}_{\mathrm{LP}} \quad \text{with} \quad \overline{Z}^{(0)} = \overline{Z}.$$

Trimming is well motivated by Lemma 8.1, 9.1, and 9.2: The regression errors are asymptotically independent zero mean variables only if the harmonic frequencies are not too close to the origin. In fact, Robinson (1995a) not only introduced trimming. He also generalized the estimation procedure to a multivariate framework, where different orders of fractional integration of several processes can be estimated simultaneously. Moreover, he introduced the so-called pooling, where I_j is replaced by a neighboring, nonoverlapping sum, e.g.

$$\overline{I}_1 = I_1 + \cdots + I_J, \overline{I}_2 = I_{J+1} + \cdots + I_{2J}, \overline{I}_3 = \cdots.$$

Here we stick to the univariate case without pooling. Pooling has not been used in the applied literature to the best of our knowledge, although Robinson (1995a, Remark 3) showed that pooling reduces the variance – asymptotically.[3] In finite samples, however, pooling would require a smaller bandwidth in order to be local to the origin, which may deteriorate the approximation to the limiting distribution. Under weaker assumptions on f_y, we have the same limiting distribution as in Proposition 9.1 under trimming, where ℓ passes to infinity, too, but even more slowly than m.

---

3 A different type of pooling, though similar, has been proposed by Shimotsu and Phillips (2002). It reduces the variance asymptotically; however, at the expense of an increase in the bias, see Shimotsu and Phillips (2002, Theorem 2).

Proposition 9.2 *Let $\{y_t - \mu\}$ be a stationary, Gaussian process satisfying Assumption 6.3 with d_0, $|d_0| < 1/2$, and some further restrictions by Robinson (1995a). If*

$$\frac{m^{1/2} \ln m}{\ell} + \frac{\ell \ln^2 T}{m} + \frac{m^{1+1/(2\beta)}}{T} \to 0 \qquad (9.6)$$

for $\beta \in (0, 2]$ from Assumption 6.3, then

$$\sqrt{m}\left(\overline{d}_{\mathrm{LP}}^{(\ell)} - d_0\right) \overset{D}{\to} \mathcal{N}\left(0, \frac{\pi^2}{24}\right)$$

as $T \to \infty$.

Proof: Robinson (1995a, Theorem 3). □

Remember that β from Assumption 6.3 controls for the degree of smoothness at the origin and that $\beta = 2$ when $\{x_t\}$ is an invertible ARMA or an EXP process; see Proposition 6.6 and the following discussion.

Note that the third ratio from Eq. (9.6) implies $m = o(T^{4/5})$ since $\beta \leq 2$, which is familiar from Eq. (9.3). The second term determines that m has to grow fast enough relative to ℓ and T, while the first term imposes the minimum rate at which ℓ has to diverge. In practical applications, the choice of ℓ being large enough but not too large relative to m (which in turn has to be large enough but not too large relative to T) is very delicate. For this reason, many practitioners prefer to apply the log-periodogram regression without trimming. A major theoretical advantage of trimming, however, is that it allows to extend the limiting normality beyond the region of stationarity. The following result is due to Velasco (1999a).

Proposition 9.3 *Let $\{\Delta y_t\}$ be a stationary, Gaussian process satisfying Assumption 6.3 with $\delta_0 = d_0 - 1$ and some further restrictions by Velasco (1999a), largely paralleling those by Robinson (1995a). The estimator $\overline{d}_{\mathrm{LP}}^{(\ell)}$ is computed from the levels of $\{y_t\}$. If the first ratio from Eq. (9.6) is replaced by*

$$\frac{m^{1/2} \ln m}{\ell^{2(1-d_0)}} \to 0$$

then

$$\overline{d}_{\mathrm{LP}}^{(\ell)} \overset{p}{\to} d_0 \quad \text{for } d_0 \in \left[\frac{1}{2}, 1\right),$$

and

$$\sqrt{m}\left(\overline{d}_{\mathrm{LP}}^{(\ell)} - d_0\right) \overset{D}{\to} \mathcal{N}\left(0, \frac{\pi^2}{24}\right) \quad \text{for } d_0 \in \left[\frac{1}{2}, \frac{3}{4}\right),$$

as $T \to \infty$.

Proof: Velasco (1999a, Theorems 2 and 3). □

Due to nonstationarity, the trimming parameter ℓ has to diverge even faster than under Eq. (9.6) for stationarity. Further, Velasco (1999a) shows that the log-periodogram regression yields estimators that are no longer consistent for true memory parameters larger than 1. The case $d_0 = 1$ deserves special attention. Assuming a type II model, Phillips (2007, Theorem 3.3) analyzed the log-periodogram estimator from Eq. (9.2) and proved that $\widehat{d}_{PR}^{(m)} \overset{p}{\to} 1$ for $d_0 = 1$. Indeed, Kim and Phillips (2006, Theorem 3.1) showed that this limit holds for $d_0 > 1$, too. Hence, consistency is only given up to $d_0 = 1$.

Just as in Proposition 8.3, the range of allowed values of d is still too limited. We would like to have consistency and limiting normality for a larger range of nonstationarity. To extend the validity of the log-periodogram regression beyond $d < 1$, Velasco (1999a) suggested to use the full cosine bell taper given in Eq. (8.5). Replacing the periodogram by the tapered periodogram from Eq. (8.6), Velasco (1999a, Theorem 5) showed that the log-periodogram regression produces estimators that are consistent and limiting normal for the wider range of $[1/2, 3/2)$, although this comes at the price of an inflated limiting variance.

All the limiting results presented in this section were obtained under the assumption of Gaussianity of the data although the log-periodogram regression is not motivated by Gaussianity. On the contrary, in the previous chapter Gaussianity was assumed to motivate the likelihood function or to show efficiency, but it was not required to establish limiting normality. Robinson (2003, p. 11) commented on this as follows: "They assumed Gaussianity, and ironically, for technical reasons, this is harder to avoid when a nonlinear function of the periodogram, such as the log, is involved, than in Whittle estimation, which was originally motivated by Gaussianity." Only Velasco (2000) got around the assumption of Gaussianity when establishing limiting normality of log-periodogram regression estimators, however, for the rather limited case of trimming and pooling and tapering at the same time, and only for $0 < d < 1/2$. Finally, note that the assumption of Gaussian data is not as harmless as it may seem at first glance. It implies that the innovations $\{\varepsilon_t\}$ behind the MA representation of x_t are independent, ruling out, e.g. conditional heteroskedasticity.

9.2   Local Whittle Estimation

For motivational reasons, let $\{y_t\}$ be again fractionally integrated in the sense of Assumption 6.2. In order to obtain a local estimator, we consider the spectrum from Proposition 6.1,

$$f_y(\lambda) = 4^{-d}\sin^{-2d}\left(\frac{\lambda}{2}\right)f_x(\lambda),$$

at m harmonic frequencies in a vicinity of the origin:

$$\lambda_1 = \frac{2\pi}{T}, \dots, \lambda_m = \frac{2\pi m}{T} \rightarrow 0 \quad \text{as } T \rightarrow \infty.$$

Then we approximate

$$f_y(\lambda_j) \sim G \, \lambda_j^{-2d}, \quad j = 1, \dots, m, \quad G = f_x(0).$$

This local behavior is plugged into the Whittle likelihood now evaluated at m instead of M frequencies; see Eq. (4.31):

$$-\sum_{j=1}^{m} \left(\ln(f_y(\lambda_j)) + \frac{I_y(\lambda_j)}{f_y(\lambda_j)} \right) \approx -\sum_{j=1}^{m} \left(\ln G - 2d \ln(\lambda_j) + \frac{I_y(\lambda_j)}{G \lambda_j^{-2d}} \right)$$

$$=: \ell_m(G, d), \quad \frac{m}{T} \rightarrow 0.$$

Next, we concentrate out G. To that end, we obtain from $\partial \ell_m(G, d)/\partial G = 0$ that

$$\widehat{G}(d) = \frac{1}{m} \sum_{j=1}^{m} \frac{I_y(\lambda_j)}{\lambda_j^{-2d}}. \tag{9.7}$$

Inserting into $\ell_m(G, d)$ yields

$$\ell_m(\widehat{G}(d), d) = -m \, \ln(\widehat{G}(d)) + 2 \, d \sum_{j=1}^{m} \ln(\lambda_j) - \frac{1}{\widehat{G}(d)} \sum_{j=1}^{m} \frac{I_y(\lambda_j)}{\lambda_j^{-2d}}$$

$$= -m \, \ln(\widehat{G}(d)) + 2 \, d \sum_{j=1}^{m} \ln(\lambda_j) - m.$$

Maximizing this criterion function amounts to minimizing

$$R(d) := \ln \left\{ \frac{1}{m} \sum_{j=1}^{m} \frac{I_y(\lambda_j)}{\lambda_j^{-2d}} \right\} - \frac{2d}{m} \sum_{j=1}^{m} \ln(\lambda_j). \tag{9.8}$$

This defines the LW estimator $\widehat{d}_{\text{LW}}$; see Robinson (1995b, p. 1633)[4]:

$$\widehat{d}_{\text{LW}} = \arg \min R(d).$$

The limiting distribution has been established by Robinson (1995b) actually under weaker assumptions than outlined here.[5]

---

4 The name *local Whittle* seems to be widely spread nowadays, although Robinson (1995b) called this estimator *Gaussian semiparametric*. In fact, Künsch (1987, p. 71) was the first to suggest this estimator, however, without establishing any statistical properties.

5 No closed form solution exists for $\widehat{d}_{\text{LW}}$, which is sometimes considered as a drawback and complicates the analytical treatment. To circumvent this problem, one may alternatively consider the recursive, the so-called k-step version of the LW estimator; see Guggenberger and Sun (2006, p. 878). It shares the same limiting properties stated in Proposition 9.4 while having a closed form expression.

Proposition 9.4 *Let* $\{y_t - \mu\}$ *be a stationary process satisfying Assumption 6.3 with* d_0, $|d_0| < 1/2$, *and some further restrictions by Robinson (1995b); in particular, the innovations behind the process have to be conditionally homoskedastic. If*

$$\frac{1}{m} + \frac{m^{1+2\beta}\ln^2 m}{T^{2\beta}} \to 0, \quad \beta \in (0, 2], \tag{9.9}$$

with β *from Assumption 6.3, then*

$$\sqrt{m}\left(\hat{d}_{\text{LW}} - d_0\right) \overset{D}{\to} \mathcal{N}\left(0, \frac{1}{4}\right)$$

as $T \to \infty$.

Proof: Robinson (1995b, Theorem 2). □

Note that Eq. (9.9) does not impose a minimum rate at which the bandwidth m has to diverge. The maximum rate depends on β. The larger β, the faster may diverge m, which is not surprising since β from Assumption 6.3 regulates the smoothness of f_x at the origin. From Proposition 6.6 we know that $\beta = 2$ for conventional ARMA or EXP processes $\{x_t\}$. In that case, we may choose the bandwidth as

$$m = cT^\alpha, \quad \alpha = 4/5 - \epsilon, \quad \epsilon > 0,$$

where c is some positive finite constant, without loss of generality $c = 1$. With this choice, it holds that

$$\frac{m^5 \ln^2 m}{T^4} = \frac{T^{4-5\epsilon}}{T^{4-4\epsilon}} \left(\frac{\alpha \ln T}{T^{2\epsilon}}\right)^2 \to 0$$

as required by Eq. (9.9).

The variance given in Proposition 9.4 shows that the LW estimator is more efficient than the log-periodogram regression; see Proposition 9.1:

$$\frac{1}{4} < \frac{\pi^2}{24} \approx 0.41.$$

Indeed, Robinson (1995b, p. 1640) conjectured that the LW estimator is efficient in the class of semiparametric estimators for a given sequence m. Further, Robinson and Henry (1999) and Shao and Wu (2007) established that the LW estimator has desirable robustness properties in the presence of conditional heteroskedasticity; we omit details here.

Quite similarly to the extension by Velasco (1999a), the LW estimator has been carried to nonstationarity by Velasco (1999b). We have the following result paralleling Proposition 9.3. We now assume that $\{y_t\}$ is $I(d)$ with $d \geq 1/2$, such that $\{\Delta y_t\}$ is stationary.

Proposition 9.5 *Let $\{\Delta y_t\}$ be a stationary process satisfying Assumption 6.3 with $\delta_0 = d_0 - 1$, and some further restrictions by Velasco (1999b), largely paralleling those by Robinson (1995b). The estimator $\widehat{d}_{LW}$ is computed from the levels of $\{y_t\}$. It then holds under Eq. (9.9) that*

$$\widehat{d}_{LW} \xrightarrow{p} d_0 \ \text{for} \ \frac{1}{2} \le d_0 < 1,$$

and

$$\sqrt{m}\left(\widehat{d}_{LW} - d_0\right) \xrightarrow{D} \mathcal{N}\left(0, \frac{1}{4}\right) \ \text{for} \ \frac{1}{2} \le d_0 < \frac{3}{4},$$

as $T \to \infty$.

Proof: Velasco (1999b, Theorems 2 and 3). □

The case $d_0 \ge 1$ deserves again separate consideration. Shao and Wu (2007, Theorem 3.3) established for type I processes that $\widehat{d}_{LW} \xrightarrow{p} 1$. In particular for $d_0 = 1$, Shao and Wu (2007, Theorem 3.4) even established the limiting distribution of $\sqrt{m}(\widehat{d}_{LW} - 1)$, which, however, turned out to be nonnormal. Using appropriate data tapers, Velasco (1999b) showed that consistent, and even asymptotically normal, estimation is possible for any degree of nonstationarity.

To obtain consistency and limiting normality beyond $d = 1$ and $d = 3/4$, respectively, Shimotsu and Phillips (2005) proposed to refine the LW estimator. We do not intend to provide a deeper motivation or justification of this so-called *exact* LW (ELW) procedure, but rather present it as a recipe. It amounts to replacing $\lambda_j^{2d} I_y(\lambda_j)$ in Eq. (9.8) by the periodogram of the differenced data, $I_{\Delta^d y}(\lambda_j)$:

$$R_E(d) := \ln\left\{\frac{1}{m}\sum_{j=1}^{m} I_{\Delta^d y}(\lambda_j)\right\} - \frac{2d}{m}\sum_{j=1}^{m}\ln(\lambda_j).$$

To compute the fractional differences, it is assumed that $\{y_t\}$ is given by a type II process from Definition 6.2. The range for the true parameter d_0 can be any interval of a maximum length $9/2$, e.g.

$$d_0 \in (-1, 7/2).$$

One then defines

$$\widehat{d}_{ELW} = \arg\min R_E(d).$$

For a zero mean type II process with assumptions paralleling those by Robinson (1995b), we have the following result.

Proposition 9.6 *Let $\{y_t\}$ be a type II process of order d_0,*

$$y_t = \Delta_+^{-d_0} x_t, \quad \Delta_1 < d_0 < \Delta_2, \quad \Delta_2 - \Delta_1 \le 9/2,$$

where the spectrum f_x meets Assumption 6.3 with $\beta \in (0, 2]$ and some further restrictions by Shimotsu and Phillips (2005); in particular, the innovations behind the process have to be conditionally homoskedastic. If

$$\frac{1}{m} + \frac{m^{1+2\beta}\ln^2 m}{T^{2\beta}} + \frac{\ln T}{m^\alpha} \to 0 \text{ for any } \alpha > 0, \tag{9.10}$$

then

$$\sqrt{m}\left(\hat{d}_{\mathrm{ELW}} - d_0\right) \overset{D}{\to} \mathcal{N}\left(0, \frac{1}{4}\right)$$

as $T \to \infty$.

Proof: Shimotsu and Phillips (2005, Theorem 2.2). □

Note that Eq. (9.10) is slightly more restrictive than (9.9) in that it requires m to diverge fast enough.

Clearly, the assumption of a zero mean process is not satisfactory for applied purposes. For means different from 0, Shimotsu (2010) suggested to demean $\{y_t\}$ with an appropriate estimator $\hat{\mu}$ and to compute the ELW estimator from the demeaned data. The objective function to be minimized becomes

$$R_{E,\mu}(d) := \ln\left\{\frac{1}{m}\sum_{j=1}^{m} I_{\Delta^d(y-\hat{\mu})}(\lambda_j)\right\} - \frac{2d}{m}\sum_{j=1}^{m}\ln(\lambda_j),$$

where $I_{\Delta^d(y-\hat{\mu})}(\lambda_j)$ is the periodogram of $\{\Delta_+^d(y_t - \hat{\mu})\}$. From Eq. (7.14) we know that the usual arithmetic mean $\bar{y}$ converges fast for small values of true d_0, while it is not a reliable estimator for larger orders of integration. In particular, for the range of nonstationarity, the first sample observation is a better estimator: $\hat{\mu} = y_1$. Assuming a type II framework,

$$y_t = \mu + \Delta_+^{-d_0}x_t,$$

we have in this case as estimation error $\hat{\mu} - \mu = y_1 - \mu = x_1$. Without prior knowledge about d_0, Shimotsu (2010) put forward the following weighted estimator:

$$\hat{\mu}(d) = v(d)\,\bar{y} + (1 - v(d))\,y_1,$$

where

$$v(d) = \begin{cases} 1, & d \leq 1/2 \\ \dfrac{1 + \cos(4\pi d)}{2}, & 1/2 < d < 3/4 \\ 0, & d \geq 3/4 \end{cases}.$$

In practice, d has to be replaced by a consistent first-step estimator $\widehat{d}_{(1)}$. First, one uses an estimator $\widehat{d}_{(1)}$ independent of μ in order to get an estimator of the constant: $\widehat{\mu} = \widehat{\mu}(\widehat{d}_{(1)})$. Then, one could obtain an ELW estimator by minimizing $R_{E,\mu}(d)$, but Shimotsu (2010, Section 4.4) advised against doing so for analytical reasons. Instead, Shimotsu (2010, Section 4.1) recommended a two-step estimator. To that end, the slope and Hessian of $R_{E,\mu}(d)$ are evaluated at $\widehat{d}_{(1)}$ to compute the feasible two-step estimator:

$$\widehat{d}_{2\text{ELW}} := \widehat{d}_{(1)} - \frac{R'_{E,\mu}(\widehat{d}_{(1)})}{R''_{E,\mu}(\widehat{d}_{(1)})}.$$

Shimotsu (2010, Theorem 3) showed that the limiting result of Proposition 9.6 carries over to this two-step procedure under certain assumptions.[6] The estimation should improve upon iterating this idea:

$$\widehat{d}_{2\text{ELW}}^{(n)} := \widehat{d}_{2\text{ELW}}^{(n-1)} - \frac{R'_{E,\mu}(\widehat{d}_{2\text{ELW}}^{(n-1)})}{R''_{E,\mu}(\widehat{d}_{2\text{ELW}}^{(n-1)})} \quad \text{with } \widehat{d}_{2\text{ELW}}^{(0)} = \widehat{d}_{(1)}.$$

In any case, the two-step estimation requires of course a first-step estimator $\widehat{d}_{(1)}$ that is consistent, preferably also for $d \geq 1$. Shimotsu (2010) proposed to employ the tapered estimator by Velasco (1999b). An alternative without tapering is discussed next.

Abadir et al. (2007) proposed an extension of the LW estimator based on the extended periodogram $I_y(\cdot; d)$ defined in Eq. (8.7). The LW objective function from Eq. (9.8) is modified accordingly, i.e. the periodogram is replaced by $I_y(\cdot; d)$[7]:

$$U(d) := \ln \left\{ \frac{1}{m} \sum_{j=1}^{m} \frac{I_y(\lambda_j; d)}{\lambda_j^{-2d}} \right\} - \frac{2d}{m} \sum_{j=1}^{m} \ln(\lambda_j).$$

Although Abadir et al. (2007) allowed for a parameter range from $-3/2$ to an arbitrary degree of integration (which requires to generalize the extended periodogram), we focus again on the empirically most relevant range:

$$d \in \left(-\frac{1}{2}, \frac{1}{2} \right) \cup \left(\frac{1}{2}, \frac{3}{2} \right).$$

---

6 Shimotsu (2010) also allowed to correct more generally for polynomial time trends.
7 Note that $U(d) = \widetilde{U}(d)$ with

$$\widetilde{U}(d) = \ln \left\{ \frac{1}{m} \sum_{j=1}^{m} \frac{I_y(\lambda_j; d)}{j^{-2d}} \right\} - \frac{2d}{m} \sum_{j=1}^{m} \ln(j),$$

where $\widetilde{U}(d)$ is the expression originally proposed in Abadir et al. (2007, Eq. (2.6)).

The minimizer of $U(d)$ is then called *fully extended local Whittle* estimator, in short, FELW:

$$\widehat{d}_{\text{FELW}} := \arg\min U(d).$$

It has the nice property that under nonstationarity $(1/2 < d < 3/2)$, the observable process may be driven by an additional time trend that does not have to be accounted for; see Abadir et al. (2007, Eq. (2.26)). Hence, the model is under nonstationarity

$$\Delta y_t = \mu + \Delta^{-\delta} x_t, \quad |\delta| < \frac{1}{2},$$

if $y_t \sim I(d)$ with $d = 1 + \delta$. The stationary case is covered by

$$y_t = \mu + \Delta^{-\delta} x_t, \quad |\delta| < \frac{1}{2}.$$

We summarize the following limiting result:

Proposition 9.7 *Let* $\{\xi_t\} = \{\Delta^{-\delta_0} x_t\}$ *be a stationary fractionally integrated process with spectrum* f_ξ, *and*

$$\Delta^D y_t = \mu + \xi_t, \quad D \in \{0, 1\}, \quad |\delta_0| < \frac{1}{2},$$

where the spectrum f_ξ *meets Assumption 6.3 with* $\beta \in (0, 2]$ *and some further restrictions by Abadir et al. (2007). If*

$$\frac{1}{m} + \frac{m^{1+2\beta}}{T^{2\beta}} \to 0, \quad \beta \in (0, 2], \tag{9.11}$$

then

$$\sqrt{m} \left(\widehat{d}_{\text{FELW}} - d_0 \right) \xrightarrow{D} \mathcal{N}\left(0, \frac{1}{4}\right)$$

as $T \to \infty$, *where* $d_0 = D + \delta_0$.

Proof: Abadir et al. (2007, Corollary 2.1). □

Remember once more from Proposition 6.6 that invertible and stationary ARMA processes $\{x_t\}$, or EXP processes, imply $\beta = 2$ in Assumption 6.3. Further, it is worthwhile to stress again that Proposition 9.7 allows for a linear trend in the data if $D = 1$:

$$y_t = \mu t + \sum_{j=1}^{t} \xi_j.$$

This is a distinct feature not held by other estimators treated here; see Abadir et al. (2007, p. 1359) for an explanation.

9.3 Finite Sample Approximation

One goal of semiparametric inference is hypothesis testing about the order of fractional integration or the computation of confidence intervals relying on approximate normality. The feasible tests from the previous sections are approximate and rely on limiting normality as the bandwidth passes to infinity. However, there are different ways of standardization, resulting in approximations to the limiting distribution of different quality. Moreover, the approximation of normality depends on the choice of m, as do the bias and the variance of the estimator. Hence, the bandwidth selection is of particular importance in applied work. Before we turn to it in Section 9.5, we discuss different ways of approximating the limiting variance.

The original proposal on how to compute standard errors for $\widehat{d}_{\mathrm{PR}}^{(m)}$ by Geweke and Porter-Hudak (1983, Theorem 2) was to use the empirical standard errors s_{emp} building on

$$
s_{\mathrm{emp}}^2 = \frac{\widehat{\sigma}^2}{4 \sum_{j=1}^{m} (X_j - \overline{X}_m)^2}, \qquad \widehat{\sigma}^2 = \frac{1}{m} \sum_{j=1}^{m} \widehat{U}_j^2,
$$

with residuals $\widehat{U}_j$ from Eq. (9.1). The asymptotic standard error is of course from Proposition 9.1:

$$
s_{\mathrm{asy}}^2 = \frac{\pi^2/24}{m} = \frac{\pi^2/6}{4\,m}.
$$

The denominators are linked by, see Hurvich and Beltrao (1994, Lemma 1, p. 301),

$$
\frac{1}{m} \sum_{j=1}^{m} (X_j - \overline{X}_m)^2 = \frac{1}{m} \sum_{j=1}^{m} v_j^2 + o(1) = 1 + o(1), \tag{9.12}
$$

where

$$
v_j := \ln j - \frac{1}{m} \sum_{\ell=1}^{m} \ln \ell,
$$

such that

$$
\sum_{j=1}^{m} (X_j - \overline{X}_m)^2 \sim m.
$$

Hence, a third standard error approximates between s_{emp}^2 and s_{asy}^2, replacing $\widehat{\sigma}^2$ by $\pi^2/6$:

$$
s_{\mathrm{app}}^2 = \frac{\pi^2/24}{\sum_{j=1}^{m} (X_j - \overline{X}_m)^2}.
$$

Hassler et al. (2006, Eq. (7)) provided experimental evidence that the approximation to the standard normal distribution is best when using a t-type statistic relying on the approximate standard error, i.e.

$$t_{\text{app}} = \frac{\widehat{d}_{\text{PR}}^{(m)} - d_0}{s_{\text{app}}}.$$

Building on regression Eq. (9.4) with $Z_j = -2 \ln \lambda_j \sim -2X_j$, one may of course analogously studentize $\widehat{d}_{\text{LP}}$ as

$$t = \frac{\widehat{d}_{\text{LP}} - d_0}{\pi / \sqrt{6}} \sqrt{\sum_{j=1}^{m} (Z_j - \overline{Z})^2},$$

which is recommended by Andrews and Guggenberger (2003, p. 686) to obtain good finite sample accuracy. Because of Eq. (9.12) we may alternatively consider

$$t = \frac{\widehat{d}_{\text{LP}} - d_0}{\pi / \sqrt{24}} \sqrt{\sum_{j=1}^{m} v_j^2},$$

with v_j from Eq. (9.12). In fact, it is simple to verify the following exact equality:

$$\widehat{d}_{\text{LP}} = -\frac{1}{2} \frac{\sum_{j=1}^{m} v_j \ln I_j}{\sum_{j=1}^{m} v_j^2}.$$

We now turn to the LW estimator, where an alternative variance approximation to $1/4m$ (stemming from Proposition 9.4) has been discussed. It can be motivated by the analog of the Fisher information in ML estimation. Consider the second derivative of the objective function from Eq. (9.8). When evaluated at $\widehat{d}_{\text{LW}}$, we learn from Robinson (1995b, Eq. (4.10)) that[8]

$$\frac{\partial^2 R(\widehat{d}_{\text{LW}})}{\partial d^2} = \frac{4}{m} \sum_{j=1}^{m} v_j^2 + o_p(1) \to 4,$$

where the limit 4 arises from Eq. (9.12). With the Fisher information analogy (see, e.g. Theorem 8.1), Hassler and Olivares (2013) suggested to build tests or confidence intervals on

$$\sqrt{m} \, (\widehat{d}_{\text{LW}} - d_0) \sim \mathcal{N}\left(0, \left(\partial^2 R(\widehat{d}_{\text{LW}})/\partial d^2\right)^{-1}\right),$$

or

$$2\sqrt{\sum_{j=1}^{m} v_j^2} \, (\widehat{d}_{\text{LW}} - d_0) \sim \mathcal{N}(0, 1). \tag{9.13}$$

---

8 The same holds true for $\partial^2 R_E / \partial d^2$ by Shimotsu and Phillips (2005, p. 1916) for the exact LW estimator.

They reported experimental evidence that Eq. (9.13) outperforms the approximation from Proposition 9.4 in terms of size distortions. A closely related proposal can be found in Hurvich and Chen (2000, p. 163) for their tapered estimator where v_j is replaced by $X_j - \overline{X}$ with $X_j \sim \ln(2\pi j/T)$ from Eq. (8.10).

9.4 Bias Approximation and Reduction

The choice of the bandwidth balances the trade-off between bias and variance as we will quantify next. Let us now maintain Assumption 6.3, i.e. $f_y(\lambda) = \lambda^{-2d}h(\lambda)$. An early bias approximation for the LW estimator was provided on heuristic grounds by Henry and Robinson (1996); see also the footnote in Andrews and Sun (2004, p. 584). Given the Taylor expansion (with $h'(0) = 0$)

$$\ln(h(\lambda_j)) = \ln(h(0)) + 0 + \frac{h''(0)}{h(0)} \frac{\lambda_j^2}{2} + o(\lambda_j^2),$$

it is not surprising that the term $h''(0)/h(0)$ shows up in the rigorous bias approximation given in the next proposition.

Proposition 9.8 *Let $\{y_t - \mu\}$ be a stationary process satisfying Assumption 6.3 for $\beta = 2$ with spectrum $f_y(\lambda) = \lambda^{-2d_0}h(\lambda)$, $|d_0| < 1/2$. Let h be twice differentiable around the origin with a bounded second derivative and*

$$b_2 := \frac{h''(0)}{h(0)}.$$

Further, assume

$$m \to \infty \quad and \quad m = O\left(T^{4/5}\right).$$

It then holds under some additional assumptions made in Andrews and Guggenberger (2003) and Andrews and Sun (2004) for the log-periodogram regression estimator from Eq. (9.4) and the LW estimator from Eq. (9.8), respectively, that $E(\widehat{d}_{\text{LP}})$ and $E(\widehat{d}_{\text{LW}})$ equal

$$d_0 - \frac{2\pi^2}{9} b_2 \frac{m^2}{T^2} + o(1).$$

Proof: By our assumptions, $h''(\lambda) - h''(0)$ is bounded in absolute value. Hence, we can choose $s = 2$ in the notation of Andrews and Guggenberger (2003) and Andrews and Sun (2004). For $r = 0$ they cover the log-periodogram regression estimator and the LW estimator, respectively. Then the bias approximation follows from Andrews and Guggenberger (2003, Theorem 1) and Andrews and Sun (2004, Corollary 1); note that Andrews and Guggenberger (2003) assume Gaussianity of the process. □

An analogous bias approximation is available for the Geweke–Porter-Hudak version of the log-periodogram regression estimator $\widehat{d}_{\mathrm{PR}}^{(m)}$ from Eq. (9.2). Let us now maintain Assumption 6.2 with spectrum f_x from $\Delta^d y_t = x_t$. Assuming $E(U_j) = 0$ for the errors from Eq. (9.1), Hassler (1993, Eq. (4)) approximated $E\left(\widehat{d}_{\mathrm{PR}}^{(m)}\right) - d$ heuristically as

$$-\frac{1}{2}\frac{\sum_{j=1}^{m}(X_j - \overline{X}_m)\ln(f_x(\lambda_j))}{\sum_{j=1}^{m}(X_j - \overline{X}_m)^2}.$$

The same heuristic formula follows from Agiakloglou et al. (1993, Eq. (2.2)), and it does a satisfactory job in explaining the finite sample bias observed in computer experiments. It is clear that a major source for the bias is the fact that $\ln(f_x(\lambda_j))$ is in general not constant for finite T. A more rigorous bias approximation was obtained by Hurvich et al. (1998, Theorem 1) under Definition 6.1 with spectrum f_x and $f_x'(0) = 0$. We wish to analyze $\widehat{d}_{\mathrm{LP}}$ and $\widehat{d}_{\mathrm{LW}}$ under this assumption, too. To this end, we simply have to equate $(4\sin^2 \lambda/2)^{-d}f_x(\lambda)$ with $\lambda^{-2d}h(\lambda)$ and combine Proposition 9.8 with 6.6.

Corollary 9.1 *Let $\{y_t - \mu\}$ be a stationary process satisfying Assumption 6.2, $y_t - \mu = (1-L)^{-d_0}x_t$, where $\{x_t\}$ possesses a three times continuously differentiable spectrum f_x at zero and in a neighborhood of the origin with $0 < f_x(0) < \infty$ and $f_x'(0) = 0$. Further, assume*

$$m \to \infty \quad \text{and} \quad m = O\left(T^{4/5}\right).$$

It then holds under some additional assumptions made in Hurvich et al. (1998), Andrews and Guggenberger (2003), and Andrews and Sun (2004) that

$$E\left(\widehat{d}_{\mathrm{PR}}^{(m)}\right) = d_0 - \left[\frac{2\pi^2}{9}\frac{f_x''(0)}{f_x(0)}\frac{m^2}{T^2} + o\left(\frac{m^2}{T^2}\right)\right] + O\left(\frac{\ln^3 m}{m}\right) \tag{9.14}$$

and $E(\widehat{d}_{\mathrm{LP}})$ and $E(\widehat{d}_{\mathrm{LW}})$ equal

$$d_0 - \frac{2\pi^2}{9}\left(\frac{f_x''(0)}{f_x(0)} + \frac{d_0}{6}\right)\frac{m^2}{T^2} + o(1).$$

Proof: The result in Eq. (9.14) is obtained by Hurvich et al. (1998, Theorem 1). The second and third results follow from Propositions 9.8 and 6.6 with

$$b_2 = \frac{f_x''(0)}{f_x(0)} + \frac{d_0}{6}.$$

$\square$

The bias in (9.14) is made up by two terms. The first one in squared brackets arises from the Taylor expansion of $\ln(f_x(\lambda_j))$ (see Hurvich et al. (1998, Lemma 1)), while the second big-O term is due to $E(U_j) \neq 0$ (see Hurvich et al.

(1998, Lemma 8)). Clearly, the larger m (relative to T), the larger is the bias as long as $f_x''(0) \neq 0$. Also the sign of the bias is rather intuitive: If f_x has a relative maximum at the origin with $f_x''(0) < 0$, then the true d_0 will be overestimated. It is noteworthy that $\widehat{d}_{PR}^{(m)}$ has a somewhat simpler bias term than the other two estimators in that it does not depend on d_0. Since the bias vanishes under the assumptions of Proposition 9.1, the limiting distribution given there for $\widehat{d}_{PR}^{(m)}$ continues to hold for $\widehat{d}_{LP}$.

We now turn to refinements of the log-periodogram regression and LW estimation, where the rate at which the bias converges to 0 is made faster, while the variance is blown up only by a constant factor at the same time. Andrews and Guggenberger (2003) pioneered this for the log-periodogram regression. Let us return to Assumption 6.3 where $f_y(\lambda) = \lambda^{-2d} h(\lambda)$. Andrews and Guggenberger (2003) suggested to treat $\ln h(\lambda_j)$ not as approximately constant, but to approximate it by a Taylor expansion. Therefore, one simply extends the regression in Eq. (9.4) as follows: Regress $\ln I_j$ on a constant intercept and Z_j, and on $\lambda_j^2, \lambda_j^4, \dots, \lambda_j^{2r}$; note that odd powers of λ_j are not required since h is an even function. The estimator associated with Z_j is now written as $\widetilde{d}_{LP}^{(r)}$:

$$\ln I_j = \widetilde{d}_{LP}^{(r)} Z_j + \sum_{s=0}^{r} \widetilde{\beta}_s \lambda_j^{2s} + \widetilde{U}_j^{(r)}, \quad j = 1, \dots, m. \tag{9.15}$$

Hence, $r = 0$ recovers the usual log-periodogram regression: $\widetilde{d}_{LP}^{(0)} = \widehat{d}_{LP}$. Andrews and Guggenberger (2003) recommended rather small values of r such as $r = 1$ or $r = 2$. The next proposition gives the approximate bias and the limiting distribution of this bias-reduced log-periodogram regression estimator; to simplify the proposition we maintain stricter assumptions on h than necessary.

Proposition 9.9 *Let $\{y_t - \mu\}$ be a stationary, Gaussian process satisfying Assumption 6.3 for $\beta = 2$ with spectrum $f_y(\lambda) = \lambda^{-2d_0} h(\lambda)$, $|d_0| < 1/2$. Let h be $(2 + 2r)$ times continuously differentiable around the origin, and*

$$m = o\left(T^{(4+4r)/(5+4r)}\right)$$

as $T \to \infty$. It then holds for the bias-reduced log-periodogram regression estimator that

$$\mathrm{E}\left(\widetilde{d}_{LP}^{(r)}\right) - d_0 = O\left(\frac{m^{2+2r}}{T^{2+2r}}\right),$$

$$\sqrt{m}\left(\widetilde{d}_{LP}^{(r)} - d_0\right) \xrightarrow{D} \mathcal{N}\left(0, \frac{\pi^2}{24} c_r\right),$$

where c_r is given in Andrews and Guggenberger (2003, Eq. (3.11)).

Proof: Andrews and Guggenberger (2003, Theorems 1 and 2). □

Of course, $c_0 = 1$, while, e.g. $c_1 = 2.25$ and $c_2 = 3.52$. This variance inflation with growing r is the price for the faster rate at which the bias vanishes. Comparing Proposition 9.9 with 9.8, we have

$$\mathrm{E}\left(\widetilde{d}_{\mathrm{LP}}^{(r)}\right) - d_0 = O\left(\mathrm{E}\left(\widehat{d}_{\mathrm{LP}}\right) - d_0\right) \frac{m^{2r}}{T^{2r}},$$

such that the bias is reduced by the factor m^{2r}/T^{2r}.

Andrews and Sun (2004) suggested a similar bias reduction for the LW estimator. The idea is again to approximate the log-spectrum of the short memory component by an even polynomial of order $2r$:

$$\ln f_x(\lambda) \approx \ln G - P_r(\lambda; g_1, \dots, g_r), \quad G = f_x(0),$$

where $P_{r,j}$ will also be short for

$$P_{r,j} = P_r(\lambda_j; g_1, \dots, g_r) = \sum_{k=1}^{r} g_k \lambda_j^{2k}.$$

For $r = 0$, we adopt the usual convention that $P_0 = 0$. Now, remember the LW likelihood $\ell_m(G, d)$ above Eq. (9.8):

$$\ell_m(G, d) = -\sum_{j=1}^{m} \left(\ln\left(\frac{G}{\lambda_j^{2d}}\right) + \frac{I_y(\lambda_j)}{G\lambda_j^{-2d}}\right).$$

Replacing G by $G \exp(-P_r(\lambda_j; g_1, \dots, g_r))$ yields

$$\ell_m^{(r)}(G, d, g_1, \dots, g_r) := -\sum_{j=1}^{m} \left(\ln\left(\frac{G}{\lambda_j^{2d}} e^{-P_{r,j}}\right) + \frac{I_y(\lambda_j)}{G\lambda_j^{-2d} e^{-P_{r,j}}}\right).$$

Following Andrews and Sun (2004), the maximization of $\ell_m^{(r)}$ amounts to a minimization of $R^{(r)} = R^{(r)}(d, g_1, \dots, g_r)$ with

$$R^{(r)} := \ln\left\{\frac{1}{m}\sum_{j=1}^{m} \frac{I_y(\lambda_j)}{\lambda_j^{-2d}} \exp(P_{r,j})\right\} - \frac{1}{m}\sum_{j=1}^{m}(P_{r,j} + 2d\ln(\lambda_j)). \tag{9.16}$$

The so-called local polynomial Whittle estimator $\widetilde{d}_{\mathrm{LW}}^{(r)}$ is then given by

$$\left(\widetilde{d}_{\mathrm{LW}}^{(r)}, \widetilde{g}_1, \dots, \widetilde{g}_r\right)' := \arg\min R^{(r)}(d, g_1, \dots, g_r).$$

For $r = 0$, we have $\widetilde{d}_{\mathrm{LW}}^{(0)} = \widehat{d}_{\mathrm{LW}}$ of course.

Without the assumption of Gaussianity of the data, Andrews and Sun (2004) established the following results paralleling Proposition 9.9.

Proposition 9.10 *Let $\{y_t - \mu\}$ be a stationary process satisfying Assumption 6.3 for $\beta = 2$ with spectrum $f_y(\lambda) = \lambda^{-2d_0} h(\lambda)$, $|d_0| < 1/2$. Let h be $(2 + 2r)$ times* *continuously differentiable around the origin, and*

$$\frac{m^{2r+1/2}}{T^{2r}} \to \infty, \quad and \ m = O(T^{(4+4r)/(5+4r)})$$

as $T \to \infty$. It then holds under some additional assumptions made in Andrews and Sun (2004) for the local polynomial Whittle estimator that

$$\mathrm{E}\left(\widetilde{d}_{\mathrm{LW}}^{(r)}\right) - d_0 = O\left(\frac{m^{2+2r}}{T^{2+2r}}\right),$$

$$\sqrt{m}\left(\widetilde{d}_{\mathrm{LW}}^{(r)} - d_0\right) \overset{D}{\to} \mathcal{N}\left(0, \frac{1}{4} c_r\right),$$

where c_r is given in Andrews and Guggenberger (2003, Eq. (3.11)).

Proof: Andrews and Sun (2004, Theorem 2). □

The numbers c_r from Propositions 9.9 and 9.10 are the same. Therefore, just as in the case $r = 0$, it generally holds that the limiting variance of $\widehat{d}_{\mathrm{LP}}^{(r)}$ exceeds that of $\widetilde{d}_{\mathrm{LW}}^{(r)}$. Further, the comments made with respect to bias and variance of the log-periodogram regression below Proposition 9.9 hold for the local polynomial Whittle estimator, too. It has been assessed for finite samples that the bias reduction of $\widetilde{d}_{\mathrm{LP}}^{(r)}$ and $\widetilde{d}_{\mathrm{LW}}^{(r)}$ is quite effective. In an extensive simulation study, Nielsen and Frederiksen (2005, p. 439) concluded that "the bias reduced log-periodogram regression and local polynomial Whittle estimators [...] even outperform the correctly specified time domain parametric methods." Furthermore, Guggenberger and Sun (2006) suggested bias-reduced estimators without variance inflation. A serious limitation of all these bias-reduced estimators is, however, that their properties are only established for $|d| < 1/2$. In practice, many time series may be more persistent. If one knew this for sure a priori, one could of course integer difference them to reach the range of $|d| < 1/2$; if, however, the true d is smaller than $1/2$, then the differenced data will be overdifferenced, falling again out of the range $|d| < 1/2$.

9.5 Bandwidth Selection

From Sections 9.1 and 9.2, we know that the variances of both, the local periodogram regression estimation and the LW estimation, are of order $1/m$. This demonstrates that a smaller bias (due to a smaller m) comes at the price of a larger variance. This makes the choice of the bandwidth so delicate and practically relevant at the same time.

In order to determine a reliable bandwidth, many empirical researchers plot semiparametric estimates such as $\widehat{d}_{\mathrm{PR}}^{(m)}$ against different values of m; see also Taqqu and Teverovsky (1996) for graphical bandwidth selection. Ideally, three regimes can be observed: First, for small values of m, the estimates are highly variable; second, the plot of the estimates becomes approximately flat for medium bandwidths; third, with m getting large, the estimates start to increase or decrease with m because of the bias due to a short memory component. In such an ideal situation, it is recommended to select m from the middle regime. If there are only two regimes observed, i.e. the estimates of d remain more or less constant with growing m approaching $T/2$, then this indicates the absence of additional short memory and the process may be modeled as fractionally integrated noise, FIN(d). In practice, however, such ideal situations will rarely be met even with large sample sizes. Therefore, some applied workers prefer the so-called automatic or data-driven bandwidth selection, where one typically uses an optimality criterion like minimizing the mean squared error (MSE) to determine m.

For the rest of this section, we assume a fractionally integrated model from Assumption 6.2 with spectrum $f_y(\lambda) = (4\sin^2 \lambda/2)^{-d} f_x(\lambda)$ meeting the additional assumptions of Proposition 9.1 on the derivatives of f_x, see also Corollary 9.1.[9] Robinson (1994c) pioneered optimal bandwidth selection in the context of long memory. In particular, he considered the so-called averaged periodogram estimator, which we do not discuss in this book.[10] Henry and Robinson (1996) determined the optimal bandwidth for the LW estimator on heuristic grounds; see also Henry (2001). Combining Corollary 9.1 with Proposition 9.4, we obtain as approximate MSE:

$$\text{MSE}\left(\hat{d}_{\text{LW}}\right) = \left[-\frac{2\pi^2}{9}\left(\frac{f_x''(0)}{f_x(0)} + \frac{d}{6}\right)\right]^2 \frac{m^4}{T^4} + \frac{1}{4\,m}.$$

The asymptotically optimal (minimizing the MSE) bandwidth choice for the LW estimator results as

$$m_{\text{LW}} = \left(\frac{3}{4\pi}\right)^{4/5}\left(\frac{1}{\tau + d/12}\right)^{2/5} T^{4/5},$$

where

$$\tau = \frac{f_x''(0)}{2\,f_x(0)}, \tag{9.17}$$

which reproduces Henry and Robinson (1996, Eq. (1.6)). Next, we consider the log-periodogram regression estimator from Eq. (9.2). Given the asymptotic bias

---

9 Alternatively, we could maintain Assumption 6.3 and use Proposition 9.8 instead of Corollary 9.1.

10 Robinson (1994d) suggested this estimator of d based on the averaged periodogram and established its consistency under adequate assumptions. Lobato and Robinson (1996), however, showed that its limiting distribution is only normal for the range of $0 < d < 1/4$ and nonnormal for $1/4 < d < 1/2$. Moreover, the estimator requires a further user-chosen parameter on top of the bandwidth selection. For these reasons the averaged periodogram estimator is not much used nowadays.

and variance, one has for the MSE:

$$\text{MSE}\left(\widehat{d}_{\text{PR}}^{(m)}\right) = \left[-\frac{2\pi^2}{9}\frac{f_x''(0)}{f_x(0)}\right]^2 \frac{m^4}{T^4} + \frac{\pi^2}{24\,m}.$$

Minimizing this expression yields as optimal bandwidth (see Hurvich et al. (1998, Eq. (9)) and also the discussion in Hurvich and Deo (1999)):

$$m_{\text{PR}} = \left(\frac{27}{512\,\pi^2}\right)^{1/5} \left(\frac{1}{\tau}\right)^{2/5} T^{4/5}.$$

Both bandwidths depend on the unknown parameter τ, and m_{LW} is further plagued by the unknown d. We can show that the latter is also true if the log-periodogram regression is computed from Eq. (9.4). Combining Corollary 9.1 with Proposition 9.1, we have

$$\text{MSE}\left(\widehat{d}_{\text{LP}}\right) = \left[-\frac{2\pi^2}{9}\left(\frac{f_x''(0)}{f_x(0)} + \frac{d}{6}\right)\right]^2 \frac{m^4}{T^4} + \frac{\pi^2}{24\,m}.$$

Minimizing this term results in

$$m_{\text{LP}} = \left(\frac{27}{512\,\pi^2}\right)^{1/5} \left(\frac{1}{\tau + d/12}\right)^{2/5} T^{4/5},$$

which depends on d again.

A feasible version for the above bandwidths requires an estimate of τ from Eq. (9.17). To that end, Hurvich and Deo (1999) suggested a log-periodogram regression similar to the bias-reduced regression from Eq. (9.15). However, they sticked to the original regressor X_j from Eq. (9.1) and suggested $r = 1$. Clearly, this is motivated by the following Taylor approximation, where we maintain $f_x'(0) = 0$,

$$\ln f_x(\lambda) \approx \ln f_x(0) + \frac{f_x''(0)}{2\,f_x(0)}\,\lambda^2,$$

which is substituted into

$$\ln f_y(\lambda) = -2d\,\ln\left(2\,\sin\,\lambda/2\right) + \ln f_x(\lambda).$$

This motivates the following regression estimated by OLS:

$$\ln I_y(\lambda_j) = \widetilde{c} - 2\widetilde{d}_{\text{PR}}^{(1)}X_j + \widetilde{\tau}\,\lambda_j^2 + \widetilde{\varepsilon}_j, \quad j = 1, \dots, L = A\,T^a, \tag{9.18}$$

where the positive constants A and $a < 1$ have to be picked adequately to estimate τ consistently; see Hurvich and Deo (1999, Lemma 1). In particular, Hurvich and Deo (1999) considered $a = 6/7$ and $A = 1/4$ in Monte Carlo experiments. For further elaborations on this approach to estimate τ, see Andrews and Guggenberger (2003, Theorem 4). Replacing τ in the optimal bandwidth formula m_{PR} by $\widetilde{\tau}$ results in

$$\widehat{m}_{\text{PR}} = \left(\frac{27}{512\,\pi^2}\right)^{1/5} \left(\frac{1}{\widetilde{\tau}}\right)^{2/5} T^{4/5}.$$

Such a data-driven or automatic bandwidth choice provides a so-called plug-in estimator $\widehat{d}_{PR}^{(\widehat{m}_{PR})}$.

The MSE-minimizing bandwidths m_{LW} and m_{LP} depend on both, τ and d. Following Henry (2001), one may employ an iterative procedure: Use $\widehat{m}^{(k)}$ to estimate $\widehat{d}^{(k)}$, which is used to determine $\widehat{m}^{(k+1)}$, until convergence,

$$\widehat{m}_{LW}^{(k+1)} = \left(\frac{3}{4\pi}\right)^{4/5} \left(\frac{1}{\widetilde{\tau} + \widehat{d}^{(k)}/12}\right)^{2/5} T^{4/5},$$

and

$$\widehat{m}_{LP}^{(k+1)} = \left(\frac{27}{512\,\pi^2}\right)^{1/5} \left(\frac{1}{\widetilde{\tau} + \widehat{d}^{(k)}/12}\right)^{2/5} T^{4/5}.$$

As initial estimate, one may choose

$$\widehat{d}^{(0)} \quad \text{with} \quad \widehat{m}^{(0)} = T^{4/5}.$$

Note that an alternative method of estimating τ from Eq. (9.17) had been pioneered by Delgado and Robinson (1996, Eq. (4.3)); we omit details.

Unfortunately, computer experiments by, e.g., Hurvich and Deo (1999), Henry (2001), or Andrews and Guggenberger (2003) have shed some doubt on the usefulness of plug-in bandwidth choice in practice.

The above plug-in methods are not fully automatic in that they require an initial choice of the bandwidth. If m is determined exclusively from the data, then the resulting estimator has been called adaptive; see, e.g. Giraitis et al. (2000), Iouditsky et al. (2001), Andrews and Sun (2004), and Guggenberger and Sun (2006).

We conclude this section with a comment on optimal bandwidth and limiting normality. Note that the MSE-minimizing m violates $m = o(T^{4/5})$, ensuring the limiting distributions according to Propositions 9.1 and 9.4. If m is of order $T^{4/5}$, and not of smaller order, the asymptotic bias does not vanish anymore. It is, however, possible to correct for this. Let us now choose

$$m = B\,T^{4/5} \tag{9.19}$$

for some positive constant B, which is typically taken from the MSE-minimizing formulae of m_{LW}, m_{PR}, or m_{LP}. Under the assumptions of Proposition 9.1, it holds by Hurvich and Deo (1999, Theorem 2) that

$$\sqrt{m}\left(\widehat{d}_{PR}^{(m)} - d_0\right) \xrightarrow{D} \mathcal{N}\left(-\frac{4\pi^2}{9}B^{5/2}\tau, \frac{\pi^2}{24}\right), \tag{9.20}$$

where τ is from Eq. (9.17) again. From that we obtain under Eq. (9.19) the following limiting result:

$$\sqrt{m}\left(\widehat{d}_{PR}^{(m)} - d_0 + \frac{4\pi^2}{9}\frac{B^{5/2}}{\sqrt{m}}\,\tau\right) \xrightarrow{D} \mathcal{N}\left(0, \frac{\pi^2}{24}\right).$$

To obtain feasible tests or confidence intervals, a consistent estimator of τ has to be plugged in. Resulting tests or confidence intervals have been labeled as bias-corrected log-periodogram regression inference. Similarly, it holds for the bias and the limiting distribution of $\widehat{d}_{\mathrm{LP}}$ under Eq. (9.19) that (see Andrews and Guggenberger (2003, p. 687), and Guggenberger and Sun (2006, Eq. (2.9)) for a typo-correction)

$$\sqrt{m}\left(\widehat{d}_{\mathrm{LP}} - d_0 + \frac{4\pi^2}{9}\frac{B^{5/2}}{\sqrt{m}}\left(\tau + \frac{d_0}{12}\right)\right) \overset{D}{\to} \mathcal{N}\left(0, \frac{\pi^2}{24}\right).$$

Note that also for $\widehat{d}_{\mathrm{LP}}$, we maintained here the model from Assumption 6.2; see also Corollary 9.1. For the LW estimator it follows from Andrews and Sun (2004, Theorem 2) under the conditions of Corollary 9.1 that

$$\sqrt{m}\left(\widehat{d}_{\mathrm{LW}} - d_0\right) \overset{D}{\to} \mathcal{N}\left(-\frac{4\pi^2}{9}B^{5/2}\left(\tau + \frac{d_0}{12}\right), \frac{1}{4}\right).$$

See also Guggenberger and Sun (2006, Eq. (2.10)). Hence, choosing the optimal rate $m = B\, T^{4/5}$ for some B combined with a consistent estimation of τ in order to remove the asymptotic bias allows for standard normal inference[11]:

$$2\sqrt{m}\left(\widehat{d}_{\mathrm{LW}} - d_0\right) + \frac{8\pi^2}{9}B^{5/2}\left(\tau + \frac{d_0}{12}\right) \overset{D}{\to} \mathcal{N}(0, 1).$$

For a given confidence level $1 - \alpha$ and $z_{1-\alpha/2}$ being a quantile of the standard normal distribution, an approximate confidence interval is readily available from $\widehat{d}_{\mathrm{LW}}$: The lower endpoint becomes

$$\mathrm{LE}_{1-\alpha} = \left(\widehat{d}_{\mathrm{LW}} + \frac{4\pi^2 B^{5/2}}{9\sqrt{m}}\tau - \frac{z_{1-\alpha/2}}{2\sqrt{m}}\right)\left(1 - \frac{\pi^2 B^{5/2}}{27\sqrt{m}}\right)^{-1},$$

where we assumed that $\pi^2 B^{5/2}/27\sqrt{m}$ is less than 1 and the upper endpoint $\mathrm{UE}_{1-\alpha}$ is obtained by simply replacing $-z_{1-\alpha/2}$ by $z_{1-\alpha/2}$:

$$\mathrm{UE}_{1-\alpha} = \left(\widehat{d}_{\mathrm{LW}} + \frac{4\pi^2 B^{5/2}}{9\sqrt{m}}\tau + \frac{z_{1-\alpha/2}}{2\sqrt{m}}\right)\left(1 - \frac{\pi^2 B^{5/2}}{27\sqrt{m}}\right)^{-1}.$$

It then holds that

$$\lim_{m\to\infty} \mathrm{P}(\mathrm{LE}_{1-\alpha} \le d_0 \le \mathrm{UE}_{1-\alpha}) = 1 - \alpha,$$

and that

$$\lim_{m\to\infty} \mathrm{P}(\mathrm{LE}_{1-\alpha} > d_0) = \lim_{m\to\infty} \mathrm{P}(\mathrm{UE}_{1-\alpha} < d_0) = \frac{\alpha}{2}.$$

---

11 A corresponding result is available for the fully extended LW estimator, too; see Abadir et al. (2007, Corollary 2.1).

Similarly, one may construct confidence intervals from $\hat{d}_{\mathrm{LP}}$ or $\hat{d}_{\mathrm{PR}}^{(m)}$. The asymptotic idea to combine the MSE-minimizing bandwidth with a bias correction to obtain limiting normality, however, may not work so well in finite samples according to the computer simulations by Hurvich and Deo (1999) and Andrews and Guggenberger (2003).

9.6 Global Estimators

So far the estimators in this chapter are *local* in that they all evaluate the periodogram only in a vicinity of the origin. Correspondingly, when assuming $\Delta^d y_t = x_t$, the above propositions require assumptions with respect to the spectrum of $\{x_t\}$ only close to the origin; see Assumption 6.3. In this section, we still refrain from parametric assumptions, but we do maintain assumptions on f_x over the full frequency range from 0 to π. This is required because the periodogram will be evaluated over this full range. Consequently, we adopt the terminology by Moulines and Soulier (2003, p. 259) and call these estimators *global*. The same feature has also been called *broadband semiparametric estimation*; see Hurvich and Brodsky (2001, p. 223) and also Moulines and Soulier (1999).

The first procedure by Bhansali et al. (2006) builds on the parametric Whittle estimator from Section 8.4. The idea is to approximate the unknown structure of $\{x_t\}$ by an AR(k) model, where the autoregressive order $k = k_T$ increases with the sample size. The approximate model for the long memory process hence is ARFI(k, d),

$$\left(1 - \sum_{j=1}^{k} a_j L^j\right) \Delta^d y_t = \varepsilon_t,$$

where we drop the index and write k instead of k_T for brevity. The Whittle estimation minimizes

$$Q_k(\varphi) = \frac{2\pi}{M} \sum_{j=1}^{M} \frac{I_y(\lambda_j)}{T_C(\lambda_j; \varphi)}, \quad M = \left\lfloor \frac{T-1}{2} \right\rfloor,$$

with respect to $\varphi' = (d, a_1, \ldots, a_k)$, where

$$T_C(\lambda_j; \varphi) = \left(2 \sin \frac{\lambda_j}{2}\right)^{-2d} \left|1 - \sum_{n=1}^{k} a_n e^{-i\lambda_j n}\right|^{-2}.$$

The minimization with respect to d has to be restricted to an interval centered around a preliminary estimator $\tilde{d}$:

$$d \in \left[\tilde{d} \pm \frac{c}{K_T}\right] \cap \left[0 \pm \frac{1}{2}\right], \quad c > 0, \quad K_T \to \infty. \tag{9.21}$$

As preliminary estimator, Bhansali et al. (2006, Remark 2.3) suggested the LW estimator from Section 9.2 with bandwidth $m = T^{0.7}$ and with $K_T = T^{0.2}$ in Eq. (9.21); but they are silent on how to choose the constant c. Let's call the resulting estimator for the fractional integration parameter $\hat{d}_W^{(k)}$.

Proposition 9.11 *Let* $\{y_t - \mu\}$ *be a stationary process with* $y_t - \mu = (1 - L)^{-d_0}x_t$, $|d_0| < 1/2$, *where* $x_t - \sum_{j=1}^{\infty} a_j x_{t-j} = \varepsilon_t \sim iid(0, \sigma^2)$ *with* $\sum_{j=1}^{\infty} |a_j| < \infty$ *and* $\sum_{j=1}^{\infty} a_j \neq 1$. *Subject to Eq. (9.21) and under some further assumptions by Bhansali et al. (2006), it holds that*

$$\sqrt{\frac{T}{k}}\left(\hat{d}_W^{(k)} - d_0\right) \xrightarrow{D} \mathcal{N}(0, 1),$$

as $T \to \infty$, $k \to \infty$ *with* $k = O(\ln T)$.

Proof: Bhansali et al. (2006, Remark 2.1, Corollary 3.1). □

Note that the previous local semiparametric estimators converge with $\sqrt{m}$ and $m = O(T^{0.8})$. Hence, the global estimator $\hat{d}_W^{(k)}$ achieves a faster rate:

$$\frac{T^{0.8}}{T/\ln T} \to 0.$$

Unfortunately, however, the result in Proposition 9.11 has not been extended to the nonstationarity region $d_0 \geq 1/2$.

The second global estimator relies on the log-periodogram regression of a FEXP process. The OLS estimator $\hat{d}_{\text{Exp}}^{(k)}$ is computed from Eq. (8.12):

$$\ln I_j = c - 2dX_j + \sum_{\ell=1}^{k} \psi_\ell \cos(\ell \lambda_j) + U_j, \quad j = 1, \dots, M = \left\lfloor \frac{T-1}{2} \right\rfloor,$$

where the abbreviations are defined in Section 8.5. Hurvich and Brodsky (2001) worked under the assumptions spelled out in Proposition 8.5 but replaced the finite order EXP model by the assumption that

$$\ln f_x(\lambda) = \sum_{\ell=0}^{\infty} \psi_\ell \cos(\ell \lambda), \quad \sum_{\ell=0}^{\infty} |\psi_\ell| < \infty.$$

Under the assumption that $k = k_T = O(\ln T)$ diverges with T when computing $\hat{d}_{\text{Exp}}^{(k)}$, Hurvich and Brodsky (2001, Theorem 2) obtained

$$\frac{T}{k} \operatorname{Var}(\hat{d}_{\text{Exp}}^{(k)}) \to \frac{\pi^2}{6}, \quad T \to \infty.$$

The issue how to determine $k = k_T$ from the data has been addressed by Moulines and Soulier (2000), and distributional results have been provided by Moulines and Soulier (1999, Theorem 1). All three papers related to $\hat{d}_{\text{Exp}}^{(k)}$

assumed $|d| < 1/2$ as in Proposition 9.11, and further they maintained the assumption of Gaussianity of the process. At the same time the limiting variance exceeds the one of the global semiparametric Whittle estimator; see again Proposition 9.11. Therefore, we omit details since $\widehat{d}_W^{(k)}$ is considered as superior to $\widehat{d}_{\mathrm{Exp}}^{(k)}$.

9.7 Technical Appendix: Proofs

Proof of Lemma 9.2

From Giraitis et al. (2012, Theorem 5.3.1), we have that

$$2 \frac{I_y(\lambda_{j_i})}{f_y(\lambda_{j_i})} \xrightarrow{D} \chi^2(2), \quad i = 1, \dots, p,$$

and that asymptotic independence applies. Hence, we are left with an application of Lemma 4.2. To that end assume $X \sim \chi^2(2)$ with density ϕ_x:

$$\phi_x(x) = \frac{1}{2} e^{-\frac{x}{2}}, \quad x \in D_x = (0, \infty).$$

Consider the transformation

$$G = -\ln\left(\frac{X}{2}\right) = h(X)$$

on the domain $D_g = (-\infty, \infty)$ with

$$h^{-1}(G) = 2\exp(-G).$$

According to Lemma 4.2, the density of G is given by

$$\phi_g(g) = \left| \frac{dh^{-1}(g)}{dg} \right| \phi_x(h^{-1}(g)).$$

Substituting h and ϕ_x, we obtain

$$\phi_g(g) = |-2\exp(-g)| \frac{1}{2} \exp\left\{ -\frac{h^{-1}(g)}{2} \right\}$$

$$= 2 e^{-g} \frac{1}{2} \exp\{-e^{-g}\} = e^{-g} \exp\{-e^{-g}\}.$$

We now compare this with a Gumbel distribution with parameters a and b,

$$Y \sim \mathrm{Gum}(a, b),$$

where $E(Y) = a - b\,\Gamma'(1)$ with $\Gamma'(1) = -\gamma = -0.5772\dots$ from Eq. (5.20), and variance $\mathrm{Var}(Y) = b^2\,\pi^2/6$. The density of Y is

$$\phi_y(y) = \frac{1}{b} \exp\left\{ -\frac{y-a}{b} \right\} \exp\left\{ -\exp\left(-\frac{y-a}{b} \right) \right\}$$

for $y \in \mathbb{R}$; see, e.g. Mood et al. (1974, p. 542). Comparing ϕ_y with ϕ_g shows that G follows a Gumbel distribution with $a = 0$ and $b = 1$, such that

$$E(G) = -\Gamma'(1) = \gamma \quad \text{and} \quad \text{Var}(G) = \frac{\pi^2}{6}.$$

Hence, the proof is complete. $\qquad\qquad\qquad\qquad\qquad\qquad\qquad\qquad\qquad$ □

10

Testing

Tests can be performed employing the limiting normality of the previous estimators. However, power gains are available when constructing efficient tests, e.g. from the Lagrange multiplier (LM) principle. Before turning to LM tests, we will review procedures that are related to the so-called rescaled range or variance. Then we recap the LM or score principle, in order to carry it to the problem of fractional integration testing. LM tests have been proposed in the frequency domain as well as in the time domain. In particular, the regression-based LM test settled in the time domain has the property of being robust with respect to conditional and unconditional heteroskedasticity.

10.1 Hypotheses on Fractional Integration

Hypotheses of applied interest on the fractional integration parameter d are, e.g. short memory versus long memory,

$$H_0 : d = 0 \text{ vs. } H_1 : d > 0,$$

or nonstationarity versus stationarity,

$$H_0 : d = 0.5 \text{ vs. } H_1 : d < 0.5.$$

Under nonstationarity we consider the type II model:

$$y_t = \sum_{j=0}^{t-1} \pi_j(-d)x_{t-j}.$$

For large j the behavior of $\{\pi_j(-d)\}$ is characterized in Eq. (5.26). We observe $\pi_j(-d) \to 0$ if and only if $d < 1$, which is sometimes called *mean reversion*, contrasting the $I(1)$ behavior of an autoregressive unit root ($d = 1$). A third pair of hypotheses of interest hence is

$$H_0 : d = 1 \text{ vs. } H_1 : d < 1,$$

which amounts to a so-called unit root test.

Time Series Analysis with Long Memory in View, First Edition. Uwe Hassler.
© 2019 John Wiley & Sons, Inc. Published 2019 by John Wiley & Sons, Inc.

Of course one may construct asymptotic tests from the previous two chapters. For an ML type estimator $\hat{d}$ from Chapter 8, we reject H_0 if $\sqrt{T}|\hat{d} - d_0|$ exceeds a critical value determined by the limiting variance and the approximating Gaussian distribution. Since the requirement of a parametric model for $\hat{d}$ may not be realistic, we studied semiparametric estimators in the last chapter. Again, we reject H_0 if $\sqrt{m}|\hat{d}_{sp} - d_0|$ is too large, where m is the bandwidth associated with the semiparametric estimator $\hat{d}_{sp}$. As discussed in Section 9.3, one may replace m by $\sum_{j=1}^{m} v_j^2$ with $v_j = \ln j - (\ln m!)/m$ to achieve tests with better size properties in finite samples. In any case, semiparametric estimators typically have a large variance converging only slowly with the bandwidth m to 0, which will not result in very powerful tests. Further, all procedures introduced previously rule out unconditional heteroskedasticity that we would like to allow for because of its empirical relevance.[1] The present chapter is devoted to procedures that explicitly address the testing problem without the primary goal of estimation.

In this chapter we will maintain the assumption of a type II process defined in Definition 6.2:

$$\Delta_+^{d_0+\theta} y_t = x_t \quad \text{or} \quad y_t = \sum_{j=0}^{t-1} \pi_j(-d_0 - \theta)x_{t-j},$$

where the $I(0)$ sequence $\{x_t\}$ satisfies Assumption 6.1. Null hypotheses about the order of integration are then recast in terms of θ:

$$H_0 : y_t \sim I(d_0) \quad \text{or} \quad \theta = 0. \tag{10.1}$$

Many procedures in the literature are designed to test for the specific value $d_0 = 0$. To extend them to general fractional hypotheses, we can apply them with the corresponding fractional differences, where the data is differenced under H_0. In accordance with the type II assumption, we define as differenced data under H_0:

$$x_{t,d} := \Delta_+^{d_0} y_t = \sum_{j=0}^{t-1} \pi_j(d_0)y_{t-j}, \quad t = 1, \dots, T. \tag{10.2}$$

For notational convenience, we do not use the symbol x_{t,d_0} but $x_{t,d}$ when differencing of order d_0 according to the null hypothesis. Hence, the null hypothesis becomes in terms of differences

$$H_0 : x_{t,d} \sim I(0),$$

and $x_{t,d} = x_t$ under the null. We summarize these assumptions as follows:

---

1 Demetrescu and Sibbertsen (2016) show that the log-periodogram regression estimator $\hat{d}_{LP}$ is not robust with respect to unconditional heteroskedasticity: While it remains unbiased, the limiting variance is affected.

Assumption 10.1 (*Null hypothesis*) *Let $\{y_t\}$ be a fractionally integrated type II process of order $d = d_0$ with $\{x_t\}$ from Assumption 6.1: $y_t = \Delta_+^{-d} x_t$. The differences $x_{t,d} := \Delta_+^{d_0} y_t$ are computed under the null hypothesis that $y_t \sim I(d_0)$.*

In practice, one may wish to allow for deterministic components beyond fractional integration. In differences, a typical assumption allows for a nonzero mean:

$$x_{t,d} = \mu + \Delta_+^{-\theta} x_t. \tag{10.3}$$

In levels this implies

$$y_t = \Delta_+^{-d_0} \mu + \Delta_+^{-d_0 - \theta} x_t.$$

For $d_0 = 1$, a linear time trend arises from Eq. (5.22): $\Delta_+^{-1} \mu = \mu\, t$. Alternatively, some papers directly model a deterministic function f of time in levels:

$$y_t = f(t) + \Delta_+^{-d_0 - \theta} x_t. \tag{10.4}$$

Most often, this function simply accounts for a linear time trend: $f(t) = a + b\, t$.

10.2 Rescaled Range or Variance

The rescaled range analysis, also called range over standard deviation (R/S) analysis, dates back to the pathbreaking paper by Hurst (1951) on historical data of the river Nile; see also Figure 1.1. Let us assume a sample of observed data, $y_1, \ldots, y_T$, which is used to construct the partial sum of demeaned observations:

$$S_t^* = \sum_{j=1}^{t} (y_j - \bar{y}), \quad \bar{y} = \frac{1}{T} \sum_{t=1}^{T} y_t.$$

If $y_t \sim I(d)$ satisfies Proposition 7.3, we have

$$T^{-d-1/2} S_{\lfloor rT \rfloor}^* \Rightarrow C_{S,d}^{II}(B_d^{II}(r) - r\, B_d^{II}(1)) = C_{S,d}^{II}\, BB_d(r), \quad \text{say,}$$

where BB_d is sometimes called a fractional Brownian bridge (of type II). The range of S_t^* is defined as

$$R = \max_{1 \le t \le T} S_t^* - \min_{1 \le t \le T} S_t^*,$$

and the following ratio has been called rescaled range (or range over standard deviation):

$$\frac{R}{s} = \frac{\displaystyle\max_{1 \le t \le T} S_t^* - \min_{1 \le t \le T} S_t^*}{s}, \quad s^2 = \frac{1}{T} \sum_{t=1}^{T} (y_t - \bar{y})^2.$$

Under stationarity we have $s^2 \xrightarrow{p} \gamma_y(0) = \text{Var}(y_t)$, such that by the continuous mapping theorem (CMT) with growing sample size

$$T^{-d-1/2}\frac{R}{s} \xrightarrow{D} \frac{C^{\text{II}}_{S,d}}{\sqrt{\gamma_y(0)}}\left(\max_{0\leq r\leq 1} \text{BB}_d(r) - \min_{0\leq r\leq 1} \text{BB}_d(r)\right) = V, \quad \text{say.}$$

For the expectation this allows to approximate

$$T^{-d-1/2}\text{E}(R/s) \approx \text{E}(V),$$

or upon taking logs,

$$\ln \text{E}(R/s) \approx \ln \text{E}(V) + (d + 1/2)\ln T.$$

To develop an estimator on heuristic grounds, one may compute a sequence of rescaled ranges from nonoverlapping subsamples of size n, R_n/s_n, and average to approximate $\text{E}(R/s)$. A regression of the logs thereof on $\ln n$ with varying n yields the rescaled range (R/S) estimator of d, or of the so-called Hurst coefficient $H = d + 1/2$. An early study into the robustness properties of the (R/S) estimator is by Mandelbrot and Wallis (1969). In this chapter, however, we are no longer interested in estimating d.

We focus on testing the $I(0)$ hypothesis for $\{x_{t,d}\}$ from Eq. (10.2). Then the scaling factor in front of $\max \text{BB}_d - \min \text{BB}_d$ becomes $C^{\text{II}}_{S,0} = \omega$, where ω^2 is the usual long-run variance of $\{x_t\}$ defined in Eq. (2.10). Hence, the limit of the classical R/S statistic depends on $\sqrt{\omega^2/\gamma(0)}$, which equals 1 if $\{x_t\}$ is free of serial correlation. In general, ω is a nuisance parameter one has to eliminate. Hence, Lo (1991) suggested a modified R/S statistic that becomes asymptotically robust against short memory of $\{x_t\}$ captured in ω^2:

$$\text{RS}_q = \frac{\max\limits_{1\leq t\leq T} S^*_{t,d} - \min\limits_{1\leq t\leq T} S^*_{t,d}}{\sqrt{T}\,\hat{\omega}_q}, \tag{10.5}$$

where $\hat{\omega}^2_q \xrightarrow{p} \omega^2$, and the partial sum is defined in terms of fractional differences from Eq. (10.2):[2]

$$S^*_{t,d} = \sum_{j=1}^{t}(x_{j,d} - \bar{x}_d), \quad \bar{x}_d = \frac{1}{T}\sum_{t=1}^{T} x_{t,d}. \tag{10.6}$$

The consistent estimation of the long-run variance is a standard problem in econometrics since Newey and West (1987), who popularized the usage of the so-called Bartlett weights,

$$w_h(q) = 1 - \frac{h}{q+1}, \quad h = 1,\ldots,q, \tag{10.7}$$

---

2 Remember that $S^*_{t,d}$ defined in terms of $x_{t,d} = \Delta^{d_0}_{+} y_t$ is computed from differences under the null: $d = d_0$.

depending on the so-called bandwidth parameter q. The estimator becomes

$$\widehat{\omega}_q^2 = \frac{1}{T}\sum_{t=1}^{T}(x_{t,d} - \overline{x}_d)^2 + 2\sum_{h=1}^{q} w_h(q)\widehat{\gamma}(h),$$

with

$$\widehat{\gamma}(h) = \frac{1}{T}\sum_{t=1}^{T-h}(x_{t,d} - \overline{x}_d)(x_{t+h,d} - \overline{x}_d),$$

see also Andrews (1991) and Hamilton (1994, Section 10.5) for a discussion on alternative weights and data-driven choices of q. Consistency requires the familiar minimum bandwidth restriction:

$$\frac{1}{q} + \frac{q}{T} \to 0 \quad \text{as} \quad T \to \infty. \tag{10.8}$$

If $\{x_t\}$ was white noise, we could choose $q = 0$ with $\widehat{\omega}_0^2 = s^2$ being the usual variance estimator.

We now give the limiting distribution of the modified R/S statistic under the null Eq. (10.1) that $\{y_t\}$ is integrated of order d_0. Under the assumption of a type II process, this translates into $\{x_{t,d}\} \sim I(0)$. A mean different from 0 does not affect the result.

Proposition 10.1 *Let $\{x_{t,d} - \mu\}$ from Assumption 10.1 satisfy some additional assumptions by Lo (1991, Theorem 3.1). It then holds under Eq. (10.8) that*

$$RS_q \xrightarrow{D} \max_{0 \le r \le 1} BB_0(r) - \min_{0 \le r \le 1} BB_0(r),$$

as long as $q/T^{1/4} \to 0$ as $T \to \infty$.

Proof: Lo (1991, Theorem 3.1). $\qquad\qquad\qquad\qquad\qquad\qquad\qquad\qquad$ □

Lo (1991, footnote 16) conjectured that the bandwidth restriction $q/T^{1/4} \to 0$ maintained in Proposition 10.1 could be relaxed to $q/T^{1/2} \to 0$ as suggested by Andrews (1991). In fact, a result corresponding to Proposition 10.1 was established by Giraitis et al. (2003, Proposition 3.1) under the minimum assumption $\frac{1}{q} + \frac{q}{T} \to 0$ from Eq. (10.8), which does not impose any restriction on q except for diverging at a slower rate than T.

Note that BB_0 is a standard Brownian bridge. The distribution function of $\max BB_0 - \min BB_0$ is known from the literature and was used by Lo (1991, Table II) to provide asymptotic critical values. When testing within the model $\{x_{t,d}\} \sim I(\theta)$ against $\theta > 0$, we reject for too large values. Consistency of the test was explored by Lo (1991, Theorem 3.3) and Giraitis et al. (2003, Proposition 3.2).

To achieve a better balance between empirical size and power than for the (modified) R/S test, Giraitis et al. (2003) introduced a so-called rescaled variance statistic, building on the sample variance of $\{S^*_{t,d}\}$ from (10.6):

$$\text{VS}_q = \frac{\sum_{t=1}^{T} \left(S^*_{t,d} - \frac{1}{T} \sum_{j=1}^{T} S^*_{j,d} \right)^2}{T^2\, \hat{\omega}^2_q}. \tag{10.9}$$

Again, $\hat{\omega}^2_q$ is the consistent estimator defined above. Under some assumptions ($\{x_t\}$ is assumed to be fourth order stationary, i.e. to have constant fourth moments) and the minimum requirement $q/T \to 0$, Giraitis et al. (2003) establish the limiting distribution under the null hypothesis in terms of the standard Brownian bridge.

Proposition 10.2 *Let $\{x_{t,d} - \mu\}$ from Assumption 10.1 satisfy some additional assumptions by Giraitis et al. (2003, Proposition 3.5). It then holds under Eq. (10.8) that*

$$\text{VS}_q \xrightarrow{D} \int_0^1 (\text{BB}_0(r))^2 dr - \left(\int_0^1 \text{BB}_0(r) dr \right)^2$$

as $T \to \infty$.

Proof: Giraitis et al. (2003, Proposition 3.5). ☐

When testing against $d > d_0$, the null hypothesis is rejected for too large values; for a proof of consistency, see Giraitis et al. (2003, Proposition 3.6). Asymptotic critical values are available from the distribution function of the limit in Proposition 10.2, which is known from the literature; see Giraitis et al. (2003, Remark 3.3).

The rescaled variance statistic is closely related to the so-called KPSS test by Kwiatkowski et al. (1992). We simply drop the mean of the partial sum in Eq. (10.9) to define

$$\text{KPSS}_q = \frac{\sum_{t=1}^{T} (S^*_{t,d})^2}{T^2\, \hat{\omega}^2_q}. \tag{10.10}$$

In view of Proposition 10.2, the limiting distribution does not come as a surprise.

Corollary 10.1 *Under the assumptions of Proposition 10.2, it holds that*

$$\text{KPSS}_q \xrightarrow{D} \int_0^1 (BB_0(r))^2 dr$$

as $T \to \infty$.

Proof: Kwiatkowski et al.(1992 Eq. (14)) or Giraitis et al. (2003, Proposition 3.4).

$\square$

Note that Kwiatkowski et al. (1992, p. 165) maintained $q/T^{1/2} \to 0$, while Giraitis et al. (2003) worked again under the more liberal assumption Eq. (10.8). Critical values have been published before Kwiatkowski et al. (1992) by Anderson and Darling (1952, Table 1), who discussed the limiting distribution of KPSS_q in the context of goodness-of-fit tests. When testing against $x_{t,d} \sim I(\theta)$ and $\theta > 0$, we reject for too large values of KPSS_q. Originally, Kwiatkowski et al. (1992) had proposed to test against a unit root in $\{x_{t,d}\}$, i.e. $\theta = 1$. Later, Lee and Schmidt (1996) and Lee and Amsler (1997) established that KPSS_q has power against fractional alternatives, too; see also Giraitis et al. (2003, Proposition 3.4).

The asymptotic behavior of all three test statistics of this section is characterized by Giraitis et al. (2003) under long memory alternatives for $\{x_{t,d}\}$, $0 < \theta < 1/2$.

Proposition 10.3   *Let*

$$x_{t,d} = \mu + (1 - L)^{-\theta} x_t, \quad 0 < \theta < 1/2,$$

be a stationary process meeting Assumption 6.2. It then holds under Eq. (10.8) and additional assumptions by Giraitis et al. (2003) that

$$\left(\frac{q}{T}\right)^{\theta} \text{RS}_q \overset{D}{\to} \mathcal{L}_1,$$

$$\left(\frac{q}{T}\right)^{2\theta} \text{VS}_q \overset{D}{\to} \mathcal{L}_2,$$

$$\left(\frac{q}{T}\right)^{2\theta} \text{KPSS}_q \overset{D}{\to} \mathcal{L}_3,$$

as $T \to \infty$. The limits $\mathcal{L}_1$ through $\mathcal{L}_3$ are given in Giraitis et al. (2003).

Proof: Giraitis et al. (2003, Propositions 3.2, 3.4, and 3.6). $\square$

The limits given in this proposition are not so interesting. More relevant is that the test statistics diverge at rate $(T/q)^{2\theta}$, or $(T/q)^{\theta}$, i.e. all the faster, the stronger the violation of H_0 with $\theta = 0$ is. The power will, however, crucially hinge on the bandwidth q. The bandwidth has to diverge to infinity for consistency of $\hat{\omega}_q$, which in turn is required to control the size of the tests under the null. At the same time there is a trade-off in terms of power under the alternative: The test statistics diverge all the more slowly, the faster q grows. In fact, Giraitis et al. (2003, Theorem 3.1) show under the alternative hypothesis that

$$\hat{\omega}_q^2 = O_p(q^{2\theta}),$$

see also Giraitis et al. (2005). This makes the choice of q a delicate problem in practice. Experimentally, Giraitis et al. (2003, p. 280) report that small bandwidths lead to massive size distortions (rejections with a probability larger than the significance level) if H_0 is true. The other way around, large values of q come at a price in terms of power in that H_0 is rarely rejected under H_1. For a related discussion focused on the RS_q statistic, see also Teverovsky et al. (1999). For applied purposes, there seems to be little guidance on how to choose the crucial tuning parameter q.

10.3 The Score Test Principle

Within the framework of maximum likelihood, the trilogy of likelihood ratio tests, Wald tests, and LM tests is well known. There is a controversy whether to call the last of these three principles LM test or (Rao's) score test. The label LM test dates back to Silvey (1959), who elaborated on the test proposed by Aitchison and Silvey (1958), and before that by Rao (1948), although Aitchison and Silvey (1958) carried out their work independently. The paper by Rao (1948) "was not 'noticed' by others for a long time even by researchers engaged in the area of testing. [...] Silvey (1959), who presented a test identical to that by Rao, made no reference to Rao's work." (quoted from Bera and Bilias, 2001, p. 10). The *Journal of Statistical Planning and Inference* dedicated a special issue to Rao's seminal contribution; see in particular Bera and Bilias (2001) and Rao (2001) on the historical development. Following, e.g. Breusch and Pagan (1980) or Engle (1984), it is common to use the LM terminology in econometrics. According to Buse (1982, p. 154), people "continue to use the term Lagrange multiplier test even though our account is based on the score formulation and despite the fact that the score form (Rao, 1948) has historical precedent over the Lagrange form (Aitchison and Silvey, 1958). The Lagrange terminology has become so firmly embedded in the econometrics literature that to press for the score terminology would be futile."

We now briefly summarize the score test for the case of a scalar parameter θ and a log-likelihood function $\ell(\theta)$. The score function and the Fisher information are defined as

$$S(\theta) = \frac{\partial \ell(\theta)}{\partial \theta} \quad \text{and} \quad F(\theta) = -\mathrm{E}\left(\frac{\partial^2 \ell(\theta)}{\partial \theta^2}\right).$$

Under standard regularity conditions, see, e.g. Mood et al. (1974, p. 320), it holds that

$$F(\theta) = \mathrm{E}(S^2(\theta)).$$

An approximation to $F(\theta)$ sometimes called *observed Fisher information* opposed to *expected Fisher information* is

$$\widehat{F}(\theta) = -\frac{\partial^2 \ell(\theta)}{\partial \theta^2},$$

see, e.g. Efron and Hinkley (1978). Rao's score statistic defined as RS in Eq. (10.11) evaluates S and F exclusively under the null hypothesis, $H_0 \colon \theta = \theta_0$. Under appropriate assumptions it follows under H_0 a χ^2 distribution with one degree of freedom asymptotically:

$$\text{RS} = \frac{S^2(\theta_0)}{F(\theta_0)} \xrightarrow{D} \chi^2(1). \tag{10.11}$$

In what follows, we will stick to the convention of much of the (econometric) literature and call Rao's score statistic LM statistic.[3]

10.4 Lagrange Multiplier (LM) Test

We assume for the moment a very special case of model Eq. (10.3) with $\mu = 0$ and a Gaussian short memory component that is iid, $x_t = \varepsilon_t \sim \text{iid}(0, \sigma^2)$. The Gaussian log-likelihood function in terms of $x_{t,d}$ is

$$\ell(\theta, \sigma^2) = -\frac{T}{2} \ln(2\pi\sigma^2) - \frac{1}{2\sigma^2} \sum_{t=1}^{T} (\Delta^\theta x_{t,d})^2.$$

The gradient thereof becomes

$$\frac{\partial \ell(\theta, \sigma^2)}{\partial \theta} = -\frac{1}{\sigma^2} \sum_{t=1}^{T} \Delta^\theta x_{t,d} \ln(\Delta) \Delta^\theta x_{t,d},$$

where a Taylor expansion provides

$$\ln(\Delta) = \ln(1 - L) = -\sum_{j=1}^{\infty} \frac{L^j}{j},$$

see Tanaka (1999, Eq. (39)). Hence, under H_0 we obtain for the score evaluated at $\theta = 0$, S, that

$$S = \frac{\partial \ell(\theta, \sigma^2)}{\partial \theta} \bigg|_{\theta=0}$$

$$= -\frac{1}{\sigma^2} \sum_{t=1}^{T} x_{t,d} \ln(\Delta) x_{t,d}$$

$$= \frac{1}{\sigma^2} \sum_{t=2}^{T} x_{t,d} x_{t-1,d}^*$$

---

[3] According to Bera and Bilias (2001, p. 10) "It is fair to say that econometricians can claim major credit (and blame) in popularizing the use of Rao's score test (and giving it a somewhat whimsical terminology)."

where we maintain the type II model from Assumption 10.1 with $0 = x_{0,d} = x_{-1,d} = \cdots$, such that

$$x_{t-1,d}^* = \sum_{j=1}^{t-1} \frac{x_{t-j,d}}{j}. \tag{10.12}$$

Under the assumption that $x_{t,d} = x_t$ is iid, the expectation of the squared score becomes

$$E(S^2) = \frac{\sum_{t=2}^{T} E((x_{t-1,d}^*)^2)}{\sigma^2}.$$

Following Eq. (10.11), the usual LM or score statistic is

$$\frac{S^2}{E(S^2)} = \frac{\left(\sum_{t=2}^{T} x_{t,d} x_{t-1,d}^*\right)^2}{\sigma^2 \sum_{t=2}^{T} E((x_{t-1,d}^*)^2)}.$$

In practice, one has to estimate the noise variance consistently, e.g.

$$\hat{\sigma}^2 = T^{-1} \sum_{t=1}^{T} x_{t,d}^2.$$

Consequently, we may write as LM statistic

$$LM = \frac{\left(\sum_{t=2}^{T} x_{t,d} x_{t-1,d}^*\right)^2}{\hat{\sigma}^2 \sum_{t=2}^{T} E((x_{t-1,d}^*)^2)}.$$

Note that

$$Var(x_{t-1,d}^*) = \sigma^2 \sum_{j=1}^{t-1} \frac{1}{j^2} \to \sigma^2 \frac{\pi^2}{6},$$

as $t \to \infty$; see Lemma 3.1 (e). Hence, it is not surprising that we will learn in Proposition 10.4 that

$$\frac{\sum_{t=2}^{T} E((x_{t-1,d}^*)^2)}{T} \to \sigma^2 \frac{\pi^2}{6}.$$

Moreover, the proposition establishes that LM converges to a χ^2 distribution with one degree of freedom, since $LM = \tau_{LM}^2$, where τ_{LM} from Proposition 10.4 (c) has a standard normal limit.

Proposition 10.4 *Let $\{x_{t,d}\}$ from Assumption 10.1 be iid$(0, \sigma^2)$. It then holds under the null hypothesis*

(a) with $x^*_{t-1,d}$ from Eq. (10.12) that

$$\frac{\sum_{t=2}^{T} \mathrm{E}((x^*_{t-1,d})^2)}{T} \to \sigma^2 \frac{\pi^2}{6},$$

$$\frac{\sum_{t=2}^{T} (x^*_{t-1,d})^2}{T} \overset{a.s.}{\to} \sigma^2 \frac{\pi^2}{6},$$

(b) and that

$$\frac{1}{\sqrt{T}} \sum_{t=2}^{T} x_{t,d} x^*_{t-1,d} \overset{D}{\to} \mathcal{N}\left(0, \sigma^4 \frac{\pi^2}{6}\right),$$

(c) such that

$$\tau_{\mathrm{LM}} := \frac{\sum_{t=2}^{T} x_{t,d} x^*_{t-1,d}}{\hat{\sigma} \sqrt{\sum_{t=2}^{T} \mathrm{E}((x^*_{t-1,d})^2)}} \overset{D}{\to} \mathcal{N}(0,\, 1),$$

as $T \to \infty$.

Proof: See Appendix to this chapter.

Because of

$$\sum_{t=2}^{T} x_{t,d} x^*_{t-1,d} = T \sum_{h=1}^{T-1} \frac{\widetilde{\gamma}_x(h)}{h} \quad \text{with} \quad \widetilde{\gamma}_x(h) = T^{-1} \sum_{t=1}^{T-h} x_{t,d} x_{t+h,d},$$

we may rewrite the LM statistic in terms of weighted sample autocorrelations:

$$\tau_{\mathrm{LM}} = \frac{\hat{\sigma} T \sum_{h=1}^{T-1} \frac{\widetilde{\rho}_x(h)}{h}}{\sqrt{\sum_{t=2}^{T} \mathrm{E}\left((x^*_{t-1,d})^2\right)}}, \quad \widetilde{\rho}_x(h) = \widetilde{\gamma}_x(h)/\hat{\sigma}^2.$$

Given these results, it is natural to define as in Robinson (1991, Eq. (8)) or Tanaka (1999, Eq. (41)) the test statistic t_{LM} as[4]

$$t_{\mathrm{LM}} := \sqrt{T} \frac{\sqrt{6}}{\pi} \sum_{h=1}^{T-1} \frac{\widetilde{\rho}_x(h)}{h} \tag{10.13}$$

with a standard normal limit under H_0; see also Robinson (1991, Theorem 2.1) and Tanaka (1999, Theorem 3.1):

$$t_{\mathrm{LM}} \overset{D}{\to} \mathcal{N}(0,\, 1).$$

---

4 See also the statistic $\overline{R}$ in Robinson (1994b, p. 1422).

In fact, Robinson (1991) and Tanaka (1999) considered this statistic in terms of demeaned data (or more generally after removing a deterministic component), i.e.

$$\hat{t}_{\text{LM}} := \sqrt{T} \frac{\sqrt{6}}{\pi} \sum_{h=1}^{T-1} \frac{\hat{\rho}_x(h)}{h},$$

with $\hat{\rho}_x(h)$ being defined in Section 7.5.

Similarly as in Chapter 8, the limiting normality in Proposition 10.4 does not require the assumption of Gaussian data; but under Gaussianity the test is locally efficient (most powerful against local fractional alternatives); see Robinson (1994b, p. 1425) and Tanaka (1999, Corollary 3.2). This is the major motivation to stick to LM type tests for the rest of this chapter.

The assumptions behind Proposition 10.4 are clearly too restrictive for applied work. In practice, one encounters deterministic components different from 0 and serial correlation and (conditional) heteroskedasticity in $\{x_t\}$. Let us turn to serial correlation first. If $\{x_t\}$ is allowed to be a stable AR(k) process, $A(L) x_t = \varepsilon_t$, one may compute t_{LM} from Eq. (10.13) with $\tilde{\rho}_x(h)$ and $\hat{\sigma}^2$ being replaced by the sample autocorrelations and the variance estimation of the AR residuals $\hat{\varepsilon}_{t,d} = x_{t,d} - \sum_{j=1}^{k} \hat{a}_j x_{t-j,d}$, respectively, which affects the limiting distribution, though. Let us define here as F the Fisher information for the AR polynomial, which essentially contains the autocovariances of $\{x_t\}$ (see Brockwell and Davis, 1991, Example 8.8.1),

$$F = (f_{ij})_{i,j=1,\dots,k} \quad \text{where } f_{ij} = \sum_{n=0}^{\infty} c_n c_{n+|i-j|}$$

with $A^{-1}(L) = \sum_{n=0}^{\infty} c_n L^n$. Further, define

$$\kappa_j = \sum_{n=j}^{\infty} \frac{c_{n-j}}{n}, \quad j = 1, \dots, k.$$

Then the limiting variance of t_{LM} computed from the AR residuals $\hat{\varepsilon}_{t,d}$ becomes

$$1 - \frac{6}{\pi^2} \kappa' F^{-1} \kappa, \quad \text{with } \kappa' = (\kappa_1, \dots, \kappa_k),$$

see Tanaka (1999, Theorem 3.3) for a proof. Although this variance can be estimated from the data, it is not straightforward to do so in applied situations. Further, it requires knowledge about the number of lags, k.

As another modification of the LM statistic in order to robustify against short memory in $\{x_t\}$, Harris et al. (2008) suggested a truncated version of t_{LM} from Eq. (10.13). They noted that the numerator

$$N = \sqrt{T} \sum_{h=1}^{T-1} \frac{\tilde{\gamma}_x(h)}{h}$$

behind t_{LM} is not necessarily centered around 0 if $\gamma_x(h) \neq 0$, since

$$E(\widetilde{\gamma}_x(h)) = \frac{T-h}{T}\,\gamma_x(h),$$

which holds true by, e.g. Fuller (1996, p. 313) as long as the true mean is known to be 0. Therefore,

$$E\left(\sum_{h=1}^{T-1} \frac{\widetilde{\gamma}_x(h)}{h}\right) = \sum_{h=1}^{T-1} \frac{\gamma_x(h)}{h} - \sum_{h=1}^{T-1} \frac{\gamma_x(h)}{T}$$

$$\rightarrow \sum_{h=1}^{\infty} \frac{\gamma_x(h)}{h} \quad \text{as } T \rightarrow \infty,$$

where the last term is in general $\neq 0$, such that $|E(N)| \rightarrow \infty$. This motivates as truncated numerator

$$N_K = \sqrt{T-K} \sum_{h=K}^{T-1} \frac{\widetilde{\gamma}_x(h)}{h-K+1},$$

which requires for a limiting standard normal distribution the denominator D_q with

$$D_q^2 = \sum_{j=-q}^{q} h_j \sum_{i=-q}^{q} \widetilde{\gamma}_x(i)\widetilde{\gamma}_x(i+j),$$

where $h_0 = \pi^2/6$, and

$$h_j = \frac{1}{|j|} \sum_{i=1}^{|j|} \frac{1}{i}.$$

We then have the following result under the null hypothesis.

Proposition 10.5 *Let* $\{x_{t,d}\}$ *from Assumption 10.1 meet additional assumptions by Harris et al. (2008, Theorem 1). Under* $K = \sqrt{cT}$ *for some constant c and*

$$\frac{1}{q} + \frac{q^2}{T} \rightarrow 0,$$

it holds that

$$\frac{N_K}{D_q} \xrightarrow{D} \mathcal{N}(0,1)$$

as $T \rightarrow \infty$.

Proof: Harris et al. (2008, Theorem 1). $\qquad\qquad\square$

The test still works if the data is demeaned or detrended: If $\widetilde{\gamma}_x(h)$ is replaced by $\widetilde{\gamma}_{\hat{x}}(h)$ computed from $\hat{x}_{t,d}$ as residuals upon projecting on a constant or a constant and a linear time trend, then Proposition 10.5 continues to hold (see Harris et al., 2008, Theorem 1). Further, consistency is established in Harris et al. (2008, Theorem 2). Under Eq. (10.3) with $\theta > 0$, N_K/D_q is shown to diverge asymptotically. The rate of divergence is related to $1/\ln q$, which shows that there will be again a trade-off between correct size (requiring large q) and power (decreasing in q). What is more, the second tuning parameter, K, has to be fixed in practice, too. Harris et al. (2008) recommended $c = 1$ with $K = \sqrt{T}$.

10.5 LM Test in the Frequency Domain

Following the time domain formulation by Robinson (1991) given in Eq. (10.13), Robinson (1994b) recast the LM statistic in the frequency domain as

$$\widetilde{LM}(0) = T \frac{6}{\pi^2} \left(\frac{2\pi}{\hat{\sigma}^2} \frac{1}{T} \sum_{j=1}^{T-1} \ln\left(2\sin\frac{\lambda_j}{2} \right) I_x(\lambda_j) \right)^2,$$

where I_x is the periodogram of $x_{t,d}$ from Eq. (10.2) computed under H_0, and as usual $\lambda_j = 2\pi j/T$. The score test nature of this statistic is shown in the Appendix to this chapter where we bridge the gap to $\hat{t}_{LM}$ from Eq. (10.13):

$$\widetilde{LM}(0) \approx T \frac{6}{\pi^2} \left(\sum_{h=1}^{T-1} \frac{\hat{\rho}_x(h)}{h} \right)^2. \tag{10.14}$$

Not surprisingly, under the assumptions of Proposition 10.4, where $x_{t,d} \sim$ iid, Robinson (1994b, Theorem 1) established that $\widetilde{LM}(0) \xrightarrow{D} \chi^2(1)$.

More generally, Robinson (1994b) maintained a parametric model for $x_{t,d} = x_t$ under H_0 as in Chapter 8 with spectrum

$$f_x(\lambda; \psi) = \frac{\sigma^2}{2\pi} g(\lambda; \psi).$$

Here, ψ is a finite-dimensional parameter vector, and $g(\lambda; \psi)$ is the power transfer function of $x_t = \sum_{j=0}^{\infty} c_j \varepsilon_{t-j}$. Define the gradient column vector $\partial \ln g(\lambda; \psi)/\partial \psi$, and

$$A = \frac{\pi^2}{6} - \frac{1}{\pi} B' \left[\int_0^{2\pi} \frac{\partial \ln g(\lambda; \psi)}{\partial \psi} \left(\frac{\partial \ln g(\lambda; \psi)}{\partial \psi} \right)' d\lambda \right]^{-1} B$$

with

$$B = \int_0^{2\pi} \ln\left(2\sin\frac{\lambda}{2} \right) \frac{\partial \ln g(\lambda; \psi)}{\partial \psi} d\lambda.$$

The LM statistic accounting for short memory via $g(\lambda; \psi)$ becomes

$$\widetilde{\text{LM}}(\psi) = T \frac{1}{A} \left(\frac{2\pi}{\hat{\sigma}^2} \frac{1}{T} \sum_{j=1}^{T-1} \frac{\ln\left(2\sin\frac{\lambda_j}{2}\right)}{g(\lambda_j; \psi)} I_x(\lambda_j) \right)^2 = (\tilde{t}(\psi))^2, \qquad (10.15)$$

where $\tilde{t}(\psi)$ has been defined implicitly. Obviously, for $x_{t,d} \sim$ iid we have $g(\lambda; \psi) = 1$ and $A = \pi^2/6$, which reproduces $\widetilde{\text{LM}}(0)$. The next proposition presents the limiting distribution under the null hypothesis and under local alternatives,

$$x_{t,d} = \Delta_+^{-\theta_T} x_t, \quad \theta_T = c/\sqrt{T}. \qquad (10.16)$$

Note that for $c \neq 0$, we have a triangular array, $t = 1, \dots, T$, $T = 1, 2, \dots$, which would strictly speaking require a further index, $\{x_{t,T,d}\}$, which we omit for notational convenience. The null hypothesis, $\theta = 0$, is of course embedded for $c = 0$.

Proposition 10.6 *Let $\{x_t\}$ behind $\{x_{t,d}\}$ from Eq. (10.16) have a spectrum $f_x(\lambda; \psi) = g(\lambda; \psi) \sigma^2/2\pi$. Under some further assumptions from Robinson (1994b), it holds that*

$$\tilde{t}(\psi) \xrightarrow{D} \mathcal{N}(c\sqrt{A/2}, 1)$$

as $T \to \infty$, where $\tilde{t}(\psi)$ is the positive square root of $\widetilde{\text{LM}}(\psi)$.

Proof: Robinson (1994b, Theorem 3) for $c = 0$, and Robinson (1994b, Theorem 4) for $c \neq 0$. □

In practice, ψ is not known and has to be estimated consistently, $\hat{\psi} \xrightarrow{p} \psi$. In fact, this is the case considered by Robinson (1994b), who proved limiting normality of $\tilde{t}(\hat{\psi})$. The procedure is further burdened by determining A. To facilitate this, Robinson (1994b, p. 1427) advocated the use of an EXP(k) model; see Eq. (4.26). Similarly to Corollary 4.2, we show in the Appendix to this chapter that for the EXP(k) model we have

$$A = \frac{\pi^2}{6} - \sum_{j=1}^{k} \frac{1}{j^2}. \qquad (10.17)$$

The assumption of an ARMA model makes the computation of A much more cumbersome.

Before moving to the next procedure, we want to present a modification relevant for practice. We now assume the model Eq. (10.4), i.e.

$$y_t = f(t) + \Delta_+^{-d_0-\theta} x_t.$$

Instead of removing the deterministic function $f(t)$ from y_t, Robinson (1994b) suggested to regress the differences under the null on the correspondingly differenced deterministic regressors. Consider the example of a linear time trend:

$$f(t) = (a, \, b) \, r_t, \quad r_t' = (r_{1,t}, \, r_{2,t}) = (1, \, t).$$

We now define the (type II) differences under H_0:

$$r_{t,d} = \Delta_+^{d_0} r_t.$$

In the example of a linear time trend, we have[5]

$$r_{t,d}' = \left(\sum_{j=0}^{t-1} \pi_j(d_0), \; \sum_{j=0}^{t-1} \pi_j(d_0)(t-j) \right).$$

An ordinary least squares (OLS) regression of $\{x_{t,d}\}$ on $\{r_{t,d}\}$ yields the estimators $\widehat{a}$ and $\widehat{b}$ used to define

$$\widehat{x}_{t,d} = x_{t,d} - (\widehat{a}, \, \widehat{b}) \, r_{t,d}. \tag{10.18}$$

Finally, $\{\widehat{x}_{t,d}\}$ is fed into the test statistic $\widetilde{t}(\widehat{\psi})$, replacing, for instance, I_x by $I_{\widehat{x}}$, where $I_{\widehat{x}}$ denotes the periodogram computed from $\widehat{x}_{t,d}$. The asymptotic result in Proposition 10.6 remains unchanged as was proven in Robinson (1994b, Theorems 3 and 4).

The frequency domain LM test requires the assumption of a spelled out parametric model for $\{x_t\}$. This is the price one pays for the $\widetilde{\mathrm{LM}}(\psi)$ test having more (local) power than, e.g. the truncated test based on N_k/D_q by Harris et al. (2008). The next procedure, in contrast, is nonparametric.

Lobato and Robinson (1998) proposed a test robust with respect to short memory without having to model it. To obtain such a property, it is kind of intuitive to apply the score principle by Rao (1948) to the local Whittle likelihood $m \, R(\theta)$, where $R(\theta)$ is from Eq. (9.8) computed with $x_{t,d}$, which is $I(\theta)$, and m is the familiar bandwidth. Just as a minimization of $R(\theta)$ yielded a robust estimator in Section 9.2, the score from $m \, R(\theta)$ will provide a robust test. As in Section 9.3, let us define the sequence:

$$v_j = \ln j - \frac{1}{m} \sum_{j=1}^{m} \ln j, \quad j = 1, \dots, m.$$

Further, $\overline{I}_x$ denotes the average periodogram computed from $x_{t,d}$,

$$\overline{I}_x = \frac{1}{m} \sum_{j=1}^{m} I_x(\lambda_j).$$

---

5 When we test for a unit root, $d_0 = 1$, this simplifies to $r_{t,d}' = (0, \, 1)$.

Then Lobato and Robinson (1998) suggested

$$t_m = \frac{\frac{1}{\sqrt{m}} \sum_{j=1}^{m} I_x(\lambda_j) \, v_j}{\bar{I}_x}. \tag{10.19}$$

The justification by the score principle is given in the Appendix to this chapter.

Proposition 10.7 *Let $\{x_{t,d} - \mu\}$ from Assumption 10.1 satisfy some further assumptions from Lobato and Robinson (1998), and*

$$\frac{1}{m} + \frac{m^5 \ln^2 m}{T^4} \to 0.$$

It then holds that

$$t_m \overset{D}{\to} \mathcal{N}(0, 1)$$

as $T \to \infty$.

Proof: Lobato and Robinson (1998, Theorem 1). □

The procedure is truly nonparametric in that no assumptions on the spectrum away from zero frequency are required. Consistency against fixed alternatives is established in Lobato and Robinson (1998, Theorem 2), while it "is expected that LM can be shown to have good power against local alternatives of the form" $\theta_m = c/\sqrt{m}$ (Lobato and Robinson, 1998, p. 478). Hence, due to the nonparametric nature, there is less power compared with Proposition 10.6.

It is worth noting that the limiting results in Proposition 10.5 through 10.7 were established under the somewhat restrictive assumption of conditional homoskedasticity of $\{\varepsilon_t\}$ driving $x_t = \sum_{j=0}^{\infty} c_j \varepsilon_{t-j}$. In the next section we will overcome this condition.

10.6 Regression-based LM Test

This section brings us back to the time domain. We return for the moment to the case where $x_{t,d} = x_t$ is iid under the null hypothesis and recall the definition of $x_{t-1,d}^*$ given in Eq. (10.12). In Proposition 10.4 (b) we observe limiting normality of the sample covariance of $x_{t,d}$ and $x_{t-1,d}^*$, while Proposition 10.4 (a) gives the limit of the sample variance of $x_{t-1,d}^*$. This suggests to compute the LM statistic from a simple regression as yet another alternative to t_{LM} from Eq. (10.13) or τ_{LM} from Proposition 10.4. Let the test statistic t_ϕ denote the conventional t statistic from the auxiliary OLS regression:

$$x_{t,d} = \hat{\phi} x_{t-1,d}^* + \hat{\varepsilon}_t, \quad t = 2, \dots, T, \tag{10.20}$$

where the null hypothesis translates into

$$H_0 : \phi = 0.$$

If spelled out explicitly, this statistic reads as

$$t_\phi = \frac{\hat{\phi}}{s} \sqrt{\sum_{t=2}^{T} (x_{t-1,d}^*)^2} = \frac{\sum_{t=2}^{T} x_{t,d} x_{t-1,d}^*}{s \sqrt{\sum_{t=2}^{T} (x_{t-1,d}^*)^2}}, \qquad s^2 = \frac{1}{T-2} \sum_{t=2}^{T} \hat{\varepsilon}_t^2,$$

and it has been suggested by Breitung and Hassler (2002, Eq. (5)). Note that it is rooted in the LM principle, although it comes in disguise of a t statistic. The variance estimator s^2 computed from regression residuals converges to σ^2 under the null hypothesis since $\hat{\phi}$ converges to 0 at the conventional rate $\sqrt{T}$. Of course, any other consistent estimator could replace s^2, a natural choice being $\hat{\sigma}^2 = T^{-1} \sum_{t=1}^{T} x_{t,d}^2$ imposing the null hypothesis $\phi = 0$, which is in line with the LM nature of the test. The limiting distribution is clear under H_0:

$$t_\phi \xrightarrow{D} \mathcal{N}(0, 1).$$

The more general goal of Breitung and Hassler (2002) was fractional cointegration testing in a multivariate framework; in the univariate case, however, they just recast the LM statistic by Robinson (1991) and Tanaka (1999) in a regression framework (which will turn out convenient if $\{x_t\}$ is serially correlated or heteroskedastic). Both test statistics are asymptotically equivalent; see Breitung and Hassler (2002, Theorem 1):

$$t_\phi = t_{\mathrm{LM}} + o_p(1).$$

Note that the regression-based test can also be one sided or two sided. In the one-sided case we reject $\theta = 0$ in favor of $\theta > 0$ at significance level α if t_ϕ is larger than the $1 - \alpha$ quantile of the standard normal distribution; see Demetrescu et al. (2008, Proposition 3); vice versa, too negative values of t_ϕ are indicative of $\theta < 0$, or $d < d_0$.

Before discussing further convenient properties of t_ϕ, we want to give a historical perspective of the regression approach to fractional integration testing. Agiakloglou and Newbold (1994) pioneered the regression framework for LM tests, however, replacing $x_{t-1,d}^*$ in Eq. (10.20) by

$$x_{t-1,d}^{(K)} := \sum_{j=1}^{K} \frac{x_{t-j,d}}{j}$$

for some finite user-chosen number K. Clearly, the smaller the K, the more this test deviates from the LM test and hence loses efficiency. Another regression approach, which does not draw on the LM principle, was advocated by Dolado et al. (2002). Finally, Lobato and Velasco (2007) suggested yet another regression-based test truly anchored in the Wald principle. It replaces $x^*_{t-1,d}$ in Eq. (10.20) by[6]

$$z_{t-1,d}(d_2) := \frac{\Delta^{d_2-1} - 1}{1 - d_2} x_{t,d},$$

where d_2 has to be chosen by the user or estimated from the data. For d_2 close to one, their Wald test statistic turns into t_ϕ, since

$$\frac{\Delta^{d_2-1} - 1}{1 - d_2} \to -\ln(1 - L) = \sum_{j=1}^{\infty} \frac{L^j}{j} \quad \text{as } d_2 \to 1.$$

Hence, in a vicinity of the null hypothesis of a unit root, the Wald test by Lobato and Velasco (2007) and the above LM test are asymptotically equivalent; against fixed alternatives, however, the Wald test will be more powerful; see also Lobato and Velasco (2008). A unified treatment of the different LM tests and regression-based tests is contained in Kew and Harris (2009).

We now address the empirically relevant issue of (un)conditional heteroskedasticity. A modified version of t_{LM} from Eq. (10.15), which is robust under conditional heteroskedasticity, has been provided by Robinson (1991, Theorem 2.2). In a regression framework such a robustification is available very easily. Demetrescu et al. (2008) suggested to modify t_ϕ by employing White standard errors, sometimes also called Eicker–White standard errors to acknowledge the independent work by Eicker (1967) and White (1980). Demetrescu et al. (2008) showed that the resulting test statistic is asymptotically robust against conditional heteroskedasticity of a certain degree. With Eicker–White standard errors we replace t_ϕ computed from the simple regression Eq. (10.20) by

$$\tilde{t}_\phi = \frac{\hat{\phi} \sqrt{\sum_{t=2}^{T} (x^*_{t-1,d})^2}}{\sqrt{\sum_{t=2}^{T} (x^*_{t-1,d})^2 \hat{\varepsilon}_t^2}}, \tag{10.21}$$

where $\{\hat{\varepsilon}_t\}$ denotes again the sequence of OLS regression residuals. In fact, $\tilde{t}_\phi$ is even robust against unconditional heteroskedasticity; see Proposition 10.8. Before presenting this result, we wish to account for short memory in the data at the same time.

The regression-based LM test easily lends itself to account for serial correlation in $\{x_t\}$ by adding lagged endogenous values $x_{t-j,d}$ as regressors on the right-hand side. Demetrescu et al. (2008) proposed the (lag) augmented

---

6 We allow for general d, although Lobato and Velasco (2007) focus on the presentation for $d = 1$ (unit root).

regression-based LM test (ALM) building on

$$x_{t,d} = \widehat{\phi}\, x_{t-1,d}^* + \sum_{j=1}^{k} \widehat{a}_j x_{t-j,d} + \widehat{\varepsilon}_t, \quad t = k+2, \dots, T, \tag{10.22}$$

where the variable of interest, $x_{t-1,d}^*$, is again from Eq. (10.12). The underlying short memory process $\{x_t\}$ is assumed to be an AR(k) process,

$$x_t = \sum_{j=1}^{k} a_j x_{t-j} + \varepsilon_t,$$

where $A(z) = 1 - \sum_{j=1}^{k} a_j z^j = 0$ has only roots outside of the unit circle; see Eq. (3.7). The white noise sequence $\{\varepsilon_t\}$ is assumed to have finite eighth moments, which allows for some mild conditional heteroskedasticity. To account for eventual conditional heteroskedasticity, Demetrescu et al. (2008) advocated the use of robust standard errors, justified by the following result.

Proposition 10.8 *Let $\{x_{t,d}\}$ from Assumption 10.1 be a stationary AR(k) process and satisfy some additional assumptions by Demetrescu et al. (2008). Let $\widetilde{t}_\phi$ stand for the t statistic testing for $\phi = 0$ computed from Eq. (10.22) with Eicker–White standard errors. It then holds under some further assumptions that*

$$\widetilde{t}_\phi \xrightarrow{D} \mathcal{N}(0,\,1)$$

as $T \to \infty$.

Proof: Demetrescu et al. (2008, Proposition 2). □

Demetrescu et al. (2008, Proposition 3) showed that the test has power against local alternatives where $\theta_T \neq 0$ is $\theta_T = c/\sqrt{T}$, $c \neq 0$. This is the same rate as in Proposition 10.6 due to a fully specified parametric model; see also Kew and Harris (2009, Theorem 1). Again, against the local alternative $c > 0$, the null hypothesis is rejected for too large values of $\widetilde{t}_\phi$ and vice versa for $c < 0$.

Finally, we have to account for deterministic components. Demetrescu et al. (2008) adopted the approach by Robinson (1994b) introduced in the previous section: regress the differences of the data on the differenced deterministic regressors and obtain residuals $\widehat{x}_{t,d}$ from Eq. (10.18). They enter the ALM procedure, i.e. $x_{t-1,d}^*$ is replaced by

$$\widehat{x}_{t-1,d}^* = \sum_{j=0}^{t-1} \frac{\widehat{x}_{t-j,d}}{j},$$

and regression Eq. (10.22) is computed in terms of $\{\hat{x}_{t,d}\}$. Demetrescu et al. (2008, Proposition 4) show that the limiting result from Proposition 10.8 continues to hold.

It is nowadays believed that many time series display unconditional heteroskedasticity, too. To accommodate such a feature, Kew and Harris (2009) relaxed the assumption that the sequence $\{\varepsilon_t\}$ behind $\{x_t\}$ is stationary white noise. They allowed for unconditional heteroskedasticity of very general form, in fact only ruling out that the variance is unbounded or degenerate:

$$0 < \underline{\sigma}^2 \leq E(\varepsilon_t^2) = \sigma_t^2 \leq \overline{\sigma}^2 < \infty. \tag{10.23}$$

Under some higher moment assumptions, Kew and Harris (2009) found that the statement of Proposition 10.8 remains true. For this to hold, it is crucial to employ the Eicker–White standard errors suggested by Demetrescu et al. (2008). We highlight this robustness result.

Proposition 10.9 *Let us maintain the assumptions of Proposition 10.8, where $\{x_{t,d}\}$ satisfies some additional assumptions by Kew and Harris (2009). It then holds under heteroskedasticity according to Eq. (10.23) for $\tilde{t}_\phi$ from Eq. (10.22) with Eicker–White standard errors that*

$$\tilde{t}_\phi \overset{D}{\to} \mathcal{N}(0, 1)$$

as $T \to \infty$.

Proof: Kew and Harris (2009, Theorem 1). □

Recently, Cavaliere et al. (2015) proposed an alternative test that is robust under conditional as well as unconditional heteroskedasticity. They considered a parametric ARFIMA model evaluated with a bootstrap implementation of the score statistic. The issue of unconditional heteroskedasticity is likely to attract increasing interest in the future, since many empirical series are believed to be driven by time-varying volatility.

We further have to address the choice of k for the ALM test from Eq. (10.22). At first glance, it is tempting to employ a data-driven device such as information criteria or sequential testing. Since the regressor $x_{t-1,d}^*$ is correlated with $x_{t-j,d}$, such a procedure is subject to the pitfalls of post-model selection testing. It can be shown that a data-driven lag length selection will result in size distortions that do not vanish asymptotically, and this effect has been quantified in computer experiments by Demetrescu et al. (2011); see Leeb and Pötscher (2005) for a general exposition to post-model selection testing. Therefore, Demetrescu et al. (2008) considered deterministic rules of thumb to determine k, e.g.

$$k_4 = \lfloor 4(T/100)^{1/4} \rfloor \quad \text{and} \quad k_{12} = \lfloor 12(T/100)^{1/4} \rfloor.$$

On experimental grounds they favored the choice according to k_4, providing reasonably good size but resulting in more power than k_{12}. Further note that the ALM test as proposed originally by Demetrescu et al. (2008) was semiparametric in that it imposed a fractionally integrated model on $\{y_t\}$ but was silent with respect to a parametric model for $\{x_t\}$. Demetrescu et al. (2008) only assumed an AR(∞) representation that an AR(k) model approximates with growing k as $T \to \infty$.

Finally, we want to appreciate the similarity of the ALM test to the well-known augmented Dickey-Fuller (ADF) test, pioneered by Dickey and Fuller (1979) and Said and Dickey (1984); see also Chang and Park (2002). It tests for $d_0 = 1$ and relies on the auxiliary lag augmented regression:

$$\Delta y_t = \widehat{\phi}\, y_{t-1} + \sum_{j=1}^{k} \widehat{a}_j \Delta y_{t-j} + \widehat{\varepsilon}_t, \quad y_{t-1} = \sum_{j=0}^{t-1} \Delta y_{t-j}, \tag{10.24}$$

where y_{t-1} equals the cumulation of the differences as long as the starting value is 0. Therefore, this parallels Eq. (10.22) for $d_0 = 1$ with one (crucial) exception: The regressors on $\widehat{\phi}$ are (asymptotically) stationary in Eq. (10.22) but nonstationary in Eq. (10.24). Due to the nonstationarity of $\{y_{t-1}\}$, this regressor is asymptotically uncorrelated with the lagged differences on the right-hand side. Consequently, data-driven lag length selection can be safely applied in Eq. (10.24) without running into post-model selection problems; see also Ng and Perron (1995). The resulting limiting distribution, however, is not normal due to nonstationarity; one rather has the celebrated Dickey–Fuller distribution asymptotically. Standard unit root tests in the tradition of Dickey and Fuller (1979) have little power against fractional alternatives; see Diebold and Rudebusch (1991), Hassler and Wolters (1994), and Krämer (1998). This has been one motivation to introduce tests against fractional alternatives.

10.7 Technical Appendix: Proofs

Proof of Proposition 10.4

We begin with a couple of auxiliary results and remarks. First, define

$$x^{**}_{t-1,d} = \sum_{j=1}^{\infty} j^{-1} x_{t-j,d}, \quad \delta_t = x^{**}_{t-1,d} - x^{*}_{t-1,d} = \sum_{j=t}^{\infty} j^{-1} x_{t-j,d},$$

where δ_t is the difference between the variable from Eq. (10.12) and the infeasible counterpart $x^{**}_{t-1,d}$. Notice that $\{x^{**}_{t-1,d}\}$ is strictly stationary and ergodic with variance:

$$\mathrm{Var}(x^{**}_{t-1,d}) = \mathrm{E}((x^{**}_{t-1,d})^2) = \sigma^2 \sum_{j=1}^{\infty} j^{-2} = \sigma^2 \frac{\pi^2}{6},$$

and

$$\text{Var}(\delta_t) = E(\delta_t^2) = E(\delta_t x_{t-1,d}^{**}) = \sigma^2 \sum_{j=t}^{\infty} j^{-2}.$$

We will use the following bound:

$$\sum_{j=t}^{\infty} j^{-2} - t^{-2} = \sum_{j=t}^{\infty} (j+1)^{-2} \leq \sum_{j=t}^{\infty} \int_{j}^{j+1} s^{-2} ds = t^{-1}. \tag{10.25}$$

Further, the following lemma will be convenient (see also White, 2001, Proposition 3.52):

Lemma 10.1 *Let $\{x_n\}$, $n \in \mathbb{N}$, be a stochastic sequence with $E(|x_n|) < \infty$ and $\sum_{n=1}^{\infty} E(|x_n|) < \infty$. Then $\sum_{n=1}^{N} x_n$ converges almost surely as $N \to \infty$.*

Proof: Lukacs (1975, Theorem 4.2.1) showed almost sure convergence of $\sum_{n=1}^{N} |x_n|$, such that $\sum_{n=N+1}^{\infty} |x_n| \overset{a.s.}{\to} 0$. Hence, $\left|\sum_{n=N+1}^{\infty} x_n\right|$ converges to 0, which proves the result. $\qquad\square$

Now, we are equipped to commence the proof of Proposition 10.4.

(a) Because of Eq. (10.25) we have

$$\sum_{t=1}^{T} \frac{E(\delta_t^2)}{t} \leq \sigma^2 \sum_{t=1}^{T} (t^{-2} + t^{-3}),$$

which converges by Lemma 3.1 (d). Therefore, we know from Kronecker's Lemma 2.1 that $T^{-1} \sum_{t=1}^{T} E(\delta_t^2) \to 0$, which shows the first claim:

$$\frac{\sum_{t=2}^{T} E\left((x_{t-1,d}^*)^2\right)}{T} = \frac{\sum_{t=2}^{T} E\left((x_{t-1,d}^{**})^2\right) - E(\delta_t^2)}{T} \to \sigma^2 \frac{\pi^2}{6}.$$

Further, Lemma 10.1 ensures that $\sum_{t=1}^{T} \delta_t^2/t$ converges almost surely, which implies again by Lemma 2.1 that $T^{-1} \sum_{t=1}^{T} \delta_t^2 \to 0$ with probability 1.[7] Since $E(\delta_t^2) = E(\delta_t x_{t-1,d}^{**})$, it equally holds that $T^{-1} \sum_{t=1}^{T} \delta_t x_{t-1,d}^{**} \to 0$ almost surely, such that the second claim follows by ergodicity of $\{x_{t-1,d}^{**}\}$.

(b) We first establish the limiting normality of $T^{-1/2} \sum_{t=2}^{T} x_{t,d} x_{t-1,d}^{**}$. Under the assumption that $\{x_t\}$ is iid, the product $\{\eta_t\} = \{x_{t,d} x_{t-1,d}^{**}\}$ forms a stationary martingale difference sequence (white noise). To see this, define the σ-field generated by its past, $\eta_{t-1}, \eta_{t-2}, \ldots$:

$$\mathcal{F}_{1,t-1} = \sigma(\eta_{t-1}, \eta_{t-2}, \ldots).$$

---

7 Note that Kronecker's Lemma 2.1 was formulated for deterministic sequences. But since almost sure convergence means conventional convergence with probability 1, Kronecker's Lemma holds almost surely, too; see also Davidson (1994, p. 307).

Similarly, consider the larger σ-field $F_{2,t-1}$, $F_{1,t-1} \subseteq F_{2,t-1}$, generated by the same information plus $x^{**}_{t-1,d}$:

$$F_{2,t-1} = \sigma(x^{**}_{t-1,d}, \eta_{t-1}, \eta_{t-2}, \ldots).$$

Then the MDS property is established as follows:[8]

$$\begin{aligned}
E(\eta_t \mid F_{1,t-1}) &= E(E(\eta_t \mid F_{2,t-1}) \mid F_{1,t-1}) \\
&= E(x^{**}_{t-1,d} E(x_{t,d} \mid F_{2,t-1}) \mid F_{1,t-1}) \\
&= E(x^{**}_{t-1,d} \cdot 0 \mid F_{1,t-1}) = 0 \, ;
\end{aligned}$$

the first equality follows by the so-called law of iterated expectations, (see, e.g. Breiman (1992, Proposition 4.20) or Davidson (1994, Theorem 10.26)); the second equality follows from Doob (1953, Theorem 8.3, p. 22) or Davidson (1994, Theorem 10.10); and the third equality holds because $x_{t,d}$ is iid, and hence a zero mean variable independent of $F_{2,t-1}$ (see, e.g. Breiman (1992, Proposition 4.20) or Davidson (1994, Theorem 10.22)). Since

$$\text{Var}\left(\sum_{t=2}^{T} x_{t,d} x^{**}_{t-1,d} \right) = \text{Var}\left(\sum_{t=2}^{T} \eta_t \right) = (T-1)\sigma^4 \frac{\pi^2}{6},$$

we can apply Theorem 7.1:

$$\frac{1}{\sqrt{T}} \sum_{t=2}^{T} x_{t,d} x^{**}_{t-1,d} \overset{D}{\rightarrow} \mathcal{N}\left(0, \sigma^4 \frac{\pi^2}{6} \right).$$

Second, we are left with showing that this result continues to hold when replacing $x^{**}_{t-1,d}$ by $x^{*}_{t-1,d}$. Consider

$$\sum_{t=2}^{T} x_{t,d} x^{**}_{t-1,d} - \sum_{t=2}^{T} x_{t,d} x^{*}_{t-1,d} = \sum_{t=2}^{T} x_{t,d} \delta_t = A_T, \quad \text{say}.$$

With the arguments given in the proof of (a), it is straightforward to show that $\text{Var}(T^{-1/2} A_T) \rightarrow 0$, such that $T^{-1/2} A_T$ converges to 0 in probability, as required.

(c) Since $\hat{\sigma}^2$ is consistent (by ergodicity), the proof is completed by applying (a) and (b). □

Derivations for Section 10.5

(a) Derivation of Eq. (10.14)

Let us for brevity define the Riemann sum:

$$\text{RS}_T = \frac{2\pi}{T} \sum_{j=1}^{T-1} \ln\left(2\sin\frac{\lambda_j}{2} \right) I_x(\lambda_j).$$

---

8 More precisely, the following equalities hold only with probability 1.

It can be approximated by the following integral for large T:

$$RS_T \approx \int_0^{2\pi} \ln\left(2\sin\frac{\lambda}{2}\right) I_x(\lambda)\, d\lambda.$$

Then we substitute the periodogram according to Eq. (4.9):

$$RS_T \approx \int_0^{2\pi} \ln\left(2\sin\frac{\lambda}{2}\right) \frac{1}{2\pi}\left(\hat{\gamma}_x(0) + 2\sum_{h=1}^{T-1}\hat{\gamma}_x(h)\,\cos(\lambda h)\right) d\lambda.$$

Adopting arguments from the proof of Corollary 8.1, we find that

$$\int_0^{2\pi} \ln\left(2\sin\frac{\lambda}{2}\right) d\lambda = 0.$$

Further,

$$\int_0^{2\pi} \ln\left(2\sin\frac{\lambda}{2}\right) \cos(\lambda h)\, d\lambda = -\frac{\pi}{h}, \tag{10.26}$$

where we used Gradshteyn and Ryzhik (2000, Eq. (4.384.8)). Therefore, we observe that

$$RS_T \approx 0 + \frac{2}{2\pi}\sum_{h=1}^{T-1}\hat{\gamma}_x(h)\int_0^{2\pi}\ln\left(2\sin\frac{\lambda}{2}\right)\cos(\lambda h)\, d\lambda$$

$$= -\sum_{h=1}^{T-1}\frac{\hat{\gamma}_x(h)}{h},$$

which provides us with

$$\widetilde{LM}(0) \approx T\,\frac{6}{\pi^2}\left(\sum_{h=1}^{T-1}\frac{\hat{\rho}_x(h)}{h}\right)^2.$$

(b) Derivation of Eq. (10.17)

For the EXP(k) model from Eq. (4.26), we have that $\ln g(\lambda;\psi) = \sum_{\ell=1}^{k}\psi_\ell\cos(\ell\lambda)$ with

$$\frac{\partial \ln g(\lambda;\psi)}{\partial\psi} = (\cos(\lambda), \cos(2\lambda), \ldots, \cos(k\lambda))'.$$

From Corollary 4.2 it follows that

$$\int_0^{2\pi}\frac{\partial \ln g(\lambda;\psi)}{\partial\psi}\left(\frac{\partial \ln g(\lambda;\psi)}{\partial\psi}\right)' d\lambda = \pi\, I_k.$$

By Eq. (10.26),

$$B' = \int_0^{2\pi}\ln\left(2\sin\frac{\lambda}{2}\right)(\cos(\lambda), \ldots, \cos(k\lambda))\, d\lambda = -\pi\left(1, \frac{1}{2}, \ldots, \frac{1}{k}\right).$$

Hence, under the EXP(k) assumption it holds that

$$A = \frac{\pi^2}{6} - \frac{B'B}{\pi^2} = \frac{\pi^2}{6} - \sum_{j=1}^{k} \frac{1}{j^2}.$$

(c) Derivation of Eq. (10.19)

The score under H_0, $S = S(0)$, of $m\,R(\theta)$ from Eq. (9.8) becomes

$$S(0) = m\,\frac{\partial R(\theta)}{\partial \theta}\Big|_{\theta=0}$$

$$= \frac{2}{\frac{1}{m}\sum_{j=1}^{m} I_x(\lambda_j)\lambda_j^{2\theta}} \sum_{j=1}^{m} I_x(\lambda_j)\lambda_j^{2\theta}\ln \lambda_j - 2\sum_{j=1}^{m} \ln \lambda_j\Big|_{\theta=0}$$

$$= 2\,\frac{\sum_{j=1}^{m}(I_x(\lambda_j)\ln \lambda_j - \bar{I}_x \ln \lambda_j)}{\bar{I}_x}.$$

With $\ln \lambda_j = \ln 2\pi + \ln j - \ln T$, it follows that

$$S = S(0) = 2\,\frac{\sum_{j=1}^{m} I_x(\lambda_j)v_j}{\bar{I}_x}.$$

Lobato and Robinson (1998, Eq. (A.4)) showed under H_0 that $\partial^2 R(0)/\partial\theta^2 \to 4$. Hence, the *observed* Fisher information of $m\,R(\theta)$ can be approximated under H_0 by $4m$. This yields as score statistic:

$$\frac{S^2(\theta)}{\frac{\partial^2 mR(\theta)}{\partial\theta^2}}\Big|_{\theta=0} \approx \frac{S^2}{4m} = t_m^2,$$

as required.

11

Further Topics

This final chapter addresses seven additional topics relevant for practice. They extend the material of the previous chapters into several directions, considering, for instance, generalizations of the models from Chapter 6 or modifications of estimation and test procedures from Chapters 8 through 10. Hence, we believe that the stage is well set for a brief and concise treatment in one section each. The final section closes with only listing, briefly commenting and providing some references on yet further topics not dealt with in detail.

11.1 Model Selection and Specification Testing

The maximum likelihood (ML) estimators discussed in Chapter 8 require the specification of a parametric model. The majority of applications employs ARFIMA(p, d, q) models. Often, the lag orders p and q are determined by means of the most commonly used information criteria: Akaike's information criterion (AIC) (see Akaike (1974)) or the Bayesian information criterion (BIC) by Schwarz (1978). In a short memory context, where information criteria are applied to determine the lag orders of ARMA(p, q) models, it is well known from Hannan (1980) that BIC determines the true p and q consistently (under some standard assumptions). For stable AR(p) models, Shibata (1976) proved that AIC is not consistent and tends to overestimate the true p; see also Tsay (1984) for the nonstationary case.

Under fractional integration there is only one theoretical contribution on information criteria that we are aware of, and this is the paper by Beran et al. (1998). Assuming an ARFI(p, d) process and the conditional sum of squares (CSS) estimation discussed in Section 8.3, they showed that BIC estimates p consistently (Beran et al., 1998, Theorem 2), while AIC does not. They do not address how well information criteria can discriminate between ARFI(p, d) and AR(p), i.e. between $d = 0$ and $d \neq 0$. The latter issue was treated by Crato and Ray (1996), however on experimental and not on theoretical grounds. Relying on ML approximations, their results are not very encouraging. Crato and

Time Series Analysis with Long Memory in View, First Edition. Uwe Hassler.
© 2019 John Wiley & Sons, Inc. Published 2019 by John Wiley & Sons, Inc.

Ray (1996, p. 114) indicated that "a short-memory component and long-range dependence are very difficult to distinguish in small samples." Further, Crato and Ray (1996, p. 122) concluded that "the most difficult problem in this model identification setting is the choice between short-memory ARMA models and long-memory ARFIMA models," which is once more a motivation for powerful tests discussed in Chapter 10. Further experimental evidence pointing at the superiority of BIC over AIC to identify ARFIMA(p, d, q) models was provided by Carbonez (2009). Baillie et al. (2014b) focused on the usage of information criteria in two-step procedures: First, estimate d (semiparametrically) and compute fractional differences. Second, apply information criteria to fit an ARMA model to the differences. They observed that even BIC (with a stronger penalty term than AIC) introduces "a large upward bias in the estimated order of the short memory component" (Baillie et al., 2014b, p. 121). Consequently, they suggested a modified information criterion with an even stronger penalty term; we omit details.

Milhøj (1981) suggested to test for the null hypothesis of a parametric model in the frequency domain. While he assumed short memory, Beran (1992) carried his idea to the framework of fractional integration. Let $f_0(\lambda)$ denote the spectrum under the null hypothesis, e.g.

$$f_0(\lambda) = \left[4\sin^2\left(\frac{\lambda}{2}\right) \right]^{-d} f_{p,q}(\lambda),$$

with $f_{p,q}$ being the spectrum of an ARMA(p, q) process; see Eq. (4.23). Let V_j denote the ratio of the periodogram and the hypothesized spectrum at the harmonic frequencies $\lambda_j = 2\pi j / T$, $V_j = I(\lambda_j)/f_0(\lambda_j)$. Then, Milhøj (1981) suggested the ratio

$$M_T = \frac{\frac{2\pi}{T} \sum_{j=1}^{T-1} V_j^2}{\left(\frac{2\pi}{T} \sum_{j=1}^{T-1} V_j \right)^2}.$$

For the short memory case, i.e. under the restriction $d = 0$, he established limiting normality of M_T under the null with mean $1/\pi$ and variance $2/(T\pi^2)$. This continues to hold when the parameters behind f_0 are estimated consistently; see Milhøj (1981, p. 181) for details. Beran (1992) extended these results for long memory, $0 < d < 1/2$, under the assumption of a Gaussian process. Deo and Chen (2000), however, argued that his proof is not complete. For the special case of Gaussian white noise, $f_0(\lambda) = \sigma^2/2\pi$, they demonstrated that the argument in Beran (1992, Appendix A) is defective and noted that "Though our results on the discrepancy in the asymptotic behaviour [...] do not cover the case when $f(\cdot)$ is unbounded at the origin, there would seem to be no reason to believe that this discrepancy does not exist in this case" (Deo and Chen, 2000, p. 161). Hence, the limiting behavior of M_T under $0 < d < 1/2$ is still an open question. Chen and Deo (2004) suggested to replace V_j in M_T by

$$\tilde{f}_e(\lambda_j) = \frac{2\pi}{T} \sum_{k=1}^{T-1} W(\lambda_j - \lambda_k) \frac{I(\lambda_k)}{f_0(\lambda_k)} ,$$

where $W(\cdot)$ is a so-called spectral window. The limiting normal distribution was provided by Chen and Deo (2004, Theorems 2 and 4) under the assumption of Gaussian processes.

Of particular interest is the null hypothesis that the model behind the data is fractionally integrated noise, FIN(d). A series of applications found such a simple model to provide a valid approximation to real dynamics; see Hassler and Wolters (1995) for inflation rates, Christensen and Nielsen (2007) for realized volatility, and for opinion poll series the papers by Byers et al. (1997, 2007) and Dolado et al. (2003). In order to test the FIN(d) hypothesis, Hassler and Wolters (1995) suggested and applied a test in the tradition of Hausman (1978): It compares an ML-type estimator computed under the null hypothesis, which is efficient under the null but inconsistent under the alternative, with an inefficient semiparametric estimator being consistent under both the null and the alternative hypothesis. Hassler and Wolters (1995, Eq. (3.18)) considered the difference of the local log-periodogram regression estimator $\widehat{d}_{PR}^{(m)}$ (Section 9.1) and the parametric Whittle estimator $\widehat{d}_W$ computed under FIN(d); see Section 8.4. Under FIN(d) it follows from Corollary 8.1(a) for any ML estimator that

$$\sqrt{T}(\widehat{d}_{ML} - d_0) \overset{D}{\to} \mathcal{N}\left(0, \frac{6}{\pi^2}\right) ,$$

while $\widehat{d}_{PR}^{(m)}$ converges more slowly with m; cf. Proposition 9.1. Hence Hassler and Wolters (1995) observed under FIN(d) for a Hausman-type statistic that

$$\sqrt{m} \frac{\sqrt{24}}{\pi} (\widehat{d}_{PR}^{(m)} - \widehat{d}_W) \overset{D}{\to} \mathcal{N}(0,1).$$

More generally, let us assume any semiparametric estimator $\widehat{d}_{SP}$ converging to the true parameter value with rate $\sqrt{m}$,

$$\sqrt{m}(\widehat{d}_{SP} - d_0) \overset{D}{\to} \mathcal{N}(0, V_{SP}),$$

with $m = o(T)$. It then holds under the null hypothesis FIN(d) for the Hausman statistic H that

$$H = \frac{\sqrt{m}}{\sqrt{V_{SP}}} (\widehat{d}_{SP} - \widehat{d}_{ML}) \overset{D}{\to} \mathcal{N}(0,1) ; \tag{11.1}$$

see also Davidson and Sibbertsen (2009).

11.2 Spurious Long Memory

Since Ding et al. (1993), ample evidence has been documented on long memory in power transformations of the absolute value of daily returns. Granger and Ding (1996) explained such evidence by means of fractional integration of estimated order around 0.47. In the previous section, we addressed the issue of selecting a specification within the class of fractionally integrated models. Now, we turn to the issue that observed long memory may arise from different processes than fractional integration. Lobato and Savin (1998) addressed this topic analyzing again squared and absolute returns. They coined the terminology of true long memory (essentially fractional integration) and spurious long memory that might be caused by breaks in parameters of otherwise short memory models. Much earlier, Klemeš (1974) addressed the potential confusion of long memory and nonstationarity in the mean. Further, the difficulty to disentangle slowly varying trends and long memory has been studied by Bhattacharya et al. (1983) and Künsch (1986); see also Giraitis et al. (2001b) and the review by Sibbertsen (2004).

Our exposition here focuses on the simple case of a time-varying, stochastic mean function $\{\mu_t\}$ superimposed by an iid process $\{\varepsilon_t\}$, such that the observable series $\{z_t\}$ becomes

$$z_t = \mu_t + \varepsilon_t, \quad \varepsilon_t \sim \text{iid}(0, \sigma_\varepsilon^2), \quad t = 1, \ldots, T. \tag{11.2}$$

The mean functions μ_t and ε_t are contemporaneously independent. Depending on the specification of $\{\mu_t\}$, it will follow that

$$\text{Var}\left(\sum_{t=2}^{T} \Delta z_t\right) = \Theta(T^{2\delta-1}), \quad \delta > 1/2, \tag{11.3}$$

where δ is determined by assumptions with respect to $\{\mu_t\}$. The notation $g(T) = \Theta(T^{2\delta-1})$ means that the (positive) function g diverges with rate $T^{2\delta-1}$, which is a stronger statement than $g(T) = O(T^{2\delta-1})$ with $O(\cdot)$ from Eq. (5.7). More precisely, we define Eq. (11.3) to mean that there exist lower and upper bounds b_ℓ and b_u such that

$$0 < b_\ell \leq \frac{\text{Var}\left(\sum_{t=2}^{T} \Delta z_t\right)}{T^{2\delta-1}} \leq b_u$$

for large enough T. Such a behavior has to be compared with the variance behavior under fractional integration. Let $\{y_t\}$ satisfy the assumptions of Proposition 7.2. It then holds that

$$\text{Var}\left(\sum_{t=1}^{T} y_t\right) = \Theta(T^{2d+1}), \quad y_t \sim I(d), \quad -1/2 < d < 1/2. \tag{11.4}$$

This trivially extends to the case of nonstationarity in terms of differences:

$$
\text{Var}\left(\sum_{t=2}^{T} \Delta y_t \right) = \Theta(T^{2d-1}), \quad y_t \sim I(d), \quad 1/2 < d < 3/2. \tag{11.5}
$$

Comparing Eq. (11.3) with (11.5), the process $\{z_t\}$ is likely to be confused with fractional integration of order δ, which amounts to spurious fractional integration.

Granger and Hyung (2004) considered the model of random level shifts pioneered by Chen and Tiao (1990). Let $\{q_t\}$ be an iid sequence following a Bernoulli distribution with probability $p \in (0, 1)$, and $\{\eta_t\}$ is an iid sequence independent thereof and independent of ε_t. Under zero starting value conditions, consider

$$
\mu_t = \mu_{t-1} + q_t \, \eta_t = \sum_{i=1}^{t} q_i \, \eta_i. \tag{11.6}
$$

For fixed $p > 0$, it is straightforward to verify for z_T from Eq. (11.2) that

$$
\text{Var}(z_T) = \text{Var}\left(\sum_{t=2}^{T} \Delta z_t \right) = \Theta(T), \tag{11.7}
$$

such that $\delta = 1$ in terms of Eq. (11.3) and, therefore, the process is likely to be confused with an integrated process of order 1. If the shift probability vanishes with the sample size, $p_T = c/T$, one obtains a model with rare shifts. Under this assumption, Granger and Hyung (2004, Proposition 3) approximated the bias of $\widehat{d}_{\text{PR}}^{(m)}$ from the local log-periodogram regression; see Section 9.1. More precisely, Perron and Qu (2010, Proposition 3) disentangled the effect of μ_t and ε_t from Eq. (11.2) on the periodogram of z_t. They observed that the level shift component is dominant only up to the jth harmonic frequency with $j = O(\sqrt{T})$. Under a slightly different parameterization of a slowly changing mean function, $\mu_t = (1 - p_T)\mu_{t-1} + \sqrt{p_T}\eta_t$, Smith (2005, Theorem 1) approximated $\widehat{d}_{\text{PR}}^{(m)}$, too.

For an alternative mean function, (Engle and Smith, 1999) suggested the so-called stochastic permanent break model given by

$$
\mu_t = \mu_{t-1} + q_{t-1}(\gamma) \, \varepsilon_{t-1} = \mu_0 + \sum_{i=1}^{t} q_{t-i} \, (\gamma)\varepsilon_{t-i}, \tag{11.8}
$$

with $q_{t-i}(\gamma)$ being nondecreasing in $|\varepsilon_{t-i}|$, in particular

$$
q_{t-i}(\gamma) = \frac{\varepsilon_{t-i}^2}{\gamma + \varepsilon_{t-i}^2}, \quad \gamma > 0. \tag{11.9}
$$

By construction, μ_t is independent of ε_t in Eq. (11.2). Diebold and Inoue (2001) modified this setup and replaced γ by $\gamma_T (b)$, assuming

$$
\gamma_T(b) = \Theta(T^b), \quad b > 0, \tag{11.10}
$$

such that $\gamma_T(b) \to \infty$ as T diverges. Under some additional assumptions, Diebold and Inoue (2001, Proposition 2) established that

$$\text{Var}\left(\sum_{t=2}^{T} \Delta z_t\right) = \Theta(T^{1-2b}). \tag{11.11}$$

Consequently, $\{\Delta z_t\}$ from Eq. (11.2) with (11.8) and (11.10) mimics for $0 < b < 1/2$ an $I(-b)$ process, and $\{z_t\}$ may be confused with an $I(1-b)$ process. For larger b, or smaller q_{t-1}, the mean function μ_t varies less, and the process $\{z_t\}$ becomes less persistent. Diebold and Inoue (2001) studied the effect again on the local log-periodogram regression experimentally. Further, they allowed for a Markov-switching mean function not considered here.

Finally, we turn to a classical unobserved components model, consisting of a random walk component, $\mu_t = \sum_{j=1}^{t} \eta_j$, disturbed by additive noise. With growing T, the random walk dominates the noise component, and the process behaves like an $I(1)$ process; see also Hassler and Kuzin (2009). Therefore, Hassler et al. (2014) dampened the random walk component as its variance increases with t:

$$\mu_t = \frac{\lambda}{t^{1-\delta}} \sum_{j=1}^{t} \eta_j, \tag{11.12}$$

where $\{\eta_t\}$ is again iid$(0,1)$ independent of $\{\varepsilon_t\}$, $\lambda > 0$ and $1/2 < \delta \leq 1$. The larger δ, the more weight is attached to the random walk component; for $\delta = 1$, the standard random walk plus noise model is reproduced. Assuming Eq. (11.12), it follows from Proposition 7.3 that $T^{1/2-\delta}z_T$ converges to a Brownian motion. With $z_T = \sum_{t=2}^{T} \Delta z_t$, we conclude that

$$\text{Var}\left(\sum_{t=2}^{T} \Delta z_t\right) = \Theta(T^{2\delta-1}). \tag{11.13}$$

Comparing Eqs. (11.13) with (11.5) leads to the suggestion that z_t might be taken for $I(\delta)$. In order to quantify such an intuition, Hassler et al. (2014) applied the regression-based LM test from Section 10.6 to series simulated from this model and tested for $I(\delta)$. When the true $\delta = 1$, such that $\{z_t\}$ is integrated (of order 1), the experimental level turned out to be close to the nominal one; for $\delta < 1$, however, Hassler et al. (2014, Table 4) reported considerable power indicating that the LM test for fractional integration may be able to discriminate between true and spurious long memory.

There is a growing literature on tests designed to discriminate between true and spurious long memory, with prominent papers by Ohanissian et al. (2008) and Qu (2011). Further, McCloskey and Perron (2013) studied the trimmed log-periodogram regression estimator ($\overline{d}_{\mathrm{LP}}^{(\ell)}$ from Section 10.6) in the presence of breaks in mean or of deterministic trends; under certain conditions and when choosing the trimming parameter ℓ appropriately, they proved that the estimation and inference is robust with respect to such contamination of the true long memory. A similar result has been obtained by Iacone (2010) when trimming the local Whittle estimator; see also the robust modification of the local Whittle estimator by Hou and Perron (2014).

11.3 Forecasting

Since the early papers by Granger and Joyeux (1980) and Geweke and Porter-Hudak (1983), forecasting under long memory was a major motivation to study fractionally integrated models; see also the example by Brockwell and Davis (1991, p. 533). We assume that the observed stretch of data $y_1, \ldots, y_T$ is used to predict h steps ahead, $h = 1, 2, \ldots$. Assuming an invertible fractionally integrated process from Assumption 6.2,

$$\Delta^d y_t = x_t = C(L)\,\varepsilon_t\,, \quad C(L) = \sum_{j=0}^{\infty} c_j\,L^j\,,$$

one obtains by convolution of Δ^d and $1/C(L)$ the AR(∞) representation

$$y_t = \sum_{j=1}^{\infty} a_j(d)\,y_{t-j} + \varepsilon_t\,,$$

where $\frac{\Delta^d}{C(L)} = 1 - \sum_{j=1}^{\infty} a_j(d)\,L^j$. Along the lines of Brockwell and Davis (1991, Theorem 5.5.1), the mean square distance minimizing predictor $\tilde{y}_T(h)$ for y_{T+h} becomes

$$\tilde{y}_T(h) = \sum_{j=1}^{h-1} a_j(d)\,\tilde{y}_T(h - j) + \sum_{j=h}^{\infty} a_j(d)\,y_{T+h-j}. \tag{11.14}$$

In practice, the infinite sum has to be truncated, $\sum_{j=h}^{T+h-1} a_j(d)\,y_{T+h-j}$. Further, d has to be replaced by a consistent estimator, and similarly $c_j = c_{j,\psi}$ will depend on a parameter vector ψ that has to be estimated. Most papers assume an ARMA process for $\{x_t\}$, such that $C(L) = \Theta(L)/\Phi(L)$, with the notable exception of Hurvich (2002) who allowed for FEXP models where $\{x_t\}$ follows Bloomfield's exponential model; see Section 4.4. Finally, one may account for means different from zero by forecasting from demeaned variables. The prediction method Eq. (11.14) comes in most naturally when assuming a parametric model as in Chapter 8.

In case of semiparametric estimation, it may be more natural to fractionally difference the data first, forecast the differences, and feed these forecasts into a recursion. With $\Delta^d = 1 + \sum_{j=1}^{\infty} \pi_j(d) \, L^j$, we have

$$y_t = - \sum_{j=1}^{\infty} \pi_j(d) \, y_{t-j} + x_t.$$

Assuming a forecast $\hat{x}_T(h)$ for x_{T+h}, this suggests for $h = 1, 2, \ldots$

$$\hat{y}_T(h) = - \sum_{j=1}^{h-1} \pi_j(d) \, \hat{y}_T(h - j) - \sum_{j=h}^{\infty} \pi_j(d) \, y_{T+h-j} + \hat{x}_T(h). \tag{11.15}$$

Indeed, it is not difficult to show that $\hat{y}_T(h)$ is equal to the mean square error minimizing prediction formula proposed by Peiris and Perera (1988, Theorem 3.1). Again, the infinite sum will be replaced in practice by $\sum_{j=h}^{T+h-1} \pi_j(d) \, y_{T+h-j}$. To obtain $\hat{x}_T(h)$, one would compute $\hat{x}_t = \Delta^{\hat{d}} y_t$ for $t = 1, 2, \ldots, T$ (or $\hat{x}_t = \Delta_{+}^{\hat{d}} y_t$, when assuming a type II process as in Definition 6.2), and fit a model to these empirical differences computed from some (possibly semiparametric) estimator $\hat{d}$; the issue of fractional differencing with empirical data has been addressed in Section 6.4. Then a sequence of forecasts $\hat{x}_T(1), \ldots, \hat{x}_T(h)$ must be produced by some standard method to use Eq. (11.15) with $\pi_j(\hat{d})$ replacing $\pi_j(d)$, of course. Alternatively, Papailias and Dias (2015) suggested to fractionally integrate $\hat{x}_T(h)$ to obtain forecasts for y_{T+h}:

$$\Delta_{+}^{-\hat{d}} \, \hat{x}_T(h) = \sum_{j=0}^{T+h-1} \pi_j(-\hat{d}) \, \hat{x}_T(h - j) \, ,$$

where $\hat{x}_T(h - j) = x_{T+h-j}$ for $j \geq h$; they called this procedure a two-stage forecasting approach.

Ray (1993) carried out a forecast comparison between ARFIMA models and so-called long autoregressions, i.e. AR models of high order. She did not find much gain from fractional modeling over the AR approximation when the true processes are stationary, $0 < d \leq 0.45$; see also Crato and Ray (1996) for further evidence.[1] Smith and Yadav (1994) found partly quite different results in the range of nonstationarity, $0.7 \leq d \leq 1.3$. Tiao and Tsay (1994) proposed to compute forecasts from ARMA(1, 1) models constructed from FIN(d) processes (with d assumed to be known). Brodsky and Hurvich (1999), however, showed that for unknown d multistep forecasts from fractional models are superior. Man (2003), on the other hand, provided simulation evidence that ARMA(2, 2) are competitive forecast models also under long memory.

---

1 Recently, Demetrescu and Hassler (2016) provided a theoretical underpinning to justify long AR for forecasting purposes under fractional integration even in the presence of additional mean shifts.

Yet another picture arises when comparing ex ante forecasts of real economic or financial series. Bhardwaj and Swanson (2006) concluded for very long absolute return series significantly better ARFIMA predictions relative to AR or ARMA models. Further, they reported a less clear-cut dominance with shorter macroeconomic series. Similarly, Choi et al. (2010) found a superior predictive ability of fractionally integrated models when applied to realized volatility, even if the true cause behind a spurious long memory behavior is structural breaks. Related evidence has been collected for realized volatility of exchange rates by Morana and Beltratti (2004, p. 629): "…neglecting the break process is not important for very short term forecasting once it is allowed for a long memory component …." Finally, Baillie et al. (2012, 2014a) stressed that the empirical forecast performance of fractionally integrated models may crucially depend on the employed estimation technique and the issue of bandwidth selection.

11.4 Cyclical and Seasonal Models

This section is dedicated to long memory models where the singularity in the spectrum does not occur at the origin but at some frequency $\omega > 0$. The corresponding cycle will have a dominant contribution to the process and its variance. Sometimes, such cycles or periods have a seasonal interpretation, if $\omega = 2\pi s/S, s = 1, \ldots, \lfloor S/2 \rfloor$, and $S = 12$ stands for the number of months per year, or $S = 5$ may denote the number of working days per week. Then $2\pi s/S$ are called seasonal frequencies.

Let us define the filter

$$\Delta_\omega^d := (1 - 2\cos\omega L + L^2)^d = |1 - e^{i\omega}L|^{2d}$$

as cyclical fractional differences for some frequency ω. In one single sentence, Hosking (1981, p. 175) suggested the process

$$\Delta_\omega^d y_t = \varepsilon_t , \quad \omega \in (0, \pi) , \tag{11.16}$$

for $|d| < 1/2$, noting that it "exhibits both long-term persistence and quasiperiodic behaviour; its correlation function resembles a hyperbolically dampened sine wave." Indeed, Anděl (1986, Theorem 5.1) proved that its autocorrelation function behaves like

$$\rho(h) \sim C_{\omega,d} \cos(h\omega) h^{2d-1} , \quad h \to \infty , \tag{11.17}$$

where $C_{\omega,d}$ depends on ω, d and $\sigma^2 = \text{Var}(\varepsilon_t)$. This result was also provided by Gray et al. (1989, Theorem 3) in connection with Gray et al. (1994). The expansion of the inverse filter, Δ_ω^{-d}, was given by Gray et al. (1989) in terms of so-called Gegenbauer polynomials $G_j(\omega, d)$,

$$(1 - 2\cos\omega L + L^2)^{-d} = \sum_{j=0}^{\infty} G_j(\omega, d)L^j ; \tag{11.18}$$

see also Gradshteyn and Ryzhik (2000, Section 8.93) for a definition of $G_j(\omega, d)$. Consequently, the process from Eq. (11.16) was called a Gegenbauer process. More generally, with $\{x_t\}$ being a stationary and invertible ARMA(p, q) process, Gray et al. (1989) defined the Gegenbauer ARMA process GARMA$(\omega, d; p, q)$:

$$y_t = \Delta_\omega^{-d} x_t , \quad 0 < \omega < \pi. \tag{11.19}$$

The process is stationary for $d < 1/2$ as long as $0 < \omega < \pi$; see Gray et al. (1989, Theorem 4). Note that the ARMA assumption is not crucial and $\{x_t\}$ might be more generally a covariance stationary process from Assumption 6.1. The spectrum of the process from Eq. (11.19) becomes

$$f_y(\lambda) = \left(4 \sin \left(\frac{\lambda - \omega}{2} \right) \sin \left(\frac{\lambda + \omega}{2} \right) \right)^{-2d} f_x(\lambda) ,$$

where f_x denotes the ARMA spectrum; see Anděl (1986, Eq. (5.1)) and Gray et al. (1989, p. 243). It has a singularity at frequency ω for $d > 0$:

$$f_y(\omega + \lambda) \sim G \lambda^{-2d} , \quad \lambda \to 0 , \tag{11.20}$$

with G appropriately defined. This implies that the GARMA process $\{y_t\}$ with $d > 0$ displays long memory in the sense of Definition 2.4 (see Anděl (1986, Theorem 5.2) or Hassler (1994, Proposition 1)); but it is not strongly persistent in the sense of Definition 2.5 as long as $\omega > 0$: $f_y(0) = (4 \sin^2(\omega/2))^{-2d} f_x(0) < \infty$. Note that for $\omega = 0$, the Gegenbauer filter reduces to the well-known fractional expansion from Chapter 5:

$$\Delta_0^{-d} = (1 - 2L + L^2)^{-d} = (1 - L)^{-2d} = \sum_{j=0}^{\infty} \pi_j(-2d) L^j .$$

Similarly, we have for $\omega = \pi$ by binomial expansion

$$\Delta_\pi^{-d} = (1 + 2L + L^2)^{-d} = (1 + L)^{-2d} = \sum_{j=0}^{\infty} \binom{-2d}{j} L^j .$$

For $0 < \omega < \pi$, Arteche and Robinson (2000) considered the so-called asymmetric cyclical model where the divergence at the singularity may be different from the left and from the right. Instead of Eq. (11.20), they assumed

$$f_y(\omega + \lambda) \sim G_1 \lambda^{-2d_1} , \quad f_y(\omega - \lambda) \sim G_2 \lambda^{-2d_2} \quad \lambda \to 0 , \quad \lambda > 0. \tag{11.21}$$

It is natural to allow for a singularity at the origin on top of a cyclical singularity, or for more than one cyclical singularity, in particular in the case of seasonal time series. With monthly data ($S = 12$), Porter-Hudak (1990) considered the fractional seasonal differences

$$(1 - L^{12})^d = (1 - L)^d (1 + L)^d \prod_{s=1}^{5} \Delta_{2\pi s/12}^d .$$

When defining $y_t = (1 - L^{12})^{-d} x_t$, it is clear that the spectrum f_y has seven singularities on $[0, \pi]$ as long as $d > 0$: one at the origin and one at each seasonal frequency. They all have the same order λ^{-2d} for $\lambda \to 0$ at $\lambda + 2\pi s/12$, $s = 0, 1, \ldots, 6$. To account for more flexibility, Hassler (1994) allowed for different orders of integration at different seasonal frequencies. The flexible seasonal filter becomes (for the monthly example of $S = 12$)

$$(1 - L)^{d_0}(1 + L)^{d_6} \prod_{s=1}^{5} \Delta_{2\pi s/12}^{d_s}.$$

This allows to define a stationary seasonal process with spectral singularities λ^{-2d_s} at $\lambda + 2\pi s/12$, $s = 0, 1, \ldots, 6$ as long as $0 < d_s < 1/2$. Without restriction to seasonal frequencies, Giraitis and Leipus (1995) and Woodward et al. (1998) defined a k-factor Gegenbauer process. Building on

$$\Delta_{\omega_1, \ldots, \omega_k}^{d_1, \ldots, d_k} := \prod_{i=1}^{k} \Delta_{\omega_i}^{d_i},$$

it is defined as

$$y_t = \Delta_{\omega_1, \ldots, \omega_k}^{-d_1, \ldots, -d_k} x_t = \prod_{i=1}^{k} \Delta_{\omega_i}^{-d_i} x_t ,$$

with $0 \leq \omega_1 < \cdots < \omega_k \leq \pi$.

When it comes to estimation, one may treat the so-called Gegenbauer frequency ω, or frequencies $\omega_1 < \cdots < \omega_k$, as known or not. Assuming the simple model Eq. (11.16), Chung (1996) discussed the joint estimation of unknown ω and d building on the CSS approximation to ML (see Section 8.3). Assuming the more general 1-factor model Eq. (11.19), Hidalgo and Soulier (2004) advocated a semiparametric joint estimation of unknown ω and d; for a parametric approach, see Giraitis et al. (2001a). For estimation purposes of the asymmetric model Eq. (11.21), Arteche and Robinson (2000) discussed trimmed versions of the log-periodogram regression estimator and the local Whittle estimator from Sections 9.1 and 9.2, respectively, where the Gegenbauer frequency ω is assumed to be known. Tapered versions of these estimators were studied by Arteche and Velasco (2005) under asymmetric cyclical or seasonal long memory. Similar to Chung (1996), Woodward et al. (1998) advocated a numerical ML procedure treating the frequencies ω_i of the k-factor Gegenbauer process as unknown parameters, too. Under the same assumptions, Giraitis and Leipus (1995) discussed the parametric Whittle estimation. Under the assumption of known frequencies, Palma and Chan (2005) established properties of exact ML estimation of k-factor Gegenbauer processes. In particular for seasonal models, where ω_i are known seasonal frequencies, estimation has been addressed by Reisen et al. (2006, 2014). Finally, one may test that the vector $(d_1, \ldots, d_k)'$ equals a vector of values under a null hypothesis, where the frequencies are again assumed to be known. Robinson (1994b) allowed for a corresponding LM test; Hassler et al. (2009) discussed a regression-based version along the lines of Section 10.6.

11.5   Long Memory in Volatility

Taylor (1982) considered "Financial returns modelled by the product of two stochastic processes," which lead to the model of stochastic volatility. The main applications of the material in this section are in the field of finance, but will discuss the model as a signal plus noise model, which is of broader interest. Let us assume that the observed data is z_t from

$$z_t = \mu + y_t + \eta_t, \quad y_t \sim I(d) , \tag{11.22}$$

where $\{y_t\}$ and $\{\eta_t\}$ are independent processes. Statistically speaking, Eq. (11.22) is nothing else but an errors-in-variables or signal plus noise model. We maintain the assumption that the signal $\{y_t\}$ is a fractionally integrated long memory process of type I, $0 < d < 1/2$; see Assumption 6.2. Note that not all papers require that $\{\eta_t\}$ is white noise and even the assumption that $\{y_t\}$ and $\{\eta_t\}$ are independent is sometimes relaxed; see also the discussion in Hurvich et al. (2005, 2008).

Now, we return to the finance motivation. Let $\{r_t\}$ denote a sequence of financial returns. We assume the stochastic volatility process

$$r_t = \sigma_t \, e_t, \quad e_t \sim \mathrm{iid}(0,1), \quad \sigma_t = e^{y_t/2} ,$$

where $\{y_t\}$ is a latent process behind the volatility. Let z_t abbreviate the logarithm of the squared returns, $z_t = \ln(r_t^2)$, such that it fits into the framework of Eq. (11.22) with

$$\mu := E(\ln(e_t^2)) \quad \text{and} \quad \eta_t := \ln(e_t^2) - E(\ln(e_t^2)) \sim \mathrm{iid}(0, \sigma_\eta^2) , \tag{11.23}$$

with adequately defined σ_η^2. The spectrum of $\{z_t\}$ becomes under independence of $\{y_t\}$ and $\{\eta_t\}$

$$f_z(\lambda) = f_y(\lambda) + \frac{\sigma_\eta^2}{2\pi}, \quad f_y(\lambda) \sim G\lambda^{-2d}, \quad \lambda \to 0 , \tag{11.24}$$

where the behavior of f_y is characterized in Proposition 6.1. What is the effect on estimators of d when computed from z_t? Not surprisingly, the noise term induces a contamination of the spectrum in Eq. (11.24). Consequently, estimators of d settled in the frequency domain are expected to be all the more biased, the stronger the noise is (i.e. the larger σ_η^2). Asymptotically, however, the perturbation by the noise should be negligible. Indeed, Breidt et al. (1998) showed that the parametric Whittle estimator from Section 8.4 remains consistent, when assuming a correctly specified ARFIMA model for the signal y_t. Deo and Hurvich (2001) investigated the log-periodogram regression estimator $\widehat{d}_{\mathrm{PR}}^{(m)}$ from Section 9.1 computed from z_t under the assumption that the signal y_t is Gaussian. For the limiting normality from Proposition 9.1 to continue to hold, the conditions on the bandwidth become much stronger:

$$\frac{m^{4d+1}}{T^{4d}}(\ln m)^2 \quad \to \quad 0, \tag{11.25}$$

which implies of course $m = o(T^{4d/(4d+1)})$. This means that m may diverge all the more slowly, the smaller $d > 0$ is; see Deo and Hurvich (2001, Theorem 2) for details. Similarly, Arteche (2004, Theorem 2) established that Proposition 9.4 continues to hold for the local Whittle estimator under Eq. (11.25), where $\{\eta_t\}$ is not even required to be iid but may be a short memory process independent of $\{y_t\}$.

Several attempts have been made to refine and improve the estimation of d under errors in variables. Sun and Phillips (2003) accounted for the effect of the noise by modifying the log-periodogram regression. To that end, they assumed that $\{\eta_t\}$ is a Gaussian short memory process independent of the Gaussian signal $\{y_t\}$. Hurvich and Ray (2003) modified the local Whittle estimator to account for $\{\eta_t\}$ that is assumed to be iid. Note from Eq. (11.24) that

$$f_z(\lambda) \sim \lambda^{-2d} G(1 + h_\theta(\lambda)), \quad \lambda \to 0,$$

where

$$h_\theta(\lambda) = \theta \lambda^{2d}, \quad \theta = \frac{\sigma_\eta^2}{2\pi G}.$$

The intuition is to replace λ^{-2d} by $\lambda^{-2d} G(1 + h_\theta(\lambda))$ in the criterion function $R(d)$ from Eq. (9.8). Concentrating G out yields $R_N(d; \theta)$:

$$R_N(d; \theta) := \ln\left(\frac{1}{m}\sum_{j=1}^{m}\frac{I_z(\lambda_j)}{\lambda^{-2d}(1 + h_\theta(\lambda_j))}\right) + \frac{1}{m}\sum_{j=1}^{m}\ln(\lambda_j^{-2d}(1 + h_\theta(\lambda_j))).$$

If there is no noise, $\sigma_\eta^2 = 0$, then the function h_θ is switched off, such that $h_0(\lambda) = 0$ and $R_N(d; 0) = R(d)$; see also Hurvich et al. (2005, Eq. (2.9)). Let $\widehat{d}_{\mathrm{LW},N}$ resulting from the minimization of $R_N(d; \theta)$ be called the noise-corrected local Whittle estimator. Hurvich and Ray (2003) studied its behavior under stationarity, and also for $1/2 < d < 1$ in terms of the pseudospectrum; see Section 8.4. To obtain limiting normality, they do not require Eq. (11.25), but rather

$$\frac{T^{4d}}{m^{4d+1}} \to 0,$$

which implies that m has to diverge faster than $T^{4d/(4d+1)}$, not more slowly. The limiting variance becomes according to Hurvich and Ray (2003, Theorem 1):

$$\mathrm{Var}(\sqrt{m}(\widehat{d}_{\mathrm{LW},N} - d)) \to \frac{(1 + 2d)^2}{(4d)^2},$$

where $d > 0$ is the true value: The smaller the d, the larger is the variance. This method of noise correction by Hurvich and Ray (2003) has been carried to the bias-reduced local Whittle estimator $\tilde{d}_{LW}^{(r)}$ (see Section 9.4) by Frederiksen and Nielsen (2008). Hurvich et al. (2005, 2008) extended the noise correction to work even under contemporaneous correlation of y_t and η_t. Similarly, Arteche (2012) extended the log-periodogram regression to the case of correlation between signal and noise.

11.6 Fractional Cointegration

The idea of (fractional) cointegration can be dated back to Granger (1981, Section 4). If $\{y_{1,t}\}$ and $\{y_{2,t}\}$ are two $I(d)$ processes, and if there exists a linear combination that is integrated of reduced order $d - b$, $b > 0$, then the two processes are called cointegrated of order d and b, shortly CI(d, b):

$$y_{1,t} - \beta\, y_{2,t} \sim I(d - b), \quad \beta \neq 0 , \quad b > 0. \tag{11.26}$$

Typically, one assumes $0 < b \leq d$, which does not require $d > 1/2$, although this case of nonstationarity is most widely considered. Clearly, the larger the b, the stronger is the link between $y_{1,t}$ and $y_{2,t}$ in that the deviation from the long-run relation $y_1 = \beta y_2$ is less persistent. The special case of $d = b = 1$ has become very popular in econometrics since Engle and Granger (1987).

The definition of Eq. (11.26) suggests a regression of $y_{1,t}$ on $y_{2,t}$:[2]

$$y_{1,t} = \hat{\alpha} + \hat{\beta}\, y_{2,t} + \hat{u}_t, \quad t = 1, \ldots, T. \tag{11.27}$$

Here, we only treat ordinary least squares (OLS) estimation although more efficient variants of least squares are available. The slope estimator is expected to converge to the true β as long as $b > 0$, but the rate of convergence will depend on the size of b, which has also been called *cointegration gap* (see, e.g. Hassler, et al. (2006, p. 182) or Hualde and Robinson (2010, p. 493)). We treat three cases. The first case is characterized by b being small relative to d such that $2d - b > 1$. In this case, it holds that

$$T^b(\hat{\beta} - \beta) \xrightarrow{D} \mathcal{L}_1, \quad d > 1/2, \quad 2d - b > 1, \quad d > b, \tag{11.28}$$

where $\mathcal{L}_1$ was given in Robinson and Marinucci (2001, Proposition 6.5) under some additional assumptions. In fact, the same rate continues to hold for $d = b \geq 1$; see Robinson and Marinucci (2001, Propositions 6.3 and 6.4). A second

---

2 Note that Eq. (11.26) may be renormalized such that $y_{2,t} - \frac{1}{\beta}\, y_{1,t} \sim I(d - b)$. Hence, instead of regressing $y_{1,t}$ on $y_{2,t}$ in order to estimate β, we could equally regress $y_{2,t}$ on $y_{1,t}$ to estimate $1/\beta$, as long as $\beta \neq 0$ under cointegration.

case arises if d is relatively small or b relatively large: $2d - b < 1$. In this case, Robinson and Marinucci (2001, Prop 6.1) characterized the limit $\mathcal{L}_2$ from

$$T^{2d-1}(\widehat{\beta} - \beta) \overset{D}{\to} \mathcal{L}_2, \quad d > 1/2, \quad 2d - b < 1, \quad d \geq b; \tag{11.29}$$

see also Robinson and Marinucci (2003, Theorem 3). Here $2d - 1 < b$, such that $T^{2d-1} < T^b$ when comparing with the first case. A third case arises for $2d - b = 1$ with two subcases: If $d = b = 1$, we have that $T(\widehat{\beta} - \beta)$ converges (see again Robinson and Marinucci, (2001, Proposition 6.3)), while $d > b$ implies an even more unconventional rate,

$$\frac{T^{2d-1}}{\ln T}(\widehat{\beta} - \beta) \overset{D}{\to} \mathcal{L}_3, \quad d > 1/2, \quad 2d - b = 1, \quad d > b, \tag{11.30}$$

with $\mathcal{L}_3$ given in Robinson and Marinucci (2001, Proposition 6.2).

According to Robinson and Marinucci (2001, p. 962), these regression results extend to the case of vectors of regressors integrated of order d. In fact, the orders of integration of the regressors may differ; see Robinson and Marinucci (2003).

Under cointegration, $b > 0$, Hassler et al. (2006, Theorems 1 and 2) showed that it is possible to estimate and test the order of integration $\delta = d - b$ of the cointegration relation:

$$u_t = y_{1,t} - \alpha - \beta\, y_{2,t} \sim I(\delta), \quad \delta = d - b.$$

To that end, one performs a log-periodogram regression from the OLS residuals $\widehat{u}_t = y_{1,t} - \widehat{\alpha} - \widehat{\beta}\, y_{2,t}$, or from the differences thereof, $\Delta \widehat{u}_t$. The values d and b do not have to be known, but it must be known a priori, whether case Eq. (11.28) or (11.29) applies (ruling out the unlikely case Eq. (11.30)). Having a consistent estimator of δ from OLS residuals and a consistent estimator of d (computed from $y_{1,t}$ and/or $y_{2,t}$), an estimator of the cointegration gap $b = d - \delta$ is obvious. The idea of Hassler et al. (2006) hence was to determine the strength of the cointegration relation under the assumption of cointegration.[3] In a similar single-equation framework, (Nielsen, 2004b) suggested to test for $\theta = 0$ in

$$y_{1,t} - \beta\, y_{2,t} \sim I(d - b + \theta).$$

He assumed d to be known, and $3/4 < b \leq d$. Then it is possible to perform an LM test (see Section 10.4) from regression residuals in order to test for $\theta = 0$, or $\delta = d - b$ for prespecified b; see Nielsen (2004b, Theorem 1). It is worth noting that superior estimators over OLS have been proposed to estimate cointegration relations. Corresponding frequency domain regression estimators have been discussed by Robinson and Marinucci (2001, 2003); see also Chen and Hurvich (2003) and Christensen and Nielsen (2006).

---

3 See also a quite different approach to estimate b without providing limiting distributions by Velasco (2003).

Without cointegration, spurious regressions will arise from the regression of two $I(d)$ series on each other. Even if $\{y_{1,t}\}$ and $\{y_{2,t}\}$ are independent with $y_{1,t} \sim I(d_1)$ and $y_{2,t} \sim I(d_2)$, the t-statistic testing for $\beta = 0$ in Eq. (11.27) will diverge asymptotically (under certain assumptions), indicating nonsense significance. For $d_1 = d_2 = 1$, this was established in the seminal paper by Phillips (1986). Marmol (1996, Theorem 1) proved that this continues to hold for the unbalanced case where $d_1 \neq d_2 = 1$ or $d_1 = 1 \neq d_2$. Tsay and Chung (2000, Theorem 2) established that diverging t-statistics may occur even under stationarity, as long as $1/4 < d_1, d_2 < 1/2$ with $d_1 + d_2 > 1/2$. In order to avoid spurious regression results, it is of ultimate interest to have tests to establish fractional cointegration. There are several routes to cointegration testing, some of which are described below. However, there is no unique definition of fractional cointegration as soon as more than two series are involved. Slightly differing concepts are discussed next.

Consider the (sub)vectors

$$Y_t' = (Y_{1,t}', Y_{2,t}'), \quad Y_{1,t}' = (y_{1,t}, \ldots, y_{r,t}), \quad Y_{2,t}' = (y_{r+1,t}, \ldots, y_{K,t}),$$

where $0 < r < K$. Each component is assumed to be integrated, $y_{k,t} \sim I(d_k)$, $k = 1, \ldots, K$. Analogously, we partition the K-dimensional vector X_t, where each component $x_{k,t}$ is $I(0)$ in the sense of Assumption 6.1:

$$X_t' = (X_{1,t}', X_{2,t}') = (x_{1,t}, \ldots, x_{r,t}, x_{r+1,t}, \ldots, x_{K,t}).$$

To begin with, we assume identical orders of integration for all observables $d_1 = \cdots = d_K = d$. The natural extension of the regression framework is the following triangular system,

$$Y_{1,t} = B' \, Y_{2,t} + \Delta^{b-d} \, X_{1,t}, \quad \Delta Y_{2,t} = \Delta^{1-d} \, X_{2,t}, \tag{11.31}$$

where B is a $(K - r) \times r$ matrix. If we assume $1/2 < d < 3/2$ and $d - b < 1/2$, then $\Delta^{1-d} \, X_{2,t}$ is $I(d-1)$, such that $Y_{2,t} \sim I(d)$, and $\Delta^{b-d} \, X_{1,t} \sim I(d-b)$, where these processes are of type I. In what follows, we will be sloppy with respect to a proper distinction between type I and type II processes, and we suppress assumptions for $X_{1,t}$, $X_{2,t}$, and X_t to be *jointly* $I(0)$, which requires more than being made up by individual $I(0)$ components. Define

$$\beta' = (I_r, -B'), \tag{11.32}$$

such that β contains r linearly independent cointegrating vectors reducing the order of integration from d to $d - b$, and r is called the cointegration rank:

$$\beta' Y_t = \Delta^{b-d} \, X_{1,t} \sim I(d-b).$$

When defining $f(\lambda)$ as spectral matrix of $\Delta^d \, Y_t$, then $f(0)$ has rank $K - r$, where this rank deficiency stems from the r linearly independent cointegrating vectors; see Nielsen (2004c, Eq. (9)).

Just as in the standard case with $d = b = 1$, fractional cointegration allows for an error-correction representation. Define $\alpha' = (-I_r, O_{r \times (K-r)})$, where $O_{r \times (K-r)}$ is a $r \times (K - r)$ matrix of zeros. With β from Eq. (11.32), formal manipulation of Eq. (11.31) yields the error-correction format

$$\Delta^d Y_t = \alpha \beta' \, \Delta^{d-b} \, (1 - \Delta^b) \, Y_t + \xi_t, \tag{11.33}$$

with $\xi_t' = (X_{1,t}' + X_{2,t}' B, \, X_{2,t}') \sim I(0)$. The filter $1 - \Delta^b$ showing up in Eq. (11.33) was defined as fractional lag operator L_b by Johansen (2008, Eq. (2)),

$$L_b := 1 - \Delta^b = - \sum_{j=1}^{\infty} \pi_j(b) \, L^j, \tag{11.34}$$

where $\pi_j(b)$ are of course from Eq. (5.22). For $b = 1$, $L_1 = L$, and Eq. (11.33) reproduces the standard vector error-correction structure $\Delta Y_t = \alpha \beta' \, Y_{t-1} + \xi_t$. Note that Eq. (11.33) has essentially been proposed by Granger (1986, Eq. (4.3)); see also Davidson (2002, Eq. (2.1)). To account for serial correlation of ξ_t, Granger (1986) or Davidson (2002) suggested to include lagged differences $\Delta^d Y_{t-j} = \Delta^d L^j Y_t$ on the right-hand side of Eq. (11.33). Johansen (2008, p. 664), however, considered fractional lags of $\Delta^d Y_t$, i.e.

$$\Delta^d Y_t = \alpha \beta' \, \Delta^{d-b} \, L_b Y_t + \sum_{j=1}^{p} \Gamma_j \, \Delta^d L_b^j \, Y_t + \varepsilon_t,$$

and discussed, and solved, this model under more general assumptions; see also Johansen and Nielsen (2012b). In particular, the assumption of identical memory in all observable series is not required.

We now briefly turn to the case of different orders of integration. Similar to Johansen (2008, Definitions 1 and 2) or Nielsen (2010, Definition 1), a vector Y_t may be called $I(d)$ when, loosely speaking, all components of $\Delta^d Y_t$ have finite long-run variances, but some of these long-run variances may be zero due to overdifferencing with d. Hence, d is the maximum order of integration, $d = \max\{d_1, \dots, d_K\}$. This allows for much richer dynamics. Consider an example for $K = 4$ with $d_1 = d_2 = 1.4$ and $d_3 = d_4 = 1.0$. There may be cointegration within each of the subgroups of equal integration, say,

$$y_{1,t} - y_{2,t} = z_t \sim I(1.0), \quad y_{3,t} - y_{4,t} \sim I(0.5).$$

Additionally, there may be cointegration between these groups:

$$y_{1,t} - y_{2,t} - a \, y_{3,t} - b \, y_{4,t} \sim I(0), \quad a \neq 0 \quad \text{or} \quad b \neq 0.$$

Hence, we have three linearly independent cointegrating vectors:

$$\beta' = \begin{pmatrix} 1 & -1 & 0 & 0 \\ 0 & 0 & 1 & -1 \\ 1 & -1 & -a & -b \end{pmatrix}.$$

Sometimes, such a link between groups integrated of different orders is ruled out; see, e.g. Robinson and Yajima (2002, p. 222).

There is a huge literature on determining and testing the rank of fractional integration. We do not present details of the different proposals and mention only a small selection. To begin with, Breitung and Hassler (2002) mimicked the procedure by Johansen (1988), replacing Y_{t-1} by $\sum_{j=1}^{t-1} j^{-1} \Delta^d Y_{t-j}$ when solving for generalized eigenvalues; see also the discussion in the last paragraph of Section 10.6. A nonparametric cointegration test was provided by Nielsen (2010) along the lines of Breitung (2002). Johansen and Nielsen (2012b) advocated a sequence of cointegration tests under the fractional error-correction model Eq. (11.33). Nielsen and Shimotsu (2007) addressed consistent estimation of r without providing formal tests, extending the work by Robinson and Yajima (2002).

11.7 R Packages

Long memory analysis with R is discussed, e.g. in Cowpertwait and Metcalfe (2009, Chapter 8), Shumway and Stoffer (2011, Section 5.2), McLeod et al. (2012, Section 3.3), and most recently Woodward et al. (2017, Chapter 11). On CRAN (Comprehensive R Archive Network), several well-documented R packages are available. We provide a brief summary. The latest versions that we refer to were accessed on July 20, 2017.

longmemo: This is the oldest package. The version 1.0-0 of June 15, 2011, is still posted. It relies on S scripts originally provided by Beran; see also Beran (1994). It allows to compute the parametric Whittle estimator from Section 8.4, including the covariance matrix. Further, there are R programs for the parametric log-periodogram regression estimator settled in the FEXP model (see Section 8.5 and Beran, (1993)) as well as for the specification test by Beran (1992).

fracdiff: The currently available version 1.4-1 is of December 02, 2012. As the name suggests, the package contains a code to fractionally difference a time series. In the past, the package was also used to simulate ARFIMA series; nowadays, the procedure by Jensen and Nielsen (2014) is faster; see also the discussion of simulation and differencing in Section 6.4. With respect to estimation, the package covers the log-periodogram regression from Section 8.2 and exact ML inference following the lines by Haslett and Raftery (1989); see Section 8.2.

arfima: This package has been updated recently, and version 1.5-2 is of July 17, 2017. It largely builds on Veenstra (2012), and it contains more functions than the two previous packages, addresses the computation of information criteria, allows to compute theoretical autocorrelograms implied by fitted models, and covers prediction including plots with prediction intervals and further issues like seasonality. Estimation and inference relies again on exact or on Whittle likelihood estimation.

fArma: This package (version 3010.79 of June 24, 2013) is designed for ARMA modeling, but it calls underlying fracdiff codes for exact ML estimation. It offers alternative ways to simulate long memory and further estimators, e.g. the range over standard deviation (R/S) estimator mentioned in Section 10.2.

forecast: This recently updated forecast package (version 8.1 of June 17, 2017; see also Hyndman and Khandakar, (2008)) includes prediction formulae from Peiris and Perera (1988) for ARFIMA models.

Finally, we wish to mention the package afmtools that has been removed from CRAN; the most recent version available from the CRAN archive is from December 28, 2012. This package is well documented in the paper by Contreras-Reyes and Palma (2013). It contains a couple of valuable features not given in other packages: The convolution of moving average coefficients for impulse responses of ARFIMA models, theoretical spectra of fitted ARFIMA models, or the computation of the variance of a sample average.

11.8 Neglected Topics

Long memory is still a very active field of time series research. The selection of the additional topics presented above is somehow arbitrary. There are many further interesting and important issues. We want to list at least some of them in this final section, being well aware that this list is far from complete.

State-Space Modeling and Bayesian Analysis: Chan and Palma (1998) paved the way for state-space modeling under long memory. A state space form for ARFIMA models can build on the AR(∞) or the MA(∞) representation. This allows computing the exact likelihood function using the Kalman filter. A modification accommodates for missing values; see also Palma and Chan (1997). A state-space representation and Kalman filtering was employed by Ray and Tsay (2002) for a Bayesian analysis. In view of rapidly growing computing power, Bayesian methods have become increasingly popular over the last decades. In the field of long memory, however, little has been published since the work by Ravishanker and Ray (1997) and Pai and Ravishanker (1998), notable exceptions being Ray and Tsay (2002), and more recently Holan et al. (2009) and Graves et al. (2015).

Breaks in Persistence: Ray and Tsay (2002) employed the Bayesian approach to detect changes in level or in persistence of a time series, allowing for potentially multiple changes. Changes in mean have been addressed above in Section 11.2. Changes in persistence were modeled by Ray and Tsay (2002) as a time-varying order of integration d_t. Using *classical* methods,

Gil-Alana (2008) provided analogous tools. Formal tests against one break in persistence have been proposed by Sibbertsen and Kruse (2009), Hassler and Scheithauer (2010), or Martins and Rodrigues (2014), while Hassler and Meller (2014) suggested tests for multiple breaks in d, where the number of breaks as well as the timing of the breakpoints is unknown; see also Hassler et al. (2014, Section 4). All procedures, except for Ray and Tsay (2002), are rooted in principles and estimators or tests presented in the previous chapters.

Wavelet-Based Estimation: An altogether different approach to estimation relies on so-called wavelet transformations. A wavelet is a function meeting certain requirements, from which one constructs a collection of functions by scaling and translating. For a given time series, each element of the collection is used to compute the inner product, resulting in so-called wavelet coefficients. Even if the time series is generated from a fractionally integrated process, the wavelet coefficients are approximately uncorrelated, though their variances are related to the order of integration d; see, e.g. Jensen (1999a, Theorem 1). This provides an approximately diagonal covariance matrix for ARFIMA processes upon a wavelet transformation, which can be exploited to evaluate the likelihood function; see Jensen (1999b). For further wavelet-based approaches to estimate d, see Moulines et al. (2007, 2008). In particular, Faÿ et al. (2009) provided a systematic comparison of wavelet transformation versus the more conventional Fourier transformation behind periodogram-based estimators. A more general survey is provided in Persival and Walden (2000, Chapter 9).

Regressions with Long Memory Errors: In Chapter 7 we allowed for a constant mean different from zero. This has been generalized to a regression equation. Let us maintain $y_t = \beta'x_t + e_t$, where the error sequence is fractionally integrated with long memory but the regressors are not. Classically, it is assumed that the regressors are deterministic. Yajima (1988, 1991) pioneered regressions under long memory errors. Local asymptotic normality was established by Hallin et al. (1999), and robust S-estimation was studied by Sibbertsen (2001). Robinson and Hidalgo (1997) even allowed for stationary fractionally integrated regressors independent of the long memory errors, which is an assumption not maintained under cointegration; see Section 11.6.

Why Long Memory? Is there a reason why long memory is such a widely observed phenomenon in a variety of empirical applications? Granger (1980) argued that cross-sectional aggregation over a large number of independent stationary AR(1) processes, where the autoregressive parameter varies according to a beta distribution, results in a hyperbolic decay of the autocorrelations as it is characteristic for fractional integration with positive d; see also Granger (1980) for generalizations. A series of different generating mechanisms has been suggested to explain the feature of long memory theoretically; see Parke (1999), Davidson and Sibbertsen (2005), and Miller and Park (2010); see also Lieberman and Phillips (2008, Section 2) for the special case of realized volatility. The major justification for long memory models and methods, however, is the ubiquity of long memory and strong persistence in empirical time series as a stylized fact from many different fields of science and practice.

Bibliography

Abadir, K.M., Distaso, W., and Giraitis, L. (2007). Nonstationarity-extended local Whittle estimation. *Journal of Econometrics* 141: 1353–1384.

Abadir, K.M., Distaso, W., and Giraitis, L. (2009). Two estimators of the long-run variance: beyond short memory. *Journal of Econometrics* 150: 56–70.

Abadir, K.M., Distaso, W., and Giraitis, L. (2011). An I(d) model with trend and cycles. *Journal of Econometrics* 163: 186–199.

Abadir, K.M., Distaso, W., Giraitis, L., and Koul, H.L. (2014). Asymptotic normality for weighted sums of linear processes. *Econometric Theory* 30: 252–284.

Abramowitz, M. and Stegun, I.A. (1984). *Pocket Book of Mathematical Functions*, abridged edition, Harri Deutsch.

Adenstedt, R.K. (1974). On large-sample estimation for the mean of a stationary random sequence. *Annals of Statistics* 2: 1095–1107.

Agiakloglou, C. and Newbold, P. (1994). Lagrange multiplier tests for fractional difference. *Journal of Time Series Analysis* 15: 253–262.

Agiakloglou, C., Newbold, P., and Wohar, M. (1993). Bias in an estimator of the fractional difference parameter. *Journal of Time Series Analysis* 14: 235–246.

Aitchison, J. and Silvey, S.D. (1958). Maximum-likelihood estimation of parameters subject to restraints. *The Annals of Mathematical Statistics* 29: 813–828.

Akaike, H. (1974). A new look at the statistical model identification. *IEEE Transactions on Automatic Control* 19: 716–723.

Andersen, T.G., Bollerslev, T., Diebold, F.X., and Ebens, H. (2001). The distribution of realized stock return volatility. *Journal of Financial Economics* 61: 43–76.

Andersen, T.G., Bollerslev, T., Diebold, F.X., and Labys, P. (2003). Modeling and forecasting realized volatility. *Econometrica* 71: 579–625.

Anderson, T.W. (1971). *The Statistical Analysis of Time Series*. Wiley.

Anderson, T.W. and Darling, D.A. (1952). Asymptotic theory of certain "goodness of fit" criteria based on stochastic processes. *The Annals of Mathematical Statistics* 23: 193–212.

Andrews, D.W.K. (1991). Heteroskedasticity and autocorrelation consistent covariance matrix estimation. *Econometrica* 59: 817–858.

Time Series Analysis with Long Memory in View, First Edition. Uwe Hassler.
© 2019 John Wiley & Sons, Inc. Published 2019 by John Wiley & Sons, Inc.

Andrews, D.W.K. and Chen, H.-Y. (1994). Approximately median-unbiased estimation of autoregressive models. *Journal of Business & Economic Statistics* 12: 187–204.

Andrews, D.W.K. and Guggenberger, P. (2003). A bias-reduced log-periodogram regression estimator for the long-memory parameter. *Econometrica* 71: 675–712.

Andrews, D.W.K. and Sun, Y. (2004). Adaptive local polynomial Whittle estimation of long-range dependence. *Econometrica* 72: 569–614.

Andĕl, J. (1986). Long memory time series models. *Kybernetika* 22: 105–123.

Arteche, J. (2004). Gaussian semiparametric estimation in long memory in stochastic volatility and signal plus noise models. *Journal of Econometrics* 119: 131–154.

Arteche, J. (2012). Semiparametric inference in correlated long memory signal plus noise models. *Econometric Reviews* 31: 440–474.

Arteche, J. and Robinson, P.M. (2000). Semiparametric inference in seasonal and cyclical long memory processes. *Journal of Time Series Analysis* 21: 1–25.

Arteche, J. and Velasco, C. (2005). Trimming and tapering semi-parametric estimates in asymmetric long memory time series. *Journal of Time Series Analysis* 26: 581–611.

Baillie, R.T. (1996). Long memory processes and fractional integration in econometrics. *Journal of Econometrics* 73: 5–59.

Baillie, R.T., Chung, C.-F., and Tieslau, M.A. (1996). Analysing inflation by the fractionally integrated ARFIMA-GARCH model. *Journal of Applied Econometrics* 11: 23–40.

Baillie, R.T., Kongcharoen, C., and Kapetanios, G. (2012). Prediction from ARFIMA models: comparisons between MLE and semiparametric estimation procedures. *International Journal of Forecasting* 28: 46–53.

Baillie, R.T., Kapetanios, G., and Papailias, F. (2014a). Modified information criteria and selection of long memory time series models. *Computational Statistics and Data Analysis* 76: 116–131.

Baillie, R.T., Kapetanios, G., and Papailias, F. (2014b). Bandwidth selection by cross-validation for forecasting long memory financial time series. *Journal of Empirical Finance* 29: 129–143.

Baum, C.F., Barkoulas, J.T., and Caglayan, M. (1999). Persistence in international inflation rates. *Southern Economic Journal* 65: 900–913.

Bera, A.K. and Bilias, Y. (2001). Rao's score, Neyman's $c(\alpha)$ and Silvey's LM tests: an essay on historical developments and some new results. *Journal of Statistical Planning and Inference* 97: 9–44.

Beran, J. (1989). A test of location for data with slowly decaying serial correlations. *Biometrika* 76: 261–269.

Beran, J. (1992). A goodness-of-fit test for time series with long range dependence. *Journal of the Royal Statistical Society. Series B: Methodological* 54: 749–760.

Beran, J. (1993). Fitting long-memory models by generalized linear regression. *Biometrika* 80: 817–822.

Beran, J. (1994). *Statistics for Long-Memory Processes*. Chapman & Hall/CRC.

Beran, J. (1995). Likelihood estimation of the differencing parameter for invertible short and long memory autoregressive integrated moving average models. *Journal of the Royal Statistical Society. Series B: Methodological* 57: 659–672.

Beran, J., Bhansali, R.J., and Ocker, D. (1998). On unified model selection for stationary and nonstationary short- and long-memory autoregressive processes. *Biometrika* 85: 921–934.

Beran, J., Feng, Y., Ghosh, S., and Kulik, R. (2013). *Long-Memory Processes: Probabilistic Properties and Statistical Methods*. Springer.

Bhansali, R.J., Giraitis, L., and Kokoszka, P.S. (2006). Estimation of the memory parameter by fitting fractionally differenced autoregressive models. *Journal of Multivariate Analysis* 97: 2101–2130.

Bhardwaj, G. and Swanson, N.R. (2006). An empirical investigation of the usefulness of ARFIMA models for predicting macroeconomic and financial time series. *Journal of Econometrics* 131: 539–578.

Bhattacharya, R.N., Gupta, V.K., and Waymire, E. (1983). The Hurst effect under trends. *Journal of Applied Probability* 20: 649–662.

Billingsley, P. (1968). *Convergence of Probability Measures*. Wiley.

Bloomfield, P. (1973). An exponential model for the spectrum of a scalar time series. *Biometrika* 60: 217–226.

Bloomfield, P. (1985). On series representations for linear predictors. *Annals of Probability* 13: 226–233.

Bloomfield, P. (2000). *Fourier Analysis of Time Series: An Introduction*, 2e. Wiley.

Bollerslev, T. (1986). Generalized autoregressive conditional heteroskedasticity. *Journal of Econometrics* 31: 307–327.

Bondon, P. and Palma, W. (2007). A class of antipersistent processes. *Journal of Time Series Analysis* 28: 261–273.

Box, G.E.P. and Jenkins, G.M. (1970). *Time Series Analysis - Forecasting and Control*. Holden-Day.

Box, G.E.P., Jenkins, G.M., Reinsel, G.C., and Ljung, G.M. (2015). *Time Series Analysis - Forecasting and Control*, 5e. Wiley.

Box-Steffensmeier, J.M. and Smith, R.M. (1996). The dynamics of aggregate partisanship. *American Political Science Review* 90: 567–580.

Box-Steffensmeier, J.M. and Tomlinson, A.R. (2000). Fractional integration methods in political science. *Electoral Studies* 19: 63–76.

Breidt, F.J., Crato, N., and de Lima, P. (1998). The detection and estimation of long memory in stochastic volatility. *Journal of Econometrics* 83: 325–348.

Breiman, L. (1992). *Probability*, 2e. Society for Industrial and Applied Mathematics.

Breitung, J. (2002). Nonparametric tests for unit roots and cointegration. *Journal of Econometrics* 108: 343–363.

Breitung, J. and Hassler, U. (2002). Inference on the cointegration rank in fractionally integrated processes. *Journal of Econometrics* 110: 167–185.

Breusch, T.S. and Pagan, A. (1980). The Lagrange multiplier test and its applications to model specification in econometrics. *Review of Economic Studies* 47: 239–253.

Brillinger, D.R. (1975). *Time Series: Data Analysis and Theory*. Holt, Rinehart, and Winston.

Brockwell, P.J. and Davis, R.A. (1991). *Time Series: Theory and Methods*, 2e. Springer.

Brodsky, J. and Hurvich, C.M. (1999). Multi-step forecasting for long-memory processes. *Journal of Forecasting* 18: 59–75.

Buse, A. (1982). The likelihood ratio, Wald, and Lagrange multiplier tests: an expository note. *The American Statistician* 36: 153–157.

Byers, D., Davidson, J., and Peel, D. (1997). Modelling political popularity: an analysis of long-range dependence in opinion poll series. *Journal of the Royal Statistical Society: Series A (Statistics in Society)* 160: 471–490.

Byers, D., Davidson, J., and Peel, D. (2000). The dynamics of aggregate political popularity: evidence from eight countries. *Electoral Studies* 19: 49–62.

Byers, D., Davidson, J., and Peel, D. (2002). Modelling political popularity: a correction. *Journal of the Royal Statistical Society: Series A (Statistics in Society)* 165: 187–189.

Byers, D., Davidson, J., and Peel, D. (2007). The long memory model of political support: some further results. *Applied Economics* 39: 2547–2552.

Carbonez, K.A.E. (2009). Model selection and estimation of long-memory time-series models. *Review of Business and Economics* LIV: 512–554.

Cavaliere, G., Nielsen, M.O., and Taylor, A.M.R. (2015). Bootstrap score tests for fractional integration in heteroskedastic ARFIMA models, with an application to price dynamics in commodity spot and futures markets. *Journal of Econometrics* 187: 557–579.

Chan, N.H. and Palma, W. (1998). State space modeling of long-memory processes. *The Annals of Statistics* 26: 719–740.

Chang, Y. and Park, J.Y. (2002). On the asymptotics of ADF tests for unit roots. *Econometric Reviews* 21: 431–447.

Chen, W.W. and Deo, R.S. (2004). A generalized portmanteau goodness-of-fit test for time series models. *Econometric Theory* 20: 382–416.

Chen, W.W. and Hurvich, C. (2003). Semiparametric estimation of multivariate fractional cointegration. *Journal of the American Statistical Association* 98: 629–642.

Chen, C. and Tiao, G.C. (1990). Random level-shift time series models, ARIMA approximations, and level-shift detection. *Journal of Business & Economic Statistics* 8: 83–97.

Cheung, Y.-W. and Diebold, F.X. (1994). On maximum likelihood estimation of the differencing parameter of fractionally-integrated noise with unknown mean. *Journal of Econometrics* 62: 301–316.

Choi, K., Yu, W.-C., and Zivot, E. (2010). Long memory versus structural breaks in modeling and forecasting realized volatility. *Journal of International Money and Finance* 29: 857–875.

Christensen, B.J. and Nielsen, M.O. (2006). Asymptotic normality of narrow-band least squares in the stationary fractional cointegration model and volatility forecasting. *Journal of Econometrics* 133: 343–371.

Christensen, B.J. and Nielsen, M.O. (2007). The effect of long memory in volatility on stock market fluctuations. *The Review of Economics and Statistics* 89: 684–700.

Chung, C.-F. (1994). A note on calculating the autocovariances of the fractionally integrated ARMA models. *Economics Letters* 45: 293–297.

Chung, C.-F. (1996). Estimating a generalized long memory process. *Journal of Econometrics* 73: 237–259.

Chung, C.-F. and Baillie, R.T. (1993). Small sample bias in conditional sum-of-squares estimators of fractionally integrated ARMA models. *Empirical Economics* 18: 791–806.

Cochrane, J.H. (1988). How big is the random walk in GNP? *Journal of Political Economy* 96: 893–920.

Cogley, T. and Sargent, T.S. (2005). Drifts and volatilities: monetary policies and outcomes in the post WWII US. *Review of Economic Dynamics* 8: 262–302.

Contreras-Reyes, J. and Palma, W. (2013). Statistical analysis of autoregressive fractionally integrated moving average models in R. *Computational Statistics* 28: 2309–2331.

Corbae, D., Ouliaris, S., and Phillips, P.C.B. (2002). Band spectral regression with trending data. *Econometrica* 70: 1067–1109.

Cowpertwait, P.S.P. and Metcalfe, A.V. (2009). *Introductory Time Series with R.* Springer.

Craigmile, P.F. (2003). Simulating a class of stationary Gaussian processes using the Davies-Harte algorithm, with application to long memory processes. *Journal of Time Series Analysis* 24: 505–511.

Crato, N. and Ray, B.K. (1996). Model selection and forecasting for long-range dependent processes. *Journal of Forecasting* 15: 107–125.

Dahlhaus, R. (1989). Efficient parameter estimation for self-similar processes. *The Annals of Statistics* 17: 1749–1766.

Dahlhaus, R. (2006). Correction: efficient parameter estimation for self-similar processes. *The Annals of Statistics* 34: 1045–1047.

Davidson, J. (1994). *Stochastic Limit Theory.* Oxford University Press.

Davidson, J. (2000). *Econometric Theory.* Blackwell.

Davidson, J. (2002). A model of fractional cointegration, and tests for cointegration using the bootstrap. *Journal of Econometrics* 110: 187–212.

Davidson, J. and de Jong, R.M. (2000). The functional central limit theorem and weak convergence to stochastic integrals II: fractionally integrated processes. *Econometric Theory* 16: 643–666.

Davidson, J. and Hashimzade, N. (2008). Alternative frequency and time domain versions of fractional Brownian motion. *Econometric Theory* 24: 256–293.

Davidson, R. and MacKinnon, J.G. (1993). *Estimation and Inference in Econometrics.* Oxford University Press.

Davidson, J. and Sibbertsen, P. (2005). Generating schemes for long memory processes: regimes, aggregation and linearity. *Journal of Econometrics* 128: 253–282.

Davidson, J. and Sibbertsen, P. (2009). Tests of bias in log-periodogram regression. *Economics Letters* 102: 83–86.

Davies, R.B. and Harte, D.S. (1987). Tests for Hurst effect. *Biometrika* 74: 95–101.

Davydov, Y.A. (1970). The invariance principle for stationary processes. *Theory of Probability and Its Applications* 15: 487–498.

Delgado, M.A. and Robinson, P.M. (1994). New methods for the analysis of long-memory time-series: application to Spanish inflation. *Journal of Forecasting* 13: 97–107.

Delgado, M.A. and Robinson, P.M. (1996). Optimal spectral bandwidth for long memory. *Statistica Sinica* 6: 97–112.

Demetrescu, M. and Hassler, U. (2016). (When) Do long autoregressions account for neglected changes in parameters? *Econometric Theory* 32: 1317–1348.

Demetrescu, M. and Sibbertsen, P. (2016). Inference on the long-memory properties of time series with non-stationary volatility. *Economics Letters* 144: 80–84.

Demetrescu, M., Kuzin, V., and Hassler, U. (2008). Long memory testing in the time domain. *Econometric Theory* 24: 176–215.

Demetrescu, M., Hassler, U., and Kuzin, V. (2011). Pitfalls of post-model-selection testing: experimental quantification. *Empirical Economics* 40: 359–372.

Deo, R.S. and Chen, W.W. (2000). On the integral of the squared periodogram. *Stochastic Processes and their Applications* 85: 159–176.

Deo, R.S. and Hurvich, C.M. (2001). On the log periodogram regression estimator of the memory parameter in long memory stochastic volatility models. *Econometric Theory* 17: 686–710.

Dhrymes, P.J. (1978). *Mathematics for Econometrics.* Springer.

Dickey, D.A. and Fuller, W.A. (1979). Distribution of the estimators for autoregressive time series with a unit root. *Journal of the American Statistical Association* 74: 427–431.

Diebold, F.X. and Inoue, A. (2001). Long memory and regime switching. *Journal of Econometrics* 105: 131–159.

Diebold, F.X. and Rudebusch, G.D. (1991). On the power of Dickey-Fuller tests against fractional alternatives. *Economics Letters* 35: 155–160.

Ding, Z., Granger, C.W.J., and Engle, R.F. (1993). A long memory property of stock market returns and a new model. *Journal of Empirical Finance* 1: 83–106.

Dolado, J.J., Gonzalo, J., and Mayoral, L. (2002). A fractional Dickey-Fuller test for unit roots. *Econometrica* 70: 1963–2006.

Dolado, J.J., Gonzalo, J., and Mayoral, L. (2003). Long-range dependence in Spanish political opinion poll series. *Journal of Applied Econometrics* 18: 137–155.

Donsker, M.D. (1951). An invariance principle for certain probability limit theorems. *Memoirs of the American Mathematical Society* 6: 1–12.

Doob, J.L. (1953). *Stochastic Processes*. Wiley.

Efron, B. and Hinkley, D.V. (1978). Assessing the accuracy of the maximum likelihood estimator: observed versus expected Fisher information. *Biometrika* 65: 457–483.

Eicker, F. (1967). Limit theorems for regressions with unequal and dependent errors. In: *Proceedings of the 5th Berkeley Symposium on Mathematical Statistics and Probability, Volume 1: Statistics* (ed. L. Le Cam and J. Neyman), 59–82. University of California Press.

Engle, R.F. (1982). Autoregressive conditional heteroskedasticity with estimates of the variance of U.K. inflation. *Econometrica* 50: 987–100.

Engle, R.F. (1984). Wald, likelihood ratio, and Lagrange multiplier tests in econometrics. In: *Handbook of Econometrics*, Chapter 13 (ed. Z. Griliches and M.D. Intriligator), 775–826. Elsevier.

Engle, R.F. and Bollerslev, T. (1986). Modelling the persistence of conditional variances. *Econometric Reviews* 5: 1–50.

Engle, R.F. and Granger, C.W.J. (1987). Co-integration and error correction: representation, estimation, and testing. *Econometrica* 55: 251–76.

Engle, R.F. and Smith, A. (1999). Stochastic permanent breaks. *The Review of Economics and Statistics* 81: 553–574.

Faÿ, G., Moulines, E., Roueff, F., and Taqqu, M.S. (2009). Estimators of long memory: Fourier versus wavelets. *Journal of Econometrics* 151: 159–177.

Fox, R. and Taqqu, M.S. (1986). Large-sample properties of parameter estimates for strongly dependent stationary Gaussian time series. *The Annals of Statistics* 14: 517–532.

Franses, P.H. and Ooms, M. (1997). A periodic long-memory model for quarterly UK inflation. *International Journal of Forecasting* 13: 117–126.

Franses, P.H., Ooms, M., and Bos, C.S. (1999). Long memory and level shifts: re-analyzing inflation rates. *Empirical Economics* 24: 427–449.

Frederiksen, P. and Nielsen, M.O. (2008). Bias-reduced estimation of long-memory stochastic volatility. *Journal of Financial Econometrics* 6: 496–512.

Fuller, W.A. (1996). *Introduction to Statistical Time Series*, 2e. Wiley.

Geweke, J. and Porter-Hudak, S. (1983). The estimation and application of long memory time series models. *Journal of Time Series Analysis* 4: 221–238.

Gil-Alana, L. (2008). Fractional integration and structural breaks at unknown periods of time. *Journal of Time Series Analysis* 29: 163–185.

Giraitis, L. and Leipus, R. (1995). A generalized fractionally differencing approach in long-memory modeling. *Lithuanian Mathematical Journal* 35: 53–65.

Giraitis, L. and Surgailis, D. (1990). A central limit theorem for quadratic forms in strongly dependent linear variables and its application to asymptotical normality of Whittle's estimate. *Probability Theory and Related Fields* 86: 87–104.

Giraitis, L., Robinson, P.M., and Samarov, A. (2000). Adaptive semiparametric estimation of the memory parameter. *Journal of Multivariate Analysis* 72: 183–207.

Giraitis, L., Hidalgo, J., and Robinson, P.M. (2001a). Gaussian estimation of parametric spectral density with unknown pole. *The Annals of Statistics* 29: 987–1023.

Giraitis, L., Kokoszka, P., and Leipus, R. (2001b). Testing for long memory in the presence of a general trend. *Journal of Applied Probability* 38: 1033–1054.

Giraitis, L., Kokoszka, P., Leipus, R., and Teyssiere, G. (2003). Rescaled variance and related tests for long memory in volatility and levels. *Journal of Econometrics* 112: 265–294.

Giraitis, L., Kokoszka, P., Leipus, R., and Teyssiere, G. (2005). Corrigendum to "Rescaled variance and related tests for long memory in volatility and levels": [Journal of Econometrics 112 (2003) 265–294]. *Journal of Econometrics* 126: 571–572.

Giraitis, L., Koul, H.L., and Surgailis, D. (2012). *Large Sample Inference for Long Memory Processes*. Imperial College Press.

Gradshteyn, I.S. and Ryzhik, I.M. (2000). *Tables of Integrals, Series, and Products*, 6e. Academic Press.

Granger, C.W.J. (1966). The typical spectral shape of an economic variable. *Econometrica* 34: 150–161.

Granger, C.W.J. (1980). Long memory relationships and the aggregation of dynamic models. *Journal of Econometrics* 14: 227–238.

Granger, C.W.J. (1981). Some properties of time series data and their use in econometric model specification. *Journal of Econometrics* 16: 121–130.

Granger, C.W.J. (1986). Developments in the study of cointegrated economic variables. *Oxford Bulletin of Economics and Statistics* 48: 213–228.

Granger, C.W.J. and Andersen, A. (1978). On the invertibility of time series models. *Stochastic Processes and their Applications* 8: 87–92.

Granger, C.W.J. and Ding, Z. (1996). Varieties of long memory models. *Journal of Econometrics* 73: 61–77.

Granger, C.W.J. and Hyung, N. (2004). Occasional structural breaks and long memory with an application to the S&P 500 absolute returns. *Journal of Empirical Finance* 11: 399–421.

Granger, C.W.J. and Joyeux, R. (1980). An introduction to long-memory time series models and fractional differencing. *Journal of Time Series Analysis* 1: 15–29.

Graves, T., Gramacy, R.B., Franzke, C.L.E., and Watkins, N.W. (2015). Efficient Bayesian inference for natural time series using ARFIMA processes. *Nonlinear Processes in Geophysics* 22: 679–700.

Gray, H.L., Zhang, N.-F., and Woodward, W.A. (1989). On generalized fractional processes. *Journal of Time Series Analysis* 10: 233–257.

Gray, H.L., Zhang, N.-F., and Woodward, W.A. (1994). On generalized fractional processes – a correction. *Journal of Time Series Analysis* 15: 561–562.

Guggenberger, P. and Sun, Y. (2006). Bias-reduced log-periodogram and Whittle estimation of the long-memory parameter without variance inflation. *Econometric Theory* 22: 863–912.

Hallin, M., Taniguchi, M., Serroukh, A., and Choy, K. (1999). Local asymptotic normality for regression models with long-memory disturbance. *The Annals of Statistics* 27: 2054–2080.

Hamilton, J.D. (1994). *Time Series Analysis*. Princeton University Press.

Hannan, E.J. (1970). *Multiple Time Series*. Wiley.

Hannan, E.J. (1973a). The asymptotic theory of linear time-series models. *Journal of Applied Probability* 10: 130–145.

Hannan, E.J. (1973b). Central limit theorems for time series regression. *Zeitschrift für Wahrscheinlichkeitstheorie und Verwandte Gebiete* 26: 157–170.

Hannan, E.J. (1980). The estimation of the order of an ARMA process. *The Annals of Statistics* 8: 1071–1081.

Harris, D., McCabe, B., and Leybourne, S. (2008). Testing for long memory. *Econometric Theory* 24: 143–175.

Haslett, J. and Raftery, A.E. (1989). Space-time modelling with long-memory dependence: assessing Ireland's wind power resource. *Applied Statistics* 38: 1–50.

Hassler, U. (1993). Unit root tests: the autoregressive approach in comparison with the periodogram regression. *Statistical Papers* 34: 67–82.

Hassler, U. (1994). (Mis)Specification of long memory in seasonal time series. *Journal of Time Series Analysis* 15: 19–30.

Hassler, U. (1997). Sample autocorrelations of nonstationary fractionally integrated processes. *Statistical Papers* 38: 43–62.

Hassler, U. (2012). Impulse responses of antipersistent processes. *Economics Letters* 116: 454–456.

Hassler, U. (2014). Persistence under temporal aggregation and differencing. *Economics Letters* 124: 318–322.

Hassler, U. (2016). *Stochastic Processes and Calculus: An Elementary Introduction with Applications*. Springer.

Hassler, U. (2017). Ergodic for the mean. *Economics Letters* 151: 75–78.

Hassler, U. and Hosseinkouchack, M. (2014). Effect of the order of fractional integration on impulse responses. *Economics Letters* 125: 311–314.

Hassler, U. and Kokoszka, P. (2010). Impulse responses of fractionally integrated processes with long memory. *Econometric Theory* 26: 1855–1861.

Hassler, U. and Kuzin, V. (2009). Cointegration analysis under measurement errors. In: *Advances in Econometrics*, Measurement Error: Consequences, Applications and Solutions, Vol. 24 (eds J.M. Binner, D.L. Edgerton, and T. Elger), 131–150. Emerald.

Hassler, U. and Meller, B. (2014). Detecting multiple breaks in long memory: the case of U.S. inflation. *Empirical Economics* 46: 653–680.

Hassler, U. and Olivares, M. (2013). Semiparametric inference and bandwidth choice under long memory: experimental evidence. *Journal of the Turkish Statistical Association* 6: 27–41.

Hassler, U. and Scheithauer, J. (2010). Detecting changes from short to long memory. *Statistical Papers* 52: 847–870.

Hassler, U. and Wolters, J. (1994). On the power of unit root tests against fractional alternatives. *Economics Letters* 45: 1–5.

Hassler, U. and Wolters, J. (1995). Long memory in inflation rates: international evidence. *Journal of Business & Economic Statistics* 13: 37–45.

Hassler, U. and Wolters, J. (2009). Hysteresis in unemployment rates? A comparison between Germany and the US. *Journal of Economics and Statistics* 229: 119–129.

Hassler, U., Marmol, F., and Velasco, C. (2006). Residual log-periodogram inference for long-run relationships. *Journal of Econometrics* 130: 165–207.

Hassler, U., Rodrigues, P.M.M., and Rubia, A. (2009). Testing for general fractional integration in the time domain. *Econometric Theory* 25: 1793–1828.

Hassler, U., Rodrigues, P.M.M., and Rubia, A. (2014). Persistence in the banking industry: fractional integration and breaks in memory. *Journal of Empirical Finance* 29: 95–112.

Hassler, U., Rodrigues, P.M.M., and Rubia, A. (2016). Quantile regression for long memory testing: a case of realized volatility. *Journal of Financial Econometrics* 14: 693–724.

Hasza, D.P. (1980). The asymptotic distribution of the sample autocorrelations for an integrated ARMA process. *Journal of the American Statistical Association* 75: 349–352.

Hauser, M.A. (1999). Maximum likelihood estimators for ARMA and ARFIMA models: a Monte Carlo study. *Journal of Statistical Planning and Inference* 80: 229–255.

Hausman, J. (1978). Specification tests in econometrics. *Econometrica* 46: 1251–1271.

Hayashi, F. (2000). *Econometrics*. Princeton University Press.

Henry, M. (2001). Robust automatic bandwidth for long memory. *Journal of Time Series Analysis* 22: 293–316.

Henry, M. and Robinson, P.M. (1996). Bandwidth choice in Gaussian semiparametric estimation of long range dependence. In: *Athens Conference on Applied Probability and Time Series*, Vol. II (ed. P.M. Robinson and M. Rosenblatt), 220–232. New York: Springer.

Hidalgo, J. and Soulier, P. (2004). Estimation of the location and exponent of the spectral singularity of a long memory process. *Journal of Time Series Analysis* 25: 55–81.

Hipel, K.W. and McLeod, A.I. (1994). *Time Series Modelling of Water Resources and Environmental Systems*. Elsevier.

Holan, S., McElroy, T., and Chakraborty, S. (2009). A Bayesian approach to estimating the long memory parameter. *Bayesian Analysis* 4: 159–190.

Horváth, L. and Kokoszka, P. (2008). Sample autocovariances of long-memory time series. *Bernoulli* 14: 405–418.

Hosking, J.R.M. (1981). Fractional differencing. *Biometrika* 68: 165–176.

Hosking, J.R.M. (1984). Modeling persistence in hydrological time series using fractional differencing. *Water Resources Research* 20: 1898–1908.

Hosking, J.R.M. (1996). Asymptotic distributions of the sample mean, autocovariances, and autocorrelations of long-memory time series. *Journal of Econometrics* 73: 261–284.

Hou, J. and Perron, P. (2014). Modified local Whittle estimator for long memory processes in the presence of low frequency (and other) contaminations. *Journal of Econometrics* 182: 309–328.

Hsu, C.-C. (2005). Long memory or structural changes: an empirical examination on inflation rates. *Economics Letters* 88: 289–294.

Hualde, J. and Robinson, P.M. (2010). Semiparametric inference in multivariate fractionally cointegrated systems. *Journal of Econometrics* 157: 492–511.

Hualde, J. and Robinson, P.M. (2011). Gaussian pseudo-maximum likelihood estimation of fractional time series models. *The Annals of Statistics* 39: 3152–3181.

Hurst, H.E. (1951). Long-term storage capacity of reservoirs. *Transactions of the American Society of Civil Engineers* 116: 770–799.

Hurvich, C.M. (2002). Multistep forecasting of long memory series using fractional exponential models. *International Journal of Forecasting* 18: 167–179.

Hurvich, C.M. and Beltrao, K.I. (1993). Asymptotics for the low-frequency ordinates of the periodogram of a long-memory time series. *Journal of Time Series Analysis* 14: 455–472.

Hurvich, C.M. and Beltrao, K.I. (1994). Automatic semiparametric estimation of the memory parameter of a long memory time series. *Journal of Time Series Analysis* 15: 285–302.

Hurvich, C.M. and Brodsky, J. (2001). Broadband semiparametric estimation of the memory parameter of a long-memory time series using fractional exponential models. *Journal of Time Series Analysis* 22: 221–249.

Hurvich, C.M. and Chen, W.W. (2000). An efficient taper for overdifferenced series. *Journal of Time Series Analysis* 21: 155–180.

Hurvich, C.M. and Deo, R.S. (1999). Plug-in selection of the number of frequencies in regression estimates of the memory parameter of a long-memory time series. *Journal of Time Series Analysis* 20: 331–341.

Hurvich, C.M. and Ray, B.K. (1995). Estimation of the memory parameter for nonstationary or noninvertible fractionally integrated processes. *Journal of Time Series Analysis* 16: 17–41.

Hurvich, C.M. and Ray, B.K. (2003). The local Whittle estimator of long-memory stochastic volatility. *Journal of Financial Econometrics* 1: 445–470.

Hurvich, C.M., Deo, R., and Brodsky, J. (1998). The mean squared error of Geweke and Porter-Hudak's estimator of the memory parameter of a long-memory time series. *Journal of Time Series Analysis* 19: 19–46.

Hurvich, C.M., Moulines, E., and Soulier, P. (2005). Estimating long memory in volatility. *Econometrica* 73: 1283–1328.

Hurvich, C.M., Moulines, E., and Soulier, P. (2008). Corrigendum to "Estimating long memory in volatility". *Econometrica* 76: 661–662.

Hyndman, R.J. and Khandakar, Y. (2008). Automatic time series forecasting: the forecast package for R. *Journal of Statistical Software* 26: 1–22.

Iacone, F. (2010). Local Whittle estimation of the memory parameter in presence of deterministic components. *Journal of Time Series Analysis* 31: 37–49.

Ibragimov, I.A. and Linnik, Y.V. (1971). *Independent and Stationary Sequences of Random Variables*. Wolters-Noordhoff Publishing.

Iouditsky, A., Moulines, E., and Soulier, P. (2001). Adaptive estimation of the fracional differencing coefficient. *Bernoulli* 7: 699–731.

Janacek, G.J. (1982). Determining the degree of differencing for time series via the log spectrum. *Journal of Time Series Analysis* 3: 177–183.

Jensen, A.N. and Nielsen, M.O. (2014). A fast fractional difference algorithm. *Journal of Time Series Analysis* 35: 428–436.

Jensen, M.J. (1999a). Using wavelets to obtain a consistent ordinary least squares estimator of the long-memory parameter. *Journal of Forecasting* 18: 17–32.

Jensen, M.J. (1999b). An approximate wavelet MLE of short- and long-memory parameters. *Studies in Nonlinear Dynamics & Econometrics* 3: 239–353.

Johansen, S. (1988). Statistical analysis of cointegration vectors. *Journal of Economic Dynamics and Control* 12: 231–254.

Johansen, S. (1995). *Likelihood-Based Inference in Cointegrated Vector Autoregressive Models*. Oxford University Press.

Johansen, S. (2008). A representation theory for a class of vector autoregressive models for fractional processes. *Econometric Theory* 24: 651–676.

Johansen, S. and Nielsen, M.Ø. (2012a). A necessary moment condition for the fractional functional central limit theorem. *Econometric Theory* 28: 671–679.

Johansen, S. and Nielsen, M.Ø. (2012b). Likelihood inference for a fractionally cointegrated vector autoregressive model. *Econometrica* 80: 2667–2732.

Johansen, S. and Nielsen, M.Ø. (2012c). The role of initial values in nonstationary fractional time series models. Discussion Paper 12–18. University Copagenhagen, Department of Economics.

Kashyap, R.L. and Eom, K.-B. (1988). Estimation in long-memory time series model. *Journal of Time Series Analysis* 9: 35–41.

Kew, H. and Harris, D. (2009). Heteroskedasticity-robust testing for a fractional unit root. *Econometric Theory* 25: 1734–1753.

Kim, C.S. and Phillips, P.C.B. (2006). Log periodogram regression: the nonstationary case. Cowles Foundation Discussion Papers 1587, Cowles Foundation for Research in Economics, Yale University.

Kirchgässner, G., Wolters, J., and Hassler, U. (2013). *Introduction to Modern Time Series Analysis*, 2e. Springer.

Klemeš, V. (1974). The Hurst phenomenon: a puzzle? *Water Resources Research* 10: 675–688.

Knopp, K. (1990). *Theory and Application of Infinite Series*. Dover Publications, republication of 1951 Blackie & Son edition.

Krämer, W. (1998). Fractional integration and the augmented Dickey-Fuller test. *Economics Letters* 61: 269–272.

Kumar, M.S. and Okimoto, T. (2007). Dynamics of persistence in international inflation rates. *Journal of Money, Credit, and Banking* 39: 1457–1479.

Künsch, H. (1986). Discrimination between monotonic trends and long-range dependence. *Journal of Applied Probability* 23: 1025–1030.

Künsch, H. (1987). Statistical aspects of self-similar processes. In: *Proceedings of the 1st World Congress of the Bernoulli Society*, Vol. 1 (ed. Y. Prohorov and V.V. Sazanov), 67–74. VNU Science Press.

Kwiatkowski, D., Phillips, P.C.B., Schmidt, P., and Shin, Y. (1992). Testing the null hypothesis of stationarity against the alternative of a unit root: how sure are we that economic time series have a unit root? *Journal of Econometrics* 54: 159–178.

Lebo, M. and Clarke, H. (2000). Modelling memory and volatility: recent advances in the analysis of political time series. Editor's introduction. *Electoral Studies* 19: 1–7.

Lee, H.S. and Amsler, C. (1997). Consistency of the KPSS unit root test against fractionally integrated alternative. *Economics Letters* 55: 151–160.

Lee, D. and Schmidt, P. (1996). On the power of the KPSS test of stationarity against fractionally-integrated alternatives. *Journal of Econometrics* 73: 285–302.

Leeb, H. and Pötscher, B.M. (2005). Model selection and inference: facts and fiction. *Econometric Theory* 21: 21–59.

Lévy, P. (1953). Random functions: General theory with special reference to Laplacian random function. *University of California Publications in Statistics* 1: 331–390.

Li, W.K. and McLeod, A.I. (1986). Fractional time series modelling. *Biometrika* 73: 217–221.

Lieberman, O. and Phillips, P.C.B. (2008). Refined inference on long memory in realized volatility. *Econometric Reviews* 27: 254–267.

Lieberman, O., Rosemarin, R., and Rousseau, J. (2012). Asymptotic theory for maximum likelihood estimation of the memory parameter in stationary Gaussian processes. *Econometric Theory* 28: 457–470.

Lo, A. (1991). Long-term memory in stock market prices. *Econometrica* 59: 1279–1313.

Lobato, I.N. and Robinson, P.M. (1996). Averaged periodogram estimation of long memory. *Journal of Econometrics* 73: 303–324.

Lobato, I.N. and Robinson, P.M. (1998). A nonparametric test for I(0). *The Review of Economic Studies* 65: 475–495.

Lobato, I.N. and Savin, N.E. (1998). Real and spurious long-memory properties of stock-market data. *Journal of Business & Economic Statistics* 16: 261–268.

Lobato, I.N. and Velasco, C. (2007). Efficient Wald tests for fractional unit roots. *Econometrica* 75: 575–589.

Lobato, I.N. and Velasco, C. (2008). Power comparison among tests for fractional unit roots. *Economics Letters* 99: 152–154.

Lukacs, E. (1975). *Stochastic Convergence*, 2e. Academic Press.

Lütkepohl, H. (2005). *New Introduction to Multiple Time Series Analysis*. Springer.

Maasoumi, E. and McAleer, M. (2008). Realized volatility and long memory: an overview. *Econometric Reviews* 27: 1–9.

McCloskey, A. and Perron, P. (2013). Memory parameter estimation in the presence of level shifts and deterministic trends. *Econometric Theory* 29: 1196–1237.

McElroy, T. and Politis, D.N. (2012). Fixed-b asymptotics for the studentized mean from time series with short, long, or negative memory. *Econometric Theory* 28: 471–481.

McLeod, A.I., Yu, H., and Mahdi, E. (2012). Time series analysis with R. In: *Handbook of Statistics*, vol. 30 (ed. T. Subba Rao, S.S. Rao, and C. Rao), 661–712. Elsevier.

Man, K.S. (2003). Long memory time series and short term forecasts. *International Journal of Forecasting* 19: 477–491.

Mandelbrot, B.B. (1969). Long-run linearity, locally Gaussian processes, H-spectra and infinite variances. *International Economic Review* 10: 82–111.

Mandelbrot, B.B. and Van Ness, J.W. (1968). Fractional Brownian motion, fractional noises and applications. *SIAM Review* 10: 422–437.

Mandelbrot, B.B. and Wallis, J. (1968). Noah, Joseph and operational hydrology. *Water Resources Research* 4: 909–918.

Mandelbrot, B.B. and Wallis, J. (1969). Robustness of the rescaled range R/S in the measurement of noncyclic long run statistical dependence. *Water Resources Research* 5: 967–988.

Marinucci, D. and Robinson, P.M. (1999). Alternative forms of fractional Brownian motion. *Journal of Statistical Planning and Inference* 80: 111–122.

Marinucci, D. and Robinson, P.M. (2000). Weak convergence of multivariate fractional processes. *Stochastic Processes and their Applications* 86: 103–120.

Marmol, F. (1996). Nonsense regressions between integrated processes of different orders. *Oxford Bulletin of Economics and Statistics* 58: 525–536.

Martins, L.F. and Rodrigues, P.M.M. (2014). Testing for persistence change in fractionally integrated models: an application to world inflation rates. *Computational Statistics and Data Analysis* 76: 502–522.

Milhøj, A. (1981). A test of fit in time series models. *Biometrika* 68: 177–187.

Miller, J.I. and Park, J.Y. (2010). Nonlinearity, nonstationarity, and thick tails: how they interact to generate persistence in memory. *Journal of Econometrics* 155: 83–89.

Montanari, A. (2003). Long-range dependence in hydrology. In: *Theory and Applications of Long-Range Dependence* (ed. P. Doukhan, G. Oppenheim, and M.S. Taqqu), 461–472. Birkhäuser.

Mood, A.M., Graybill, F.A., and Boes, D.C. (1974). *Introduction to the Theory of Statistics*, 3e. McGraw-Hill.

Morana, C. and Beltratti, A. (2004). Structural change and long-range dependence in volatility of exchange rates: either, neither or both? *Journal of Empirical Finance* 11: 629–658.

Moulines, E. and Soulier, P. (1999). Broadband log-periodogram regression of time series with long-range dependence. *The Annals of Statistics* 27: 1415–1439.

Moulines, E. and Soulier, P. (2000). Data driven order selection for projection estimator of the spectral density of time series with long range dependence. *Journal of Time Series Analysis* 21: 193–218.

Moulines, E. and Soulier, P. (2003). Semiparametric spectral estimation for fractional processes. In: *Theory and Applications of Long-Range Dependence* (ed. P. Doukhan, G. Oppenheim, and M.S. Taqqu), 251–301. Birkhäuser.

Moulines, E., Roueff, F., and Taqqu, M.S. (2007). On the spectral density of the wavelet coefficients of long-memory time series with application to the log-regression estimation of the memory parameter. *Journal of Time Series Analysis* 28: 155–187.

Moulines, E., Roueff, F., and Taqqu, M.S. (2008). A wavelet Whittle estimator of the memory parameter of a nonstationary Gaussian time series. *The Annals of Statistics* 36: 1925–1956.

Nelson, D.B. (1990). Stationarity and persistence in the GARCH(1,1) model. *Econometric Theory* 3: 318–334.

Newbold, P. and Agiakloglou, C. (1993). Bias in the sample autocorrelations of fractional noise. *Biometrika* 80: 698–702.

Newey, W.K. and West, K.D. (1987). A simple, positive semi-definite, heteroskedasticity and autocorrelation consistent covariance matrix. *Econometrica* 55: 703–708.

Ng, S. and Perron, P. (1995). Unit root tests in ARMA models with data-dependent methods for the selection of the truncation lag. *Journal of the American Statistical Association* 90: 268–281.

Nielsen, M.O. (2004a). Efficient likelihood inference in nonstationary univariate models. *Econometric Theory* 20: 116–146.

Nielsen, M.O. (2004b). Optimal residual-based tests for fractional cointegration and exchange rate dynamics. *Journal of Business & Economic Statistics* 22: 331–345.

Nielsen, M.O. (2004c). Spectral analysis of fractionally cointegrated systems. *Economics Letters* 83: 225–231.

Nielsen, M.O. (2010). Nonparametric cointegration analysis of fractional systems with unknown integration orders. *Journal of Econometrics* 155: 170–187.

Nielsen, M.O. (2011). Asymptotics for the Conditional-Sum-of-Squares Estimator in Fractional Time Series Models. Technical Report 1259. Queen's University, Department of Economics.

Nielsen, M.O. (2015). Asymptotics for the conditional-sum-of-squares estimator in fractional time-series models. *Journal of Time Series Analysis* 36: 154–188.

Nielsen, M.O. and Frederiksen, P.H. (2005). Finite sample comparison of parametric, semiparametric, and wavelet estimators of fractional integration. *Econometric Reviews* 24: 405–443.

Nielsen, M.O. and Shimotsu, K. (2007). Determining the cointegrating rank in nonstationary fractional systems by the exact local Whittle approach. *Journal of Econometrics* 141: 574–596.

Odaki, M. (1993). On the invertibility of fractionally differenced ARIMA processes. *Biometrika* 80: 703–709.

Ohanissian, A., Russell, J.R., and Tsay, R.S. (2008). True or spurious long memory? A new test. *Journal of Business & Economic Statistics* 26: 161–175.

Pai, J.S. and Ravishanker, N. (1998). Bayesian analysis of autoregressive fractionally integrated moving-average processes. *Journal of Time Series Analysis* 19: 99–112.

Palma, W. (2007). *Long-Memory Time Series: Theory and Methods*. Wiley.

Palma, W. (2016). *Time Series Analysis*. Wiley.

Palma, W. and Chan, N.H. (1997). Estimation and forecasting of long-memory processes with missing values. *Journal of Forecasting* 16: 395–410.

Palma, W. and Chan, N.H. (2005). Efficient estimation of seasonal long-range-dependent processes. *Journal of Time Series Analysis* 26: 863–892.

Papailias, F. and Dias, G.F. (2015). Forecasting long memory series subject to structural change: a two-stage approach. *International Journal of Forecasting* 31: 1056–1066.

Parke, W.R. (1999). What is fractional integration? *The Review of Economics and Statistics* 81: 632–638.

Paya, I., Duarte, A., and Holden, K. (2007). On the relationship between inflation persistence and temporal aggregation. *Journal of Money, Credit, and Banking* 39: 1521–1531.

Peiris, M.S. and Perera, B.J.C. (1988). On prediction with fractionally differenced ARIMA models. *Journal of Time Series Analysis* 9: 215–220.

Peligrad, M. and Utev, S. (2006). Central limit theorem for stationary linear processes. *Annals of Probability* 34: 1608–1622.

Perron, P. and Qu, Z. (2010). Long-memory and level shifts in the volatility of stock market return indices. *Journal of Business & Economic Statistics* 28: 275–290.

Persival, D.B. and Walden, A.T. (2000). *Wavelet Methods for Time Series Analysis*. Cambridge University Press.

Pesaran, M.H. (2015). *Time Series and Panel Data Econometrics*. Oxford University Press.

Phillips, P.C.B. (1986). Understanding spurious regressions in econometrics. *Journal of Econometrics* 33: 311–340.

Phillips, P.C.B. (1999). Discrete Fourier transforms of fractional processes. Cowles Foundation Discussion Papers 1243, Cowles Foundation for Research in Economics, Yale University.

Phillips, P.C.B. (2007). Unit root log periodogram regression. *Journal of Econometrics* 138: 104–124.

Phillips, P.C.B. and Solo, V. (1992). Asymptotics for linear processes. *The Annals of Statistics* 20: 971–1001.

Pipiras, V. and Taqqu, M.S. (2003). Fractional calculus and its connections to fractional Brownian motion. In: *Theory and Applications of Long-Range Dependence* (ed. P. Doukhan, G. Oppenheim, and M.S. Taqqu), 165–201. Birkhäuser.

Porter-Hudak, S. (1990). An application of the seasonal fractionally differenced model to the monetary aggregates. *Journal of the American Statistical Association* 85: 338–344.

Pötscher, B.M. and Prucha, I.R. (2001). Basic elements of asymptotic theory. In *A Companion to Theoretical Econometrics* (ed. B.H. Baltagi), 201–229. Wiley-Blackwell.

Pourahmadi, M. (1983). Exact factorization of the spectral density and its application to forecasting and time series analysis. *Communications in Statistics - Theory and Methods* 12: 2085–2094.

Priestley, M.B. (1981). *Spectral Analysis and Time Series*, Vol. 1. Academic Press.

Qu, Z. (2011). A test against spurious long memory. *Journal of Business & Economic Statistics* 29: 423–438.

Rao, C.R. (1948). Large sample tests of statistical hypotheses concerning several parameters with application to problems of estimation. *Proceedings of the Cambridge Philosophical Society* 44: 50–57.

Rao, C.R. (2001). Two score and 10 years of score tests. *Journal of Statistical Planning and Inference* 97: 3–7.

Ravishanker, N. and Ray, B.K. (1997). Bayesian analysis of vector ARFIMA processes. *Australian & New Zealand Journal of Statistics* 39: 295–311.

Ray, B.K. (1993). Modeling long-memory processes for optimal long-range prediction. *Journal of Time Series Analysis* 14: 511–525.

Ray, B.K. and Tsay, R.S. (2002). Bayesian methods for change-point detection in long-range dependent processes. *Journal of Time Series Analysis* 23: 687–705.

Reisen, V.A., Rodrigues, A.L., and Palma, W. (2006). Estimation of seasonal fractionally integrated processes. *Computational Statistics and Data Analysis* 50: 568–582.

Reisen, V.A., Zamprogno, B., Palma, W., and Arteche, J. (2014). A semiparametric approach to estimate two seasonal fractional parameters in the SARFIMA model. *Mathematics and Computers in Simulation* 98: 1–17.

Robinson, P.M. (1991). Testing for strong serial correlation and dynamic conditional heteroskedasticity in multiple regression. *Journal of Econometrics* 47: 67–84.

Robinson, P.M. (1994a). Time series with strong dependence. In: *Advances in Econometrics: Sixth World Congress*, Vol. 1 (ed. C.A. Sims), 47–95. Cambridge University Press.

Robinson, P.M. (1994b). Efficient tests of nonstationary hypotheses. *Journal of the American Statistical Association* 89: 1420–1437.

Robinson, P.M. (1994c). Rates of convergence and optimal bandwidth in spectral analysis of processes with long range dependence. *Probability Theory and Related Fields* 99: 443–473.

Robinson, P.M. (1994d). Semiparametric analysis of long-memory time series. *Annals of Statistics* 22: 515–539.

Robinson, P.M. (1995a). Log-periodogram regression of time series with long range dependence. *Annals of Statistics* 23: 1048–1072.

Robinson, P.M. (1995b). Gaussian semiparametric estimation of long range dependence. *Annals of Statistics* 23: 1630–1661.

Robinson, P.M. (2003). Long-memory time series. In: *Time Series with Long Memory* (ed. P.M. Robinson), 4–32. Oxford University Press.

Robinson, P.M. (2005a). Robust covariance matrix estimation: HAC estimates with long memory/antipersistence correction. *Econometric Theory* 21: 171–180.

Robinson, P.M. (2005b). The distance between rival nonstationary fractional processes. *Journal of Econometrics* 128: 283–399.

Robinson, P.M. (2006). Conditional-sum-of-squares estimation of models for stationary time series with long memory. In: *Time Series and Related Topics* (ed. H.-C. Ho, C.-K. Ing, and T.L. Lai), 130–137. Institute of Mathematical Statistics.

Robinson, P.M. and Henry, M. (1999). Long and short memory conditional heteroskedasticity in estimating the memory parameter of levels. *Econometric Theory* 15: 299–336.

Robinson, P.M. and Hidalgo, F.J. (1997). Time series regression with long-range dependence. *The Annals of Statistics* 25: 77–104.

Robinson, P.M. and Marinucci, D. (2001). Narrow-band analysis of nonstationary processes. *Annals of Statistics* 29: 947–986.

Robinson, P.M. and Marinucci, D. (2003). Semiparametric frequency domain analysis of fractional cointegration. In: *Time Series with Long Memory* (ed. P.M. Robinson), 334–373. Oxford University Press.

Robinson, P.M. and Yajima, Y. (2002). Determination of cointegrating rank in fractional systems. *Journal of Econometrics* 106: 217–241.

Said, S.E. and Dickey, D.A. (1984). Testing for unit roots in autoregressive-moving average models of unknown order. *Biometrika* 71: 599–607.

Samarov, A. and Taqqu, M.S. (1988). On the efficiency of the sample mean in long-memory noise. *Journal of Time Series Analysis* 9: 191–200.

Schwarz, G. (1978). Estimating the dimension of a model. *The Annals of Statistics* 6: 461–464.

Shao, X. (2010). Nonstationarity-extended Whittle estimation. *Econometric Theory* 26: 1060–1087.

Shao, X. and Wu, W.B. (2007). Local Whittle estimation of fractional integration for nonlinear processes. *Econometric Theory* 23: 899–951.

Shibata, R. (1976). Selection of the order of an autoregressive model by Akaike's information criterion. *Biometrika* 63: 117–126.

Shimotsu, K. (2010). Exact local Whittle estimation of fractional integration with unknown mean and trend. *Econometric Theory* 26: 501–540.

Shimotsu, K. and Phillips, P.C.B. (2002). Pooled log periodogram regression. *Journal of Time Series Analysis* 23: 57–93.

Shimotsu, K. and Phillips, P.C.B. (2005). Exact local Whittle estimation of fractional integration. *The Annals of Statistics* 33: 1890–1933.

Shumway, R.H. and Stoffer, D.S. (2011). *Time Series Analysis and Its Applications: With R Examples*, 3e. Springer.

Sibbertsen, P. (2001). S-estimation in the linear regression model with long-memory error terms under trend. *Journal of Time Series Analysis* 22: 353–363.

Sibbertsen, P. (2004). Long memory versus structural breaks: an overview. *Statistical Papers* 45: 465–515.

Sibbertsen, P. and Kruse, R. (2009). Testing for a break in persistence under long-range dependencies. *Journal of Time Series Analysis* 30: 263–285.

Silvey, S.D. (1959). The Lagrangian multiplier test. *The Annals of Mathematical Statistics* 30: 389–407.

Smith, A. (2005). Level shifts and the illusion of long memory in economic time series. *Journal of Business & Economic Statistics* 23: 321–335.

Smith, J. and Yadav, S. (1994). Forecasting costs incurred from unit differencing fractionally integrated processes. *International Journal of Forecasting* 10: 507–514.

Solo, V. (1992). Intrinsic random functions and the paradox of $1/f$ noise. *SIAM Journal of Applied Mathematics* 52: 270–291.

Sowell, F. (1992). Maximum likelihood estimation of stationary univariate fractionally integrated time series models. *Journal of Econometrics* 53: 165–188.

Stout, W.F. (1974). *Almost Sure Convergence*. Academic Press.

Sun, Y. and Phillips, P.C.B. (2003). Nonlinear log-periodogram regression for perturbed fractional processes. *Journal of Econometrics* 115: 355–389.

Tanaka, K. (1999). The nonstationary fractional unit root. *Econometric Theory* 15: 549–582.

Taniguchi, M. and Kakizawa, Y. (2000). *Asymptotic Theory of Statistical Inference for Time Series*. Springer.

Taqqu, M.S. (1975). Weak convergence to fractional Brownian motion and to the Rosenblatt process. *Zeitschrift für Wahrscheinlichkeitstheorie und verwandte Gebiete* 31: 287–302.

Taqqu, M.S. (2003). Fractional Brownian motion and long-range dependence. In: *Theory and Applications of Long-Range Dependence* (ed. P. Doukhan, G. Oppenheim, and M.S. Taqqu), 5–38. Birkhäuser.

Taqqu, M.S. and Teverovsky, V. (1996). Semi-parametric graphical estimation techniques for long-memory data. In: *Athens Conference on Applied Probability and Time Series*, Vol. II (ed. P.M. Robinson and M. Rosenblatt), 420–432. New York: Springer.

Taylor, S.J. (1982). Financial returns modelled by the product of two stochastic processes - a study of daily sugar prices 1961–1979. In: *Time Series Analysis: Theory and Practice 1* (ed. O.D. Anderson), 203–226. Amsterdam: North-Holland.

Teverovsky, V., Taqqu, M.S., and Willinger, W. (1999). A critical look at Lo's modified R/S statistic. *Journal of Statistical Planning and Inference* 80: 211–227.

Tiao, G.C. and Tsay, R.S. (1994). Some advances in non-linear and adaptive modelling in time-series. *Journal of Forecasting* 13: 109–131.

Tsay, R.S. (1984). Order selection in nonstationary autoregressive models. *The Annals of Statistics* 12: 1425–1433.

Tsay, W.-J. and Chung, C.-F. (2000). The spurious regression of fractionally integrated processes. *Journal of Econometrics* 96: 155–182.

Veenstra, J.Q. (2012). Persistence and anti-persistence: theory and software. PhD thesis. Western University.

Velasco, C. (1999a). Non-stationary log-periodogram regression. *Journal of Econometrics* 91: 325–372.

Velasco, C. (1999b). Gaussian semiparametric estimation of non-stationary time series. *Journal of Time Series Analysis* 20: 87–127.

Velasco, C. (2000). Non-Gaussian log-periodogram regression. *Econometric Theory* 16: 44–79.

Velasco, C. (2003). Nonparametric frequency domain analysis of nonstationary multivariate time series. *Journal of Statistical Planning and Inference* 116: 209–247.

Velasco, C. and Robinson, P.M. (2000). Whittle pseudo-maximum likelihood estimation for nonstationary time series. *Journal of the American Statistical Association* 95: 1229–1243.

White, H. (1980). A heteroskedasticity-consistent covariance matrix estimator and a direct test for heteroskedasticity. *Econometrica* 48: 817–38.

White, H. (2001). *Asymptotic Theory for Econometricians*, 2e. Academic Press.

Whittle, P. (1953a). Estimation and information in stationary time series. *Arkiv för Matematik* 2: 423–434.

Whittle, P. (1953b). The analysis of multiple stationary time series. *Journal of the Royal Statistical Society. Series B: Methodological* 15: 125–139.

Wichern, D.W. (1973). The behavior of the sample autocorrelation function for an integrated moving average process. *Biometrika* 60: 235–239.

Willinger, W., Paxson, V., Riedi, R.H., and Taqqu, M.S. (2003). Long-range dependence and data network trafic. In: *Theory and Applications of Long-range Dependence* (ed. P. Doukhan, G. Oppenheim, and M.S. Taqqu), 373–407. Birkhäuser.

Wold, H.O.A. (1938). *A Study in the Analyis of Stationary Time Series*. Almqvist and Wiksell.

Woodward, W.A., Cheng, Q.C., and Gray, H.L. (1998). A k-factor GARMA long-memory model. *Journal of Time Series Analysis* 19: 485–504.

Woodward, W.A., Gray, H.L., and Elliott, A.C. (2017). *Applied Time Series Analysis with R*, 2e. Taylor & Francis Group.

Yaglom, A.M. (1962). *An Introduction to the Theory of Stationary Random Functions.* Prentice-Hall.

Yajima, Y. (1985). On estimation of long-memory time series models. *Australian Journal of Statistics* 27: 303–320.

Yajima, Y. (1988). On estimation of a regression model with long-memory stationary errors. *The Annals of Statistics* 16: 791–807.

Yajima, Y. (1989). A central limit theorem of Fourier transforms of strongly dependent stationary processes. *Journal of Time Series Analysis* 10: 375–383.

Yajima, Y. (1991). Asymptotic properties of the LSE in a regression model with long-memory stationary errors. *The Annals of Statistics* 19: 158–177.

Index

Time Series Analysis with Long Memory in View, First Edition. Uwe Hassler.
© 2019 John Wiley & Sons, Inc. Published 2019 by John Wiley & Sons, Inc.